SCHAUM'S
outlines

Mathematica

SCHAUM'S
outlines

Mathematica

―――――――――――Second Edition

Eugene Don, Ph.D.

Professor of Mathematics
Queens College, CUNY

Schaum's Outline Series

McGraw Hill

New York   Chicago   San Francisco   Lisbon   London
Madrid   Mexico City   Milan   New Delhi   San Juan
Seoul   Singapore   Sydney   Toronto

EUGENE DON received his B.S. from Queens College, M.S. from Columbia University, and Ph.D. in Applied Mathematics and Statistics from the State University of New York at Stony Brook. He has been a member of the Faculty of Mathematics at Queens College of the City University of New York for over 30 years, where he has taught a variety of mathematics courses including a course devoted to learning and applying *Mathematica*. His area of research is numerical analysis and he is interested in the application of computers to solving mathematics problems.

Schaum's Outline of MATHEMATICA

3 4 5 6 7 8 9   CUS/CUS   1 5 4 3 2 1

ISBN 978-0-07-160828-2
MHID 0-07-160828-1

Library of Congress Cataloging-in-Publication Data

Don, Eugene.
    Schaum's outline of theory and problems of Mathematica / Eugene Don.
        —2nd ed.
            p.   cm.—(Schaum's outline series)
    Includes bibliographical references and index.
    ISBN-13: 978-0-07-160828-2 (alk. paper)
    ISBN-10: 0-07-160828-1 (alk. paper)
    1. Mathematica (Computer program language)—Problems, exercises, etc.
2. Mathematics—Data processing—Problems, exercises, etc.
3. Mathematica (Computer file)   I. Title.   II. Title: Theory and problems
of Mathematica.
QA76.95.D66   2009
510.285'536—dc22
                                                                    2009007878

To my wife, Benay,
whose patience and understanding
made this book possible.

Preface to the First Edition

This book is designed to help students and professionals who use mathematics in their daily routine to learn *Mathematica*, a computer system designed to perform complex mathematical calculations. My approach is simple: learn by example. Along with easy to read descriptions of the most widely used commands, I have included a collection of over 750 examples and solved problems, each specifically designed to illustrate an important feature of the *Mathematica* software.

I have included those commands and options that are most commonly used in algebra, trigonometry, calculus, differential equations, and linear algebra. Most examples and solved problems are short and to the point. Comments have been included, where appropriate, to clarify what might be confusing to the reader.

The reader is encouraged not only to replicate the output shown in the text, but to make modifications and investigate the resulting effect upon the output. I have found this to be the most effective way to learn the syntax and capabilities of this truly unique program.

The first three chapters serve as an introduction to the syntax and style of *Mathematica*. The structure of the remainder of the book is such that the reader need only be concerned with those chapters of interest to him or her. If, on occasion, a command is encountered that has been discussed in a previous chapter, the Index may be used to conveniently locate the command's description.

Without a doubt you will be impressed with *Mathematica*'s capabilities. It is my sincere hope that you will use the power built into this software to investigate the wonders of mathematics in a way that would have been impossible just a few years ago.

I would like to take this opportunity to thank the staff at McGraw-Hill for their help in the preparation of this book and to give a special note of thanks to Mr. Joel Lerner for his encouragement and support of this project.

EUGENE DON

Preface to the Second Edition

The recent introduction of *Mathematica* 6 and *Mathematica* 7 has brought significant changes to many of the commands that comprise the language. A complete listing of all the changes can be found in the Documentation Center that is included with your program. Most notably:

- Some of the menus and dialog boxes have changed. These changes are mostly cosmetic and should not cause any confusion.

- The BasicInput palette has been renamed Basic Math Input.

- Graphics output was enhanced in version 6. Consequently plots, particularly three-dimensional plots, may look slightly different from those in previous versions.

- In versions 4 and 5 a semicolon (;) was used merely to suppress an annoying line of output when executing graphics commands. In versions 6 and 7, the semicolon suppresses graphics output *completely* and must therefore be deleted when using commands such as `Plot`, `Plot3D`, `Show`, etc. Furthermore, since the semicolon may now be used to suppress graphics, `DisplayFunction` $\rightarrow$ `Identity` and `DisplayFunction` $\rightarrow$ `$DisplayFunction` are no longer needed.

- Some of the commands that had previously been supplied in packages (and had to be loaded prior to use) are now included in the kernel and may be used without invoking `Needs` or `<<`. Some of the commands are located in different packages, and some of them are available by download from the Wolfram website.

- Some of the commands in version 5 have been eliminated and put into "legacy" packages, included with *Mathematica* 6 and 7. They will have to be loaded prior to using them.

- Some of the commands (e.g., `ImplicitPlot`) have been eliminated and their functionality has been incorporated into other commands (e.g., `ContourPlot`).

- Animation has been significantly enhanced with the introduction of `Animate` and `Manipulate`.

A tool has been incorporated into *Mathematica* that will scan notebooks written using older versions of the software. Any incompatibilities are flagged and suggestions for correcting them are automatically generated.

This second edition incorporates all of these changes in the command descriptions, examples, and solved problems. In addition a comprehensive list of commands used in the book, together with their descriptions, is conveniently located in the appendix.

The manuscript for this book was proofread several times and all the examples and solved problems have been checked for accuracy. If you should come across a mistake that has not been caught, or would like to share your thoughts about the book, please feel free to send an e-mail to

mathematica.corrections@gmail.com

I hope you will find this book helpful in navigating through *Mathematica*. I would like to thank Professor John-Tones Amenyo of York College for his help in highlighting those parts of the text that required modification.

EUGENE DON

Contents

Included with this book is a free 30 day trial of the Wolfram *Mathematica*® software. To access your free download, simply go to

http://www.wolfram.com/books/resources

and enter license number L3262-2112. You will be guided to download and install the latest version of *Mathematica*.

CHAPTER 1

Getting Acquainted

1.1 Notation and Conventions

Mathematica is a language that is best learned by experimentation. Therefore, the reader is urged to try as many examples and problems as possible and experiment by changing options and parameters. In fact, this chapter may be considered a tutorial for those readers who want to get their hands on *Mathematica* right away.

New commands are introduced with a ▪ bullet, and options associated with them are bulleted with a • symbol for easy reference.

In keeping with *Mathematica*'s conventions, all commands and instructions will be written in Courier bold face type and *Mathematica* output in Courier light face type.

This line is written in Courier bold face type.

This line is written in Courier light face type.

Menu commands in this text are described using double arrows (⇒). For example, Format ⇒ Style ⇒ Input, written in Arial font, means go to the "Format" menu, then to the "Style" submenu, and then click on "Input."

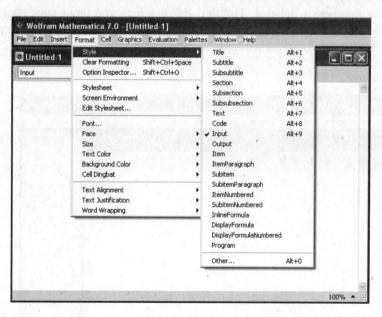

Mathematica occasionally uses a special symbol, `, which we call a backquote. Do not confuse this with an apostrophe.

Finally, most *Mathematica* commands use an arrow, →, to specify options within the command. You may use -> (– followed by >) as an alternate, if you wish. *Mathematica* will automatically convert this sequence to →. In a similar manner, the sequence != is automatically converted to ≠, <= is replaced by ≤, and >= is changed to ≥.

The examples used in this book were executed using *Mathematica* versions 6 and 7. You may notice some differences on your computer if you are using earlier versions of the software. Most noticeably, graphics, particularly three-dimensional graphics, have been enhanced in the later version and many computational algorithms have been improved, resulting in greater efficiency and speed.

1.2   The Kernel and the Front End

The *kernel* is the computational engine of *Mathematica*. You input instructions and the kernel responds with answers in the form of numbers, graphs, matrices, and other appropriate displays. The kernel works silently in the background and, for the most part, is invisible.

The interface between the user and the kernel is called the *front end* and the medium of the front end is the *Mathematica* notebook. The notebook not only enables you to communicate with the kernel, but is a convenient tool for documenting your work.

To execute an instruction, type the instruction and then press [ENTER]. Most PCs have two [ENTER] keys, but only the [ENTER] key to the *far right* of the keyboard will execute instructions. The other [ENTER] key must be pressed with the [SHIFT] key held down; otherwise you will merely get a new line. This is especially important if you are using a laptop. If you are using a Macintosh computer, do not confuse the [ENTER] key with the [RETURN] key.

The picture in Example 1 shows the standard *Mathematica* display. The symbols on the right-hand side form the Basic Math Input palette and allow access by mouse-click to the most common mathematical symbols. (If you don't see the palette on your screen, click on Palettes ⇒ BasicMathInput or Palettes ⇒ Other ⇒ Basic Math Input and it should appear.) Other palettes such as Basic Math Assistant and Classroom Assistant (version 7 and above) are available for specialized purposes and can be accessed via the Palettes menu.

Each symbol is accessed by clicking on the palette. If you use the palette, your notebooks will look like pages from a math textbook. Most examples in this book take full advantage of the Basic Math Input palette. However, each *Mathematica* symbol has an alternative descriptive format that can be typed "manually." For example, π can be represented as `Pi` and $\sqrt{5}$ can be written `Sqrt[5]`. These representations are useful for experienced *Mathematica* users who prefer not to use the mouse.

The notebook in Example 1, labeled "Untitled–1," is where you input your commands and where *Mathematica* places the result of its calculations. The picture shows the input and output of Example 1. (The display on a Macintosh computer will look slightly different.)

EXAMPLE 1   Add 2 and 3.

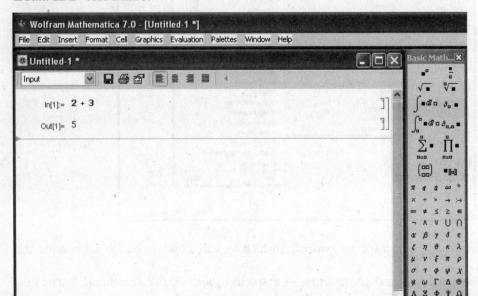

Notice that the kernel has assigned "In[1]" to the input expression and "Out[1]" to the output. This enables you to keep track of the order in which the kernel evaluates instructions. These labels are important because the order of evaluation does not always correspond to the physical position of the instruction within the notebook. In this book, however, we shall not include "In" and "Out" labels in our examples.

In working out the examples and problems in this book, you may find that your answers do not agree with the answers given in the text. This may occur if you have defined a symbol to have a specific value. For example, if x has been defined as 3, all occurrences of x will be replaced by 3. You should clear the symbol (see Section 1.5) and try the problem again. All examples and problems assume that symbols have been cleared prior to execution.

You can work on several different notebooks in a single *Mathematica* session. However, if you are using only one kernel, changes to symbols in one notebook will affect identical symbols in all notebooks.

There are times when you may wish to evaluate only part of an expression. To do this, select the portion of the expression you wish to evaluate. Then press [CTRL] + [SHIFT] + [ENTER] on a PC or [COMMAND] + [RETURN] on a Mac.

EXAMPLE 2 Suppose we wish only to perform the multiplication in the expression $2 * 3 + 5$.
First select **2 * 3**:

2 * 3 + 5

Then press [CTRL] + [SHIFT] + [ENTER] (PC) or [COMMAND] + [RETURN] (Mac).

6 + 5

A semicolon (;) at the end of a *Mathematica* command will suppress output. This is useful in long sequences of calculations when only the final answer is important.

EXAMPLE 3 Suppose we wish to define $a = 1$, $b = 2$, $c = 3$ and then display their sum. Here are two ways to write this problem.

```
a = 1              a = 1;
b = 2              b = 2;
c = 3              c = 3;
a + b + c          a + b + c
1                  6
2
3
6
```

Occasionally you may introduce an instruction that takes an excessively long time to execute, or you may inadvertently create an infinite loop. To abort a calculation, go to Evaluation ⇒ Abort Evaluation. Alternatively, you may press [ALT] + [.] to abort ([COMMAND] + [.] on the Macintosh). On the rare occasion when this does not work, you will have to terminate the kernel by going to Evaluation ⇒ Quit Kernel ⇒ Local. However, by doing so, you will lose all your defined symbols and values. Your *Mathematica* notebook will not be lost, however, so they can easily be restored.

As with all computer software, there are times when *Mathematica* will crash completely. The only remedy is to close *Mathematica* and reload it. On rare occasions, you may have to reboot your computer. In either event, your notebook changes will be lost. It is therefore extremely important to **back up your notebook often!**

Finally, there may be times when you wish to include comments within your *Mathematica* commands. Anything written within (* and *) is ignored by the *Mathematica* kernel.

EXAMPLE 4

12 + (* these words will be ignored by the kernel *) 3
15

SOLVED PROBLEMS

1.1 Multiply 12 by 17 and then add 9.

SOLUTION

```
12 * 17 + 9
```
213

1.2 Multiply the 12 by 17 in Problem 1.1, but do not add the 9.

SOLUTION

`12 * 17` + 9   ← Select `12 * 17` with the mouse.

Press [CTRL] + [SHIFT] + [ENTER]   or   [COMMAND] + [RETURN] on a Mac.

204 + 9

1.3 The following program is an infinite loop. Execute it and then abort the evaluation.

$$x = 1;$$
$$\text{While}[x > 0, x = x + 1]$$

SOLUTION

```
x = 1;
While[x > 0, x = x + 1]
```
[ALT] + .

$Aborted

1.4 Multiply 17.2 by 16.3 and then add 4.7.

SOLUTION

```
17.2 * 16.3 + 4.7
```
285.06

1.5 Multiply 17.2 by the sum of 16.3 and 4.7.

SOLUTION

```
17.2 * (16.3 + 4.7)
```
361.2

1.6 Compute the sum of $2x + 3$, $5x + 9$, and $4x + 2$.

SOLUTION

```
(2x + 3) + (5x + 9) + (4x + 2)
```
14 + 11x

1.3   *Mathematica* Quirks

Mathematica is case sensitive.

For example, **Integrate** and **integrate** are different. All *Mathematica*-defined symbols, commands and functions begin with a capital letter. Some symbols, such as **FindRoot**, use more than one capital letter. To avoid conflicts, it is a good idea for all user-defined symbols to begin with a lowercase letter.

Different brackets are used for different purposes.

- Square brackets are used for function arguments: **Sin[x]** *not* **Sin(x)**.

- Round brackets are used for grouping: **(2 + 3) * 4** means add 2 + 3 first, then multiply by 4. Never type **[2 + 3] * 4**.
- Curly brackets are used for lists: **{1, 2, 3, 4}**. More about lists in Chapter 3.

Use **E**, not **e**, for the base of the natural logarithm.

Since every *Mathematica* symbol begins with a capital letter, the base of the natural logarithm is **E**. This causes a bit of confusion, so be careful. Similarly, **I** (not **i**) is the imaginary unit. The symbols **e** and **i** from the Basic Math Input palette may be freely used if desired.

Polynomials are not written in "standard" form.

Mathematica writes polynomials with the constant term first and increasing powers from left to right. Thus, the polynomial $x^2 + 2x - 3$ would be converted to $-3 + 2x + x^2$. To see the expression in a more conventional format, the command **TraditionalForm** may be used.

- **TraditionalForm[*expression*]** prints *expression* in a traditional mathematical format.

EXAMPLE 5 Evaluate the sum of $x^2 + 3$, $2x + 5$, and $x^3 + 2$ and express the answer using **TraditionalForm**.

(x² + 3) + (2 x + 5) + (x³ + 2)

$10 + 2x + x^2 + x^3$

TraditionalForm[(x² + 3) + (2 x + 5) + (x³ + 2)]

$x^3 + x^2 + 2x + 10$

SOLVED PROBLEMS

1.7 Compute $\sqrt{81}$ using the **Sqrt** function. What happens if you do not use a capital "S"?

SOLUTION

Sqrt[81]

9

sqrt[81]

sqrt[81]   ← *Mathematica* does not recognize the (undefined) symbol sqrt.

1.8 Use parentheses to multiply the sum of 2 and 3 by the sum of 5 and 7. What happens if you use square brackets?

SOLUTION

(2 + 3) (5 + 7)

60

[2 + 3] [5 + 7]

Syntax::sntxb : Expression cannot begin with "[2+3][5+7]".

Syntax::tsntxi : "[2+3]" is incomplete; more input is needed.

Syntax::sntxi : Incomplete expression; more input is needed.

Click on the + to reveal the error.

1.9 Use the **Sin** function to compute $\sin(\pi/2)$. What happens if you use round parentheses?

SOLUTION

Sin[Pi/2] or **Sin[π/2]**

1

Sin(Pi/2)

$\dfrac{\pi\, \text{Sin}}{2}$

Mathematica thinks you want to multiply the symbol **Sin** by π and divide by 2.

1.10 Alexis typed **[4 + 1] * [6 + 2]** during a *Mathematica* session. Why didn't she get an answer of 40?

SOLUTION

Square brackets cannot be used for grouping. Round parentheses must be used.

1.11 Why didn't Ariel get an answer of 3 when she typed **sqrt[9]**?

SOLUTION

Mathematica functions must begin with a capital letter.

1.12 Why didn't Lauren get an answer of 1 when she typed **Cos(0)**?

SOLUTION

Square brackets, not round parentheses, must be used to contain arguments of functions.

1.4   *Mathematica* Gives Exact Answers

Mathematica is designed to work as a mathematician works: with 100% precision. You do not get the 10- or 12-digit numerical approximation a calculator would give, but instead get a symbolic mathematical expression.

EXAMPLE 6

$\sqrt{12}$

$2\sqrt{3}$

EXAMPLE 7

1/3 + 3/5 − 5/7 + 2/11

$\frac{463}{1155}$

EXAMPLE 8

π + π

2π

EXAMPLE 9

$\sqrt{-1}$

i

SOLVED PROBLEMS

1.13 Simplify $\sqrt{2} + \sqrt{8} + \sqrt{18}$.

SOLUTION

$\sqrt{2} + \sqrt{8} + \sqrt{18}$ or   **Sqrt[2] + Sqrt[8] + Sqrt[18]**

$6\sqrt{2}$

1.14 Compute the sum of the reciprocals of 3, 5, 7, 9, and 11.

SOLUTION

$$\frac{1}{3} + \frac{1}{5} + \frac{1}{7} + \frac{1}{9} + \frac{1}{11} \quad \text{or} \quad 1/3 + 1/5 + 1/7 + 1/9 + 1/11$$

$$\frac{3043}{3465}$$

1.15 Compute the square root of π *exactly* using the **Sqrt** function.

SOLUTION

`Sqrt[Pi]`

$\sqrt{\pi}$ ← This is the only way to represent the square root of π *exactly*.

1.16 Multiply $\sqrt{8}$ by $\sqrt{2}$.

SOLUTION

$\sqrt{8}\,\sqrt{2}$ or   `Sqrt[8] * Sqrt[2]`

4

1.17 Simplify $\sqrt{3} + \sqrt{12} + \sqrt{27} + \sqrt{48}$ leaving your answer in radical form.

SOLUTION

$\sqrt{3} + \sqrt{12} + \sqrt{27} + \sqrt{48}$

$10\sqrt{3}$

1.5 *Mathematica* Basics

In this section we discuss some of the simpler concepts within *Mathematica*. Each will be explained in greater detail in a subsequent chapter.

Symbols are defined using any sequence of alphanumeric characters (letters, digits, and certain special characters) not beginning with a digit. Once defined, a symbol retains its value until it is changed, cleared, or removed.

Arithmetic operations are performed in the obvious manner using the symbols +, –, *, and /. Exponentiation is represented by a caret, ^, so x^y means x^y. Just as in algebra, a missing symbol implies multiplication, so 2a is the same as 2*a. Be careful, however, when multiplying two symbols, since ab represents the *single* symbol beginning with a and ending with b. To multiply a by b you *must* separate the two letters with * or × (on the **Basic Math Input** palette) or a space: a*b, a×b, or a b.

EXAMPLE 10

```
a = 2
b = 3
c = a + b

2

3

5
```

Notice that the result of *each* calculation is displayed. This is sometimes annoying, and can be suppressed by using a semicolon (;) to the right of the instruction.

EXAMPLE 11

```
a = 2;
b = 3;
c = a + b
5
```

Operations are performed in the following order: (*a*) exponentiation, (*b*) multiplication and division, (*c*) addition and subtraction. If the order of operations is to be modified, parentheses, ( ), must be used. Be careful not to use [ ] or { } for this purpose.

EXAMPLE 12

```
2 + 3 * 5
17
(2 + 3) * 5
25
```

Each symbol in *Mathematica* represents something. Perhaps it is the result of a simple numerical calculation or it may be a complicated mathematical expression.

EXAMPLE 13

```
a = 3;
```
$$b = \sqrt{\frac{x^2 + 1}{2x + 3}};$$

Here, **a** is a symbol representing the numerical value 3 and **b** is a symbol representing an algebraic expression.

If you ever forget what a symbol represents, simply type **?** followed by the symbol name to recall its definition.

EXAMPLE 14   (continuation of Example 13)

```
?a
```

```
Global`a
```

```
a = 3
```

```
?b
```

```
Global`b
```

$$b = \sqrt{\frac{1+x^2}{3+2x}}$$

To delete a symbol so that it can be used for a different purpose, the **Clear** or the **Remove** command can be used.

- **Clear [*symbol*]** clears *symbol*'s definition and values, but does not clear its attributes, messages, or defaults. *symbol* remains in *Mathematica*'s symbol list. Typing ***symbol = .*** will also clear the definition of *symbol*.
- **Remove [*symbol*]** removes *symbol* completely. *symbol* will no longer be recognized unless it is redefined.

You may have noticed that when you begin to type the name of a symbol, it appears with a blue font until it is recognized as a *Mathematica* command or symbol (possibly user-defined) having some value. Then it turns black. If the symbol is cleared or removed, all instances of the symbol turn blue once again.

Parentheses, brackets, and braces remain purple until completed with a matching mate. Errors caused by having two left parentheses, but only one right parenthesis, for example, can be conveniently spotted.

EXAMPLE 15 (continuation of Example 13)

```
Clear[a]
```
> **?a**   ← **?a** recalls information about the symbol **a**.

> Global`a

```
Remove[b]
```
> **?b**

Information :: notfound : Symbol b not found. ≫

    (Clicking on ≫ gives more information about the error.)

The **N** command allows you to compute a numerical approximation.

- **N[*expression*]** gives the numerical approximation of *expression* to six significant digits (*Mathematica*'s default).
- **N[*expression*, n]** attempts to give an approximation accurate to n significant digits.

A convenient shortcut is to use **//N** to the right of the expression being approximated. Thus, *expression***//N** is equivalent to **N[*expression*]**. **//** can be used for other *Mathematica* commands as well.

- *expression* **//***Command* is equivalent to *Command*[*expression*].

Another shortcut is to type a decimal point anywhere in the expression. This will cause *Mathematica* to evaluate the expression numerically.

EXAMPLE 16

$$\frac{1}{2}+\frac{1}{3}-\frac{1}{5}$$
$$\frac{19}{30}$$

$$\frac{1}{2}+\frac{1}{3}-\frac{1}{5\,.}$$ ←Note the decimal point after the 5.
0.633333

EXAMPLE 17

N[π] or π //N
3.14159

N[π, 50]
3.1415926535897932384626433832795028841971693993751

The *Mathematica* kernel keeps track of the results of previous calculations. The symbol **%** returns the result of the previous calculation, **%%** gives the result of the calculation before that, **%%%** gives the result of the calculation before that and so forth. Using **%** wisely can save a lot of typing time.

EXAMPLE 18 To construct $\sqrt{\pi+\sqrt{\pi+\sqrt{\pi}}}$, we could type: **Sqrt[Pi+Sqrt[Pi+Sqrt[Pi]]]**. A less confusing way of accomplishing this is to type

```
Sqrt[Pi];          ← The semicolon suppresses the output of the intermediate calculations.
Sqrt[Pi + %];
Sqrt[Pi + %]
```

$$\sqrt{\pi+\sqrt{\sqrt{\pi}+\pi}}$$

Using the Basic Math Input palette, we can type

$$\sqrt{\pi} \; ;$$

$$\sqrt{\pi} + \% \; ;$$

$$\sqrt{\pi} + \%$$

$$\sqrt{\pi + \sqrt{\sqrt{\pi} + \pi}}$$

SOLVED PROBLEMS

1.18 Define $a = 3$, $b = 4$, and $c = 5$. Then multiply the sum of a and b by the sum of b and c. Print only the final answer.

SOLUTION

```
a = 3;
b = 4;
c = 5;
(a + b) * (b + c)
63
```

1.19 Let $a = 1$, $b = 2$, and $c = 3$ and add a, b, and c. Then clear a, b, and c from the kernel's memory and add again.

SOLUTION

```
a = 1;
b = 2;
c = 3;
a + b + c
6

Clear[a,b,c]
a + b + c
a + b + c
```

1.20 Obtain a 25-decimal approximation of e, the base of the natural logarithm.

SOLUTION

```
N[E, 26]   or   N[e, 26]          ← 26 significant digits gives 25 decimal places.
2.7182818284590452353602875
```

1.21 (a) Express $\dfrac{1}{7} + \dfrac{2}{13} - \dfrac{3}{19} + \dfrac{1}{23}$ as a single fraction.

(b) Obtain an approximation accurate to 15 decimal places.

SOLUTION

```
1/7 + 2/13 - 3/19 + 1/23

7249
─────
39767

N[%, 15]
0.182286820730757
```

1.22 Compute $\sqrt{968}$ (a) exactly and (b) approximately to 25 significant digits.

SOLUTION

$\sqrt{968}$ or `Sqrt[968]`

$22\sqrt{2}$

`N[%, 25]`

`31.11269837220809107363715`

1.23 Multiply 12 by 6. Then multiply 15 by 7. Then use `%` and `%%` to add the two products.

SOLUTION

`12 * 6`

`72`

`15 * 7`

`105`

`% + %%`

`177`

1.24 Compute $1 + \dfrac{1}{1 + \dfrac{1}{1 + \dfrac{1}{1 + \dfrac{1}{2}}}}$

SOLUTION

$1 + \dfrac{1}{2}$

$\dfrac{3}{2}$

$1 + \dfrac{1}{\%}$

$\dfrac{5}{3}$

$1 + \dfrac{1}{\%}$

$\dfrac{8}{5}$

$1 + \dfrac{1}{\%}$

$\dfrac{13}{8}$

1.25 Compute the value of $1 + (1 + (1 + (1 + (1+1)^2)^2)^2)^2$.

SOLUTION

`1 + 1`

`2`

`1 + %^2`

`5`

`1 + %^2`

`26`

`1 + %^2`

`677`

`1 + %^2`

`458 330`

1.6 Cells

Cells are the building blocks of a *Mathematica* notebook. Cells are indicated by brackets at the right-hand side of the notebook. (Most likely you have already noticed these brackets and were wondering what they meant.) Cells can contain sub-cells, which may in turn contain sub-sub-cells, and so forth.

The kernel evaluates a notebook on a cell-by-cell basis, so if you have several instructions within a single cell, they will all be executed with a single press of the [ENTER] key.

EXAMPLE 19

```
a = 1 + 2
b = 2 + 7      ←All three lines are contained within a single cell. [ENTER] is pressed only once.
c = a + b
```

```
3
9
12
```

A new cell can be formed by moving the mouse until the cursor becomes horizontal, and then clicking. A horizontal line will appear across the screen to mark the beginning of the new cell. Existing cells can be divided by clicking on the menu Cell ⇒ Divide Cell. The cell will be divided into two cells, the break occurring at the point where the cursor is positioned. As a shortcut, you can divide a cell by pressing (simultaneously) [SHIFT] + [CTRL] + [D].

Cells can be combined (merged) by selecting the appropriate cell brackets (a vertical black line should appear) and then clicking on Cell ⇒ Merge Cells. Alternatively, you can press [SHIFT] + [CTRL] + [M].

To avoid extremely long notebooks, cells can be closed (or compressed) by double-clicking on the cell bracket. The bracket will change appearance, looking something like a fish hook. Double-clicking a second time will open the cell.

There are different types of cells for different purposes. Only input cells can be fed to the kernel for evaluation. Text cells are used for descriptive purposes. Other cell types such as Title, Subtitle, Section, Subsection, etc. can be found by clicking on the menu Format ⇒ Style. The cell type can also be seen and changed using a drop-down box located in a toolbar at the top of your notebook. If you do not see the toolbar, go to Window ⇒ Show Toolbar to display it.

SOLVED PROBLEMS

1.26 Let $a = 2x + 3$ and $b = 5x + 6$. Then compute $a + b$.

    (a) Place each instruction in a separate cell and execute them individually.

    (b) Place all three instructions in a single cell and execute them simultaneously.

SOLUTION

This is what the output looks like *after* execution:

(a)
```
a = 2 x + 3
3 + 2 x
b = 5 x + 6
6 + 5 x
a + b
9 + 7 x
```

(b)
```
a = 2 x + 3
b = 5 x + 6
a + b
3 + 2 x
6 + 5 x
9 + 7 x
```

1.27 Let $a = 2x + 3y + 4z$, $b = x + 3y + 5z$, and $c = 3x + y + z$. Compute the sum of a, b, and c. Place four lines within a single cell and execute, printing only the final result.

SOLUTION

```
a = 2x + 3y + 4z;

b = x + 3y + 5z;

c = 3x + y + z;

a + b + c

6 x + 7 y + 10 z
```

1.7 Getting Help

There are many sources of help in *Mathematica*. First and foremost is the Documentation Center (as shown in the following figure) available from the Help menu. There you will find all available commands grouped by topic, or you can search for the help you need by typing in a few keywords. The Function Navigator contains a listing of all the functions available in *Mathematica* arranged by topic, and the entire *Mathematica* manual may be accessed by going to the Virtual Book.

The help files contain numerous examples that you may want to explore. Feel free to make any changes in the help files without fear of modifying their content. These files are protected and your changes will not be permanent.

If you know the name of the command you want, you can use a question mark, **?**, followed by the name of the command to determine its syntax. More extensive information about the command, including attributes and options, can be obtained using **??**. Or you can type the name of the command, place the cursor within its name, and then press F1. You will be taken to a page with a complete description and illustrative examples.

Occasionally, when you make an error, *Mathematica* will beep or the cell will change color. If you are not sure what you did to cause this, you can get a clue by going to Help ⇒ Why The Beep? or Help ⇒ Why The Coloring?

EXAMPLE 20   Suppose you know that the command **Plot** graphs a function, but you cannot remember its syntax.

```
?Plot
```

Plot[f, {x, x_{min}, x_{max}}] generates a plot of f as a function of x from x_{min} to x_{max}.
Plot[{f_1, f_2,...}, {x, x_{min}, x_{max}}] plots several functions f_i. ≫

If information is needed about attributes or optional settings (and their defaults), **??** can be used.

```
??Plot
```

Plot[f, {x, x_{min}, x_{max}}] generates a plot of f as a function of x from x_{min} to x_{max}.
Plot[{ f_1, f_2,...}, {x, x_{min}, x_{max}}] plots several functions f_i. ≫

```
Attributes[Plot]={HoldAll,Protected}

Options[Plot]={AlignmentPoint → Center, AspectRatio → 1/GoldenRatio,
  Axes → True, AxesLabel → None, AxesOrigin → Automatic, AxesStyle → {},
  Background → None, BaselinePosition → Automatic, BaseStyle → {},
  ClippingStyle → None, ColorFunction → Automatic, ColorFunctionScaling → True,
  ColorOutput → Automatic, ContentSelectable → Automatic,
  DisplayFunction :→ $DisplayFunction, Epilog → {},
  Evaluated → System`Private`$Evaluated, EvaluationMonitor → None,
  Exclusions → Automatic, ExclusionsStyle → None, Filling → None,
  FillingStyle → Automatic, FormatType :→ TraditionalForm, Frame → False,
  FrameLabel → None, FrameStyle → {}, FrameTicks → Automatic,
  FrameTicksStyle → {}, GridLines → None, GridLinesStyle → {},
  Imagemargins → 0., ImagePadding → All, ImageSize → Automatic,
  LabelStyle → {}, MaxRecursion → Automatic, Mesh → None,
  MeshFunctions → {#1 &}, MeshShading → None, MeshStyle → Automatic,
  Method → Automatic, PerformanceGoal :→ $PerformanceGoal,
  PlotLabel → None, PlotPoints → Automatic, PlotRange → {Full, Automatic},
  PlotRangeClipping → True, PlotRangePadding → Automatic,
  PlotRegion → Automatic, PlotStyle → Automatic,
  PreserveImageOptions → Automatic, Prolog → {}, RegionFunction → (True &),
  RotateLabel → True, Ticks → Automatic, TicksStyle → {},
  WorkingPrecision → MachinePrecision}
```

Options can also be obtained using the Options command. This is useful if you want to specify an option but cannot remember its name.

EXAMPLE 21

```
Options[Solve]
{InverseFunctions → Automatic, MakeRules → False, Method → 3, Mode → Generic,
  Sort → True, VerifySolutions → Automatic, WorkingPrecision → ∞}
```

Very often you may remember part of a symbol name, but not the whole name. If you know the beginning is "Arc," for example, type in the part you know and then press **[CTRL]** + **[K]**. This will generate a menu of all commands and functions beginning with Arc. Then click on the one you want. If you are using a Macintosh computer, use **[COMMAND]** + **[K]**. (The **[COMMAND]** key is the key with the apple on it.)

EXAMPLE 22 Type `Arc` and then press [CTRL] + [K] or [COMMAND] + [K].

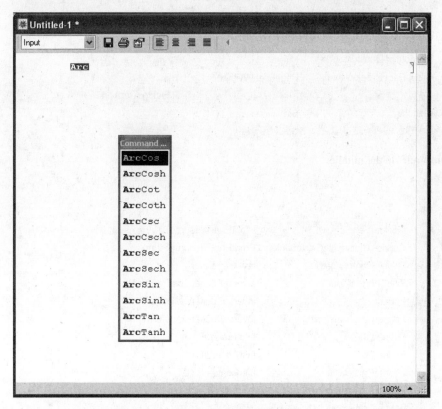

Another way of determining symbol names is to use **?** together with wildcards. The character " **∗** " acts as a "wildcard" and takes the place of any sequence of characters. Wildcards can be used anywhere, at the beginning, middle, or end of a symbol.

EXAMPLE 23 Output may vary depending upon your version of *Mathematica*

(a) Find all commands beginning with "Inv."

`?Inv*`

▼ **System`**

Inverse	InverseFunctions	InverseJacobiNS
InverseBetaRegularized	InverseGammaRegularized	InverseJacobiSC
InverseCDF	InverseGaussianDistribution	InverseJacobiSD
InverseEllipticNomeQ	InverseJacobiCD	InverseJacobiSN
InverseErf	InverseJacobiCN	InverseLaplaceTransform
InverseErfc	InverseJacobiCS	InverseSeries
InverseFourier	InverseJacobiDC	InverseWeierstrassP
InverseFourierCosTransform	InverseJacobiDN	InverseZTransform
InverseFourierSinTransform	InverseJacobiDS	Invisible
InverseFourierTransform	InverseJacobiNC	InvisibleApplication
InverseFunction	InverseJacobiND	InvisibleTimes

▼ **WebServices`**

InvokeServiceOperation

(b) Find all commands ending with "in."

`? *in`

▼ System`

ArcSin	DistributionDomain	Min	StringJoin
AxesOrigin	GroupPageBreakWithin	PageBreakWithin	Thin
Begin	Join	Plain	$MachineDomain
CoefficientDomain	Khinchin	Sin	
ConstrainedMin	LineBreakWithin	StackBegin	

(c) Find all commands with "our" in the middle.

`? *our*`

▼ System`

ButtonSource	ContourStyle	FrontEndResource
ClockwiseContourIntegral	CounterClockwiseContourIntegr	FrontEndResourceString
ContourGraphics	DoubleContourIntegral	InverseFourier
ContourIntegral	FindShortestTour	InverseFourierCosTransform
ContourLabels	Fourier	InverseFourierSinTransform
ContourLines	FourierCosTransform	InverseFourierTransform
ContourPlot	FourierDCT	LightSources
ContourPlot3D	FourierDST	ListContourPlot
Contours	FourierParameters	ListContourPlot3D
ContourShading	FourierSinTransform	$FinancialDataSource
ContourSmoothing	FourierTransform	

▼ PacletManager`

PacletResource

▼ ResourceLocator`

ResourceAdd	ResourcesLocate

Wildcards can also be used to determine which symbols have been used thus far by the kernel. Typing `? ` *` returns a list of all symbols that have been defined during your *Mathematica* session. The character ` (backquote) stands for global—you want a list of all *global* symbols. (See the appendix for a discussion of global symbols.)

EXAMPLE 24   Note: The results of this example may be slightly different on your computer, depending upon the symbols you have defined.

```
a = 3;
b2xy = 4;
xyz7 = 5;
? ` *
```

▶ Global`

a	b2xy	xyz7

`Clear["` * "]` will clear all global symbols. `Remove["`*"]` will remove all global symbols.

EXAMPLE 25

```
Remove["`*"]
? ` *              ← Check to see if any symbols remain.
```

Remove :: rmnsm : There are no symbols matching "`*". ≫

SOLVED PROBLEMS

1.28 Obtain basic information about the *Mathematica* command `Simplify`.

SOLUTION

`? Simplify`

Simplify[*expr*] performs a sequence of algebraic and other transformations on *expr*, and returns the simplest form it finds.
Simplify[*expr*, *assum*] does simplification using assumptions. »

1.29 Obtain extended information about the *Mathematica* command `Simplify` including default settings for options.

SOLUTION

`?? Simplify`

Simplify[*expr*] performs a sequence of algebraic and other transformations on *expr*, and returns the simplest form it finds.
Simplify[*expr*, *assum*] does simplification using assumptions. »

```
Attributes[Simplify]={Protected}

Options[Simplify]={Assumptions:→$Assumptions,
ComplexityFunction→Automatic,ExcludedForms→{},TimeConstraint→300,
TransformationFunctions→Automatic,Trig→True}
```

1.30 Obtain help on the *Mathematica* command `Factor` and then factor $x^3 - 6x^2 + 11x - 6$.

SOLUTION

`?Factor`

Factor[*poly*] factors a polynomial over the integers.
Factor[*poly*, Modulus → *p*] factors a polynomial modulo a prime *p*.
Factor[*poly*, Extension → {$a_1, a_2,...$}] factors a polynomial allowing coefficients that are rational combinations of the algebraic numbers a_i. »

```
Factor[x³ – 6x² + 11x – 6]
(–3 + x) (–2 + x) (–1 + x)
```

1.31 Find all *Mathematica* commands beginning with "Abs."

SOLUTION

`?Abs*`

▼ System`

Abs	AbsoluteOptions	AbsoluteTime
AbsoluteCurrentValue	AbsolutePointSize	AbsoluteTiming
AbsoluteDashing	AbsoluteThickness	

1.32 Find all *Mathematica* commands beginning with "Si" and ending with "al."

SOLUTION

`?Si*al`

▼ System`

SinhIntegral	SinIntegral

1.33 Find all *Mathematica* commands beginning with "Fi."

SOLUTION

`?Fi*`

▼ System`

Fibonacci	FileNameSetter	FindMaximum
FieldMasked	FilePrint	FindMinimum
FieldSize	FileType	FindRoot
File	Filling	FindSettings
FileByteCount	FillingStyle	FindShortestTour
FileDate	FilterRules	FinishDynamic
FileFormat	FinancialData	First
FileHash	Find	Fit
FileInformation	FindClusters	FitAll
FileName	FindFit	FixedPoint
FileNameDialogSettings	FindInstance	FixedPointList
FileNames	FindList	

▼ JLink`

FieldFunction	Fields

1.34 Find all *Mathematica* commands beginning with "Fi" and ending with "t."

SOLUTION

`?Fi*t`

▼ System`

FileByteCount	FindFit	First	FixedPointList
FileFormat	FindList	Fit	
FilePrint	FindRoot	FixedPoint	

1.8   Packages

There are many specialized functions and procedures that are not loaded when *Mathematica* is initially invoked. Rather, they must be loaded separately from files in the *Mathematica* directory on the hard drive. These files are of the form *filename*.m.

EXAMPLE 26   A map of the world can be obtained from the command **WorldPlot** which is located in the package **WorldPlot`**. To load this command, simply type (note the ` at the end)

    ≪WorldPlot`**   or   **Needs["WorldPlot`"]

The appropriate command can then be accessed.

`WorldPlot[World]`

Once a package is loaded you can get a list of the functions it contains by using the **Names** command.

EXAMPLE 27 (Continuation of Example 26)

Names["WorldPlot`*"]

{Africa, Albers, Asia, ContiguousUSStates, Equirectangular, Europe,
LambertAzimuthal, LambertCylindrical, Mercator, MiddleEast, Mollweide,
NorthAmerica, Oceania, Orthographic, RandomColors, RandomGrays, ShowTooltips,
Simple, Sinusoidal, SouthAmerica, ToMinutes, USData, USStates, World,
WorldBackground, WorldBorders, WorldClipping, WorldCountries, WorldData,
WorldDatabase, WorldFrame, WorldFrameParts, WorldGraphics, WorldGrid,
WorldGridBehind, WorldGridStyle, WorldPlot, WorldPoints, WorldProjection,
WorldRange, WorldRotatedRange, WorldRotation, WorldToGraphics}

EXAMPLE 28 The package **Calendar`** includes some interesting calendar functions.

≪Calendar`

Names["Calendar`*"]

{Calendar, CalendarChange, DateQ, DayOfWeek, DaysBetween, DaysPlus,
EasterSunday, EasterSundayGreekOrthodox, Friday, Gregorian, Islamic, Jewish,
JewishNewYear, Julian, Monday, Saturday, Sunday, Thursday, Tuesday, Wednesday}

?DaysBetween

DaysBetween[{$year_1$, $month_1$, day_1}, {$year_2$, $month_2$, day_2}] gives the number of days between the dates {$year_1$, $month_1$, day_1} and {$year_2$, $month_2$, day_2}.
DaysBetween[{$year_1$, $month_1$, day_1, $hour_1$, $minute_1$, $second_1$}, {$year_2$, $month_2$, day_2, $hour_2$, $minute_2$, $second_2$}] gives the number of days between the given dates. ≫

DaysBetween[{2007,8,3},{2008,12,5}]

490

SOLVED PROBLEMS

1.35 The function **DayOfWeek** appears in the package **Calendar`** and gives the day of the week of any date in the calendar. Load the package, obtain help to determine its syntax, and then determine which day of the week January 1, 2000, was.

SOLUTION

≪Calendar`

? DayOfWeek

DayOfWeek[{$year$, $month$, day}] gives the day of the week on which the given date {$year$, $month$, day} occurred.
DayOfWeek[{$year$, $month$, day, $hour$, $minute$, $second$}] gives the day of theweek for the given date. ≫

DayOfWeek[{2000, 1, 1}]

Saturday

1.36 The package **Combinatorica`** contains functions in combinatorics and graph theory. One of these is **KSubsets**, which lists all subsets of size *k* of a given set. Load the package and execute **Ksubsets[{1,2,3,4,5},3]**.

SOLUTION

≪Combinatorica`

KSubsets[{1, 2, 3, 4, 5}, 3]

{{1, 2, 3}, {1, 2, 4}, {1, 2, 5}, {1, 3, 4}, {1, 3, 5},
{1, 4, 5}, {2, 3, 4}, {2, 3, 5}, {2, 4, 5}, {3, 4, 5}}

1.9   A Preview of What Is to Come

If you have just purchased your copy of *Mathematica*, you probably cannot wait to give it a test run. The following examples are a collection of problems for you to try. What follows are some basic commands. To keep things simple, options have been omitted and *Mathematica*'s defaults are used exclusively. We will discuss modifications to these commands in subsequent chapters, but for now, just have fun!

EXAMPLE 29   Obtain a 50 significant digit approximation to $\sqrt{\pi}$.

$\mathbf{N[\sqrt{\pi}, 50]}$ or $\mathbf{N[Sqrt[Pi], 50]}$

1.7724538509055160272981674833411451827975494561224

EXAMPLE 30   Solve the algebraic equation $x^3 - 2x + 1 = 0$.

$\mathbf{Solve[x^3 - 2x + 1 == 0]}$ or $\mathbf{Solve[x^3 - 2x + 1 == 0]}$

$$\left\{ \{x \to 1\}, \left\{x \to \tfrac{1}{2}\left(-1 - \sqrt{5}\right)\right\}, \left\{x \to \tfrac{1}{2}\left(1 + \sqrt{5}\right)\right\} \right\}$$

EXAMPLE 31   Express $(x + 1)^{10}$ in traditional polynomial form.

$\mathbf{Expand[(x + 1)^{10}] \,//TraditionalForm}$

$x^{10} + 10x^9 + 45x^8 + 120x^7 + 210x^6 + 252x^5 + 210x^4 + 120x^3 + 45x^2 + 10x + 1$

EXAMPLE 32   What is the 1000th prime?

$\mathbf{Prime[1000]}$

7919

EXAMPLE 33   The function **ElementData** gives values of chemical and physical properties of elements. Among the properties included are **AtomicWeight** and **AtomicNumber**, whose definitions are self-explanatory. Compute the atomic weight and atomic number of titanium. (Note the quotation marks.)

$\mathbf{ElementData["Titanium","AtomicWeight"]}$

47.867

$\mathbf{ElementData["Titanium","AtomicNumber"]}$

22

EXAMPLE 34   Plot the graph of $y = \sin x$ from 0 to 2π.

$\mathbf{Plot[Sin[x], \{x, 0, 2\pi\}]}$

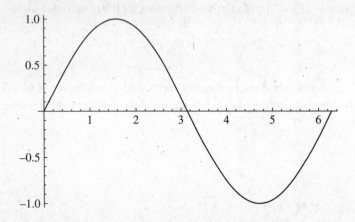

EXAMPLE 35 Sketch the graphs of $y = \sin x$, $y = \sin 2x$, and $y = \sin 3x$, $0 \le x \le 2\pi$, on one set of axes.

```
Plot[{Sin[x],Sin[2x],Sin[3x]},{x,0,2π}]
```

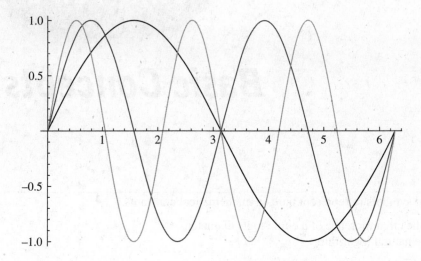

EXAMPLE 36 Sketch the three-dimensional surface defined by $z = (x^2 + 3y^2)e^{-(x^2+y^2)}$.

```
Plot3D[(x² + 3y²)e^(-(x²+y²)),{x,-3,3},{y,-3,3}]  or
Plot3D[(x^2 + 3y^2) * Exp[-(x^2 + y^2)],{x,-3,3},{y,-3,3}]
```

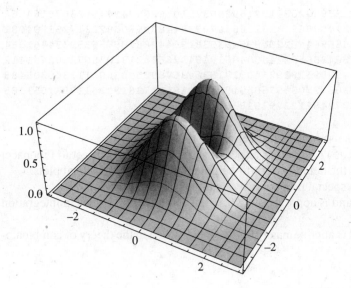

Click on the graph and drag the mouse to view the graph from any viewpoint.

CHAPTER 2

Basic Concepts

2.1 Constants

Mathematica uses predefined symbols to represent built-in mathematical constants.

- **Pi** or π is the ratio of the circumference of a circle to its diameter.
- **E** or **e** is the base of the natural logarithm.

Both **Pi** and **E** are treated symbolically and do not have values, as such. However, they may be approximated to any degree of precision.

EXAMPLE 1   **N[π, 500]** will produce a 500 significant digit approximation to π (499 decimal places).

```
N[π,500]
```

```
3.14159265358979323846264338327950288419716939937510582097494459230781640628620899862803482534211706798214808651328230664709384460955058223172535940812848111745028410270193852110555964462294895493038196442881097566593344612847564823337867831652712019091456485669234603486104543266482133936072602491412737245870066063155881748815209209628292540917153643678925903600113305305488204665213841469519415116094330572703657595919530921861173819326117931051185480744623799627495673518857527248912279381830119491
```

- **Degree** is equal to **Pi/180** and is used to convert degrees to radians.
- **GoldenRatio** has the value $(1 + \sqrt{5})/2$ and has a special significance with respect to Fibonacci series. It is used in *Mathematica* as the default width-to-height ratio of two-dimensional plots.
- **Infinity** or ∞ is a constant with special properties. For example, ∞ + 1 = ∞.
- **EulerGamma** is Euler's constant and is approximately 0.577216. It has applications in integration and in asymptotic expansions.
- **Catalan** is Catalan's constant and is approximately 0.915966. It is used in the theory of combinatorial functions.

EXAMPLE 2   How much is ∞ + ∞ ?

```
∞ + ∞
∞
```

SOLVED PROBLEMS

2.1 Approximately how many radians are in 90 degrees?

SOLUTION

```
90 Degree //N        ← expression //N is the same as N[expression]
1.5708
```

2.2 Show that `GoldenRatio` satisfies the algebraic equation $x^2 - x - 1 = 0$.

SOLUTION

```
x = GoldenRatio;
x^2 - x - 1 //N
0.
```

2.3 What happens if Zachary tries to subtract ∞ from ∞?

SOLUTION

```
∞ - ∞
```

∞::indet : Indeterminate expression $-\infty + \infty$ encountered. $\gg$

```
Indeterminate
```

2.4 Compute a 20 decimal place approximation to e, the base of the natural logarithm.

SOLUTION

```
N[E,21]   or   N[e,21]
2.71828182845904523536
```

2.2 "Built-In" Functions

In this section we discuss some of the more commonly used functions *Mathematica* offers. Because of the vast number of functions available, no attempt is made toward completeness. Additional functions are discussed in detail in later chapters.

Standard mathematical functions can be accessed by name or by clicking on their symbol in a *Mathematica* palette. For example, the square root of a number can be obtained using either the function `Sqrt` or, alternatively, by using the $\sqrt{\ }$ symbol from the Basic Math Input palette. Remember that the argument of a function must be contained within square brackets, [ ].

- `Sqrt[x]` or $\sqrt{x}$ gives the non-negative square root of x.

EXAMPLE 3

```
Sqrt[1521]   or   √1521
39
```

Higher order roots can be computed by recalling that $\sqrt[n]{x} = x^{\frac{1}{n}}$. The symbol $\sqrt[\square]{\ }$ on the Basic Math Input palette may also be used. Notice that higher order roots of negative numbers are given in a special format.

EXAMPLE 4   The cube root of 8 is given directly, but the cube root of –8 is given in terms of $\sqrt[3]{-1}$.

```
8^(1/3)   or   ∛8
2
(-8)^(1/3)   or   ∛-8
2 (-1)^(1/3)
```

EXAMPLE 5

```
N[√2]
1.41421
N[√2,50]
1.4142135623730950488016887242096980785696718753769
```

The function that returns the absolute value of *x*, |*x*|, is **Abs**.

- **Abs[x]** returns x if x ≥ 0 and −x if x < 0.

The function **Abs** can also be applied to complex numbers. If **z** is the complex number x + y i, **Abs[z]** returns its modulus, $\sqrt{x^2 + y^2}$.

EXAMPLE 6

```
Abs[5]
5
Abs[-5]
5
Abs[5 + 12i]
13
```

It is sometimes useful to have a function that determines the sign of a number.

- **Sign[x]** returns the values −1, 0, 1 depending upon whether x is negative, 0, or positive, respectively.

EXAMPLE 7

```
Sign[-27.5]
-1
Sign[0]
0
Sign[6.254]
1
```

The factorial of a positive integer, *n*, represented *n*! in mathematical literature, is the product of the integers 1, 2, 3, ... , *n*. By definition, 0! = 1. For non-integer values of *n*, *n*! is defined by Γ(*n*+1) where Γ is Euler's gamma function.

- **Factorial[n]** or **n!** gives the factorial of n if n is a positive integer and Γ(n + 1) if n has a non-integer positive value.

EXAMPLE 8

```
5!
120
0!
1

Factorial[3.5]
11.6317
```

Mathematica has a built-in random number generator. This is a useful function in probability theory and statistical analysis, e.g., random walks and Monte Carlo methods.

- **Random[ ]** gives a uniformly distributed real pseudorandom number in the interval [0, 1].
- **Random[*type*]** returns a uniformly distributed pseudorandom number of type *type*, which is either Integer, Real, or Complex. Its values are between 0 and 1, in the case of Integer or Real, and are contained within the square determined by 0 and 1+i, if *type* is Complex.
- **Random[*type, range*]** gives a uniformly distributed pseudorandom number in the interval or rectangle determined by *range*. *range* can be either a single number or a list of two numbers such as {a,b} or {a+bI, c+dI}. A single number, m, is equivalent to {0,m}.
- **Random[*type, range, n*]** gives a uniformly distributed pseudorandom number to n significant digits in the interval or rectangle determined by *range*.

Mathematica also offers the functions **RandomReal**, **RandomInteger**, and **RandomComplex** to generate pseudorandom numbers.

- **RandomReal[ ]** returns a pseudorandom real number between 0 and 1.
- **RandomReal[xmax]** returns a pseudorandom real number between 0 and xmax.
- **RandomReal[{xmin, xmax}]** returns a pseudorandom real number between xmin and xmax.
- **RandomReal[{xmin, xmax}, n]** returns a list of n pseudorandom real numbers between xmin and xmax.
- **RandomReal[{xmin, xmax}, {m, n}]** returns an $m \times n$ list of pseudorandom numbers between xmin and xmax. This extends in a natural way to lists of higher dimension. (See Chapter 3 for a complete discussion of lists.)

The definitions of **RandomInteger** and **RandomComplex** are similar to **RandomReal** and may be looked up in the Documentation Center.

- **RandomSample[{$e_1, e_2, \ldots, e_n$}, k]** gives a pseudorandom sample of k of the e_i.
- **RandomSample[{$e_1, e_2, \ldots, e_n$}]** gives a pseudorandom permutation of the list of e_i.

Any random number generator produces its output from an algorithm based upon an initial value, called a *seed*. *Mathematica* allows you to introduce a seed using the function **SeedRandom**.

- **SeedRandom[n]** initializes the random number generator using n as a seed. This guarantees that sequences of random numbers generated with the same seed will be identical.
- **SeedRandom[ ]** initializes the random number generator using the time of day and other attributes of the current *Mathematica* session.

EXAMPLE 9   (Your answers will be different from those shown.)

`Random[Integer]` `0`	← Returns 0 or 1 with equal probability.
`Random[Real]` `0.386297`	← Returns a 6 significant digit real number between 0 and 1.
`Random[Complex]` `0.420851 + 0.382187i`	← Returns a complex number in the square whose opposite vertices are 0 and 1+i.
`Random[Real,5]` `1.83872`	← Returns a real number uniformly distributed in the interval [0,5].
`Random[Real,{3,5}]` `3.95386`	← Returns a real number uniformly distributed in the interval [3,5].
`Random[Real,{3,5},10]` `4.014673296`	← Returns a real number uniformly distributed in the interval [3,5] to 10 significant digits.
`Random[Integer,{1,10}]` `7`	← Returns an integer between 1 and 10 with equal probability 1/10.
`Random[Complex,{2+I,5+6 I}]` `2.61319 + 4.30869i`	← Returns a complex number in the rectangle whose opposite vertices are the complex numbers 2+i and 5+6i.

```
RandomReal[{3, 5}]
3.62039
RandomInteger[{3, 10}, 20]
{6, 5, 7, 5, 3, 7, 10, 4, 9, 7, 5, 9, 8, 5, 4, 10, 4, 3, 9, 3}
RandomSample[{1, 2, 3, 4, 5, 6, 7, 8, 9, 10}, 5]
{2, 8, 3, 1, 10}
```

A positive integer is prime if it is divisible only by itself and 1. For technical reasons, 1 is not considered prime; the smallest prime is 2.

- `Prime[n]` returns the nth prime.
- `RandomPrime[n]` returns a pseudorandom prime number between 2 and n.
- `RandomPrime[{m, n}]` returns a pseudorandom prime number between m and n.
- `RandomPrime[{m, n}, k]` returns a list of k pseudorandom primes, each between m and n.

EXAMPLE 10   Find the 7th prime.

```
Prime[7]
17

RandomPrime[{7, 47}]
29

RandomPrime[{7, 47}, 10]
{31, 29, 41, 47, 43, 13, 31, 17, 37, 7}
```

The Fibonacci numbers are defined by

$$f_1 = 1,$$
$$f_2 = 1,$$
$$f_n = f_{n-2} + f_{n-1} \qquad n \geq 3$$

Thus, the first few Fibonacci numbers are 1, 1, 2, 3, 5, 8, 13, 21, ...

- `Fibonacci[n]` returns the nth Fibonacci number.

EXAMPLE 11

```
Fibonacci[7]
13
```

There are three *Mathematica* functions that convert real numbers to nearby integers.

- `Round[x]` returns the integer closest to x. If x lies exactly between two integers (e.g., 5.5), Round returns the nearest even integer.
- `Floor[x]` returns the greatest integer which does not exceed x. This is sometimes known as the "greatest integer function" and is represented in many textbooks by $\lfloor x \rfloor$.
- `Ceiling[x]` returns the smallest integer not less than x. Many textbooks represent this by $\lceil x \rceil$.

EXAMPLE 12

```
Round[5.75]
6

Floor[5.75]
5

Ceiling[5.75]
6
```

A decimal number can be broken up into two parts, the integer portion (number to the left of the decimal point) and the fractional portion.

- `IntegerPart[x]` gives the integer portion of x (decimal point excluded).
- `FractionalPart[x]` gives the fractional portion of x (decimal point included).

Observe that `IntegerPart[x] + FractionalPart[x] = x`.

EXAMPLE 13

```
IntegerPart[4.67]
4
FractionalPart[4.67]
0.67
IntegerPart[4.67] + FractionalPart[4.67]
4.67
```

If *m* and *n* are positive integers, there exist unique integers *q* and *r* such that

$$m = qn + r \quad \text{with} \quad 0 \leq r < n$$

This result is known as the *Division Algorithm*. *q* is called the *quotient* and *r* is the *remainder*. The *Mathematica* functions **Quotient** and **Mod** return the quotient and remainder, respectively.

- **Quotient[m, n]** returns the quotient when m is divided by n.
- **Mod[m, n]** returns the remainder when m is divided by n.

EXAMPLE 14

```
Quotient[17, 3]
5
Mod[17, 3]
2
```

Suppose *a* and *b* are two integers. If there exists an integer, *k*, such that *a = kb*, we say that *b* divides *a*. Alternatively, *a* is a multiple of *b*.

Let *m* and *n* be two integers. If *b* divides both *m* and *n*, we say that *b* is a common divisor of *m* and *n*. The largest common divisor of *m* and *n* is called their *greatest common divisor* (GCD).

If *a* is a multiple of *both m* and *n*, we say *a* is a common multiple of *m* and *n*. The smallest common multiple of *m* and *n* is called their *least common multiple* (LCM).

- **GCD[m, n]** returns the greatest common divisor of m and n.
- **LCM[m, n]** returns the least common multiple of m and n.

The functions **GCD** and **LCM** extend to more than two arguments.

EXAMPLE 15 Find the greatest common divisor and least common multiple of 24, 40, and 48.

```
GCD[24, 40, 48]
8
LCM[24, 40, 48]
240
```

The *Fundamental Theorem of Arithmetic* guarantees that every positive integer can be factored into primes in a unique way.

- The function **FactorInteger[n]** gives the prime factors of n together with their respective exponents.

EXAMPLE 16

```
FactorInteger[2 381 400]
{{2, 3}, {3, 5}, {5, 2}, {7, 2}}
```

The prime factors of 2,381,400 are 2, 3, 5, and 7 with exponents, respectively, 3, 5, 2, 2. In other words, $2,381,400 = 2^3 3^5 5^2 7^2$. The result of this operation produces a nested sequence of *lists*. (A list is a *Mathematica* object, enclosed within braces, { }, which will be discussed in detail in Chapter 3.)

In order to estimate computational efficiency, it is useful to be able to determine how long an operation or sequence of operations takes to execute.

- **Timing [*expression*]** evaluates *expression*, and returns a list of time used, in seconds, together with the result obtained.

Timing counts only the CPU time spent in the *Mathematica* kernel. It does not include overhead time spent in the front end.

EXAMPLE 17   How long does it take the kernel to compute the ten billionth prime?

Timing [Prime [10 000 000 000]]

{2.953,252 097 800 623}

Of course, the actual time taken will vary, depending upon the speed of the CPU.

Logarithms and exponential functions to any base can be computed using the function **Log**.

- **Log [x]** represents the natural logarithm. If a base, *b*, other than *e* is required, the appropriate form is **Log [b, x]**.
- The function **Exp [x]** is the natural exponential function. Other equivalent forms are **E^x** and **E^x**. Lowercase e cannot be used, but the special symbol e from the Basic Math Input palette may be used instead. Exponential functions to the base b are computed by **b^x** or **b^x**.

EXAMPLE 18   Compute ln 100, the natural logarithm of 100.

Log [100]
Log [100]

> Observe that *Mathematica* always gives *exact* answers. Approximations are supplied only when requested.

Log [100] //N
4.60517

EXAMPLE 19   Compute $\log_2 100$.

Log [2, 100]

$$\frac{\text{Log [100]}}{\text{Log [2]}}$$ ← This is the exact value of $\log_2 100$, expressed in terms of natural logarithms.

Log [2, 100] //N
6.64386

EXAMPLE 20   To compute a numerical approximation of e^2, we can write

Exp [2] //N   or   **E^2//N**   or   **e^2//N**
7.38906

- The six basic trigonometric functions, sine, cosine, tangent, secant, cosecant, and cotangent, are represented in *Mathematica* by **Sin**, **Cos**, **Tan**, **Sec**, **Csc**, and **Cot**, respectively.

Mathematica assumes the arguments of trigonometric functions to be in radians. Problems involving degrees must first be converted to radians if trigonometric functions are involved. For this purpose, one can use the built-in constant, **Degree**, whose value is $\pi/180$. The symbol °, located on the Basic Math Input palette, may be used as well.

EXAMPLE 21   60° is equivalent to $\pi/3$ radians. To compute its *sin* using radian measure, we write

Sin$\left[\frac{\pi}{3}\right]$ or   **Sin [Pi/3]**

$$\frac{\sqrt{3}}{2}$$

If we wish to compute its *sin* using degree measure, we can type

`Sin[60 Degree]`   or   `Sin[60°]`

$$\frac{\sqrt{3}}{2}$$

Care must be taken with trigonometric powers. The square of sin *x* in trigonometric form is traditionally written sin² *x*, but *Mathematica* will accept only `Sin[x]²` or `Sin[x]^2`.

EXAMPLE 22   Compute the square of sin 60°.

`Sin[60°]²`   or   `Sin[60 Degree]^2`

$$\frac{3}{4}$$

- The inverse trigonometric functions are `ArcSin`, `ArcCos`, `ArcTan`, `ArcSec`, `ArcCsc`, and `ArcCot`. However only the *principal values*, expressed in radians, are returned by these functions.

EXAMPLE 23

`ArcSin[1]`

$$\frac{\pi}{2}$$

`ArcCos[Cos[3π]]`

π

`Cos[3π] = -1` but the principal value of `ArcCos[-1]` is π.

Hyperbolic functions are combinations of exponential functions which have interesting mathematical properties. There are six hyperbolic functions. The three basic ones are

$$\sinh x = \frac{e^x - e^{-x}}{2} \qquad \cosh x = \frac{e^x + e^{-x}}{2} \qquad \tanh x = \frac{e^x - e^{-x}}{e^x + e^{-x}}$$

The other three, sech *x*, csch *x*, and coth *x*, are reciprocals, respectively, of cosh *x*, sinh *x*, and tanh *x*.

- The *Mathematica* representations of the six hyperbolic functions are `Sinh`, `Cosh`, `Tanh`, `Sech`, `Csch`, and `Coth`.

EXAMPLE 24   Compute a numerical approximation to sinh 2.

`Sinh[2]//N`

3.62686

- The inverse hyperbolic functions are represented by `ArcSinh`, `ArcCosh`, `ArcTanh`, `ArcSech`, `ArcCsch`, and `ArcCoth`.

Because `Cosh` and `Sech` are not one-to-one, `ArcCosh` and `ArcSech` return only positive values for real arguments.

EXAMPLE 25

`ArcSinh[-2] //N`

−1.44364

`ArcCosh[2] //N`

1.31696

One special command is worthy of mention at this time:

- `Print[expression]` prints *expression*, followed by a line feed.
- `Print[expression1, expression2, ...]` prints *expression1, expression2, ...* followed by a single line feed.

At first glance it may seem that **Print** is a redundant command, as simply typing the name of any object will reveal its value. However, it has a useful purpose (e.g., see loops in Section 2.8).

EXAMPLE 26

```
Print["This prints a line of text."]
This prints a line of text.
```

EXAMPLE 27

```
a = 1; b = 2; c = 3; d = 4; e = 5;
Print[a + b, b + c, c + d, d + e, e + a]
35796
```

Mathematica includes a class of functions ending in the letter Q:

AlgebraicIntegerQ	LegendreQ	PositiveDefiniteMatrixQ
AlgebraicUnitQ	LetterQ	PossibleZeroQ
ArgumentCountQ	LinkConnectedQ	PrimePowerQ
ArrayQ	LinkReadyQ	PrimeQ
AtomQ	ListQ	QuadraticIrrationalQ
CoprimeQ	LowerCaseQ	RootOfUnityQ
DigitQ	MachineNumberQ	SameQ
DistributionDomainQ	MatchLocalNameQ	SquareFreeQ
DistributionParameterQ	MatchQ	StringFreeQ
EllipticNomeQ	MatrixQ	StringMatchQ
EvenQ	MemberQ	StringQ
ExactNumberQ	NameQ	SyntaxQ
FreeQ	NumberQ	TensorQ
HermitianMatrixQ	NumericQ	TrueQ
HypergeometricPFQ	OddQ	UnsameQ
InexactNumberQ	OptionQ	UpperCaseQ
IntegerQ	OrderedQ	ValueQ
IntervalMemberQ	PartitionsQ	VectorQ
InverseEllipticNomeQ	PolynomialQ	

These functions are used to test for certain conditions and return a value of **True** or **False**. Their precise syntax can be determined from the Help menu or by using **?** as illustrated in the next examples.

EXAMPLE 28

```
?PrimeQ
```

PrimeQ[*expr*] yields True if *expr* is a prime number, and yields False otherwise. »

```
PrimeQ[5]
True
PrimeQ[6]
False
```

EXAMPLE 29

```
?PolynomialQ
```

PolynomialQ[*expr, var*] yields True if *expr* is a polynomial in *var*, and yields False otherwise.
PolynomialQ[*expr*, {*var1*,...}] tests whether *expr* is a polynomial in the *var$_i$*. »

```
PolynomialQ[x² y + x + √y, x]
```
True

```
PolynomialQ[x² y + x + √y, y]
```
False

SOLVED PROBLEMS

2.5 Compute numerical approximations to the square root and cube root of 10.

SOLUTION

```
√10 //N   or   Sqrt[10] //N
```
3.16228
```
∛10 //N   or   10^(1/3) //N
```
2.15443

2.6 Compute numerical approximations to the square root and cube root of 10 accurate to 20 significant digits.

SOLUTION

$$N\left[\sqrt{10}, 20\right]$$
3.1622776601683793320

$$N\left[\sqrt[3]{10}, 20\right]$$
2.1544346900318837218

2.7 Compute $\sqrt{3} + \sqrt{2}$ and $\sqrt{3} - \sqrt{2}$ to 50 significant digits. Then compute their product.

SOLUTION

$$a = N\left[\sqrt{3} + \sqrt{2}, 50\right]$$
3.1462643699419723423291350657155704455124771291873

$$b = N\left[\sqrt{3} - \sqrt{2}, 50\right]$$
0.31783724519578224472575761729617428837313337843343

```
a*b
```
1.0000000000000000000000000000000000000000000000000

2.8 The binomial coefficient $C(n,k) = \dfrac{n!}{k!(n-k)!}$. Use this definition to compute $C(10,4)$.

SOLUTION

$$\frac{10!}{4!(10-4)!}$$ or   `Factorial[10]/(Factorial[4] * Factorial[10 - 4])`

210

2.9 A fair die has six faces, numbered 1 through 6, and each occurs with equal probability. Simulate four tosses of a fair die.

SOLUTION

(Your answers will be different from those shown here.)

```
Random[Integer, {1, 6}]
Random[Integer, {1, 6}]
Random[Integer, {1, 6}]
```

```
Random[Integer, {1, 6}]
6
1
5
3
```

2.10 Find a 15 significant digit pseudorandom real number between π and 2π.

SOLUTION

(Your answer will be different from that shown here.)

```
Random[Real, {π, 2π}, 15]
4.13129131207734
```

2.11 What is the 27th Fibonacci number?

SOLUTION

```
Fibonacci[27]
196 418
```

2.12 Show that there is no prime between 157 and 163.

SOLUTION

```
Prime[37]        ←We determine this by experimentation.
157
Prime[38]
163
```

Since 157 and 163 are consecutive primes, there is no prime between them.

2.13 What is the integer closest to $\sqrt{159}$?

SOLUTION

```
Round[Sqrt[159]]   or   √159 //Round
13
```

2.14 Between what two consecutive integers does $(\pi^2 + 1)^5$ lie?

SOLUTION

```
Floor[(π² + 1)⁵]
151 729
Ceiling[(π² + 1)⁵]
151 730
```

The number $(\pi^2 + 1)^5$ lies between 151,729 and 151,730.

2.15 Compute the value of $\lceil x \rceil - \lfloor x \rfloor$ first using $x = 17$ and then using $x = \pi$.

SOLUTION

```
x=17;
Ceiling[x] - Floor[x]
0
x = Pi;
Ceiling[x] - Floor[x]
1
```

$\lceil x \rceil - \lfloor x \rfloor$ always equals 0 when x is an integer and 1 when x is not an integer.

2.16 What are the greatest common divisor and least common multiple of 5355 and 40425?

SOLUTION

```
GCD[5355, 40425]
```
105
```
LCM[5355, 40425]
```
2 061 675

2.17 Show that 15, 16, and 30 are relatively prime (integers are relatively prime if they have no common factor other than 1).

SOLUTION

```
GCD[15, 16, 30]
```
1

Since their GCD = 1, their only common factor is 1. Therefore, they are relatively prime.

2.18 A theorem from number theory says that the product of the GCD and LCM of two numbers is always equal to the product of the numbers. Verify this using the numbers 74613 and 85085.

SOLUTION

```
a = 74 613;
b = 85 085;
GCD[a, b] * LCM[a, b]
```
6 348 447 105
```
a * b
```
6 348 447 105

Obviously, the products are identical.

2.19 Show that 156,875,438,767 is not prime and factor.

SOLUTION

```
PrimeQ[156 875 438 767]
```
False
```
FactorInteger[156 875 438 767]
```
{{53,1},{2 959 913 939,1}}

156,875,438,767 is equal to the product of primes 53 and 2,959,913,939.

2.20 How long did it take *Mathematica* to factor 156,875,438,767 in the previous problem?

SOLUTION

```
Timing[FactorInteger[156 875 438 767]]
```
{0.011 ,{{53, 1}, {2 959 913 939, 1}}}

It took approximately 0.011 seconds. (This time will vary from computer to computer.)

2.21 Compute the natural logarithm of e^5.

SOLUTION

```
Log[e⁵]  or  Log[E^5]  or  Log[Exp[5]]
```
5

2.22 Compute the common logarithm (base 10) of e^5. What is its numerical approximation?

SOLUTION

```
Log[10, e⁵]  or  Log[10, E^5]  or  Log[10, Exp[5]]
```

$$\frac{5}{\text{Log}[10]}$$

```
% //N
2.17147
```

2.23 If Jacob starts with one cent and his money doubles every day, how much money will he have, to the penny, after 30 days?

SOLUTION

$$N\left[2^{30}/100\right]$$

```
1.07374 × 10⁷
```

If we want to get the amount to the penny, we will need 10 significant digits.

```
amount = N[2³⁰/100, 10]
```

```
1.073741824 × 10⁷
```

To see this in a more traditional format, the function **AccountingForm** can be used.

```
AccountingForm[amount]
```

```
10737418.24
```

We can group the digits into blocks of 3 and separate them with commas using the option **DigitBlock**

```
AccountingForm[amount, DigitBlock → 3]
```

```
10,737,418.24
```

2.24 What is the *exact* value of sin 15°? Compute a 20 decimal place approximation.

SOLUTION

```
Sin[15 Degree]   or   Sin[15°]
```

$$\frac{-1 + \sqrt{3}}{2\sqrt{2}}$$

```
N[%, 20]
```

```
0.25881904510252076235
```

2.25 Select a random number, x, between 0 and 1 and compute $\sin^2 x + \cos^2 x$.

SOLUTION   (Your value of x will be different from that shown here.)

```
x = Random[ ]
```

```
0.427468
```

```
Sin[x]² + Cos[x]²
```

```
1.
```

> Recall from trigonometry that
> $\sin^2 x + \cos^2 x = 1$ for all x.

2.26 Find a number between $-\pi/2$ and $\pi/2$ whose sin is 1/2.

SOLUTION

```
ArcSin[1/2]
```

$$\frac{\pi}{6}$$

2.27 Select a random number, x, between 0 and 1 and compute $\cosh^2 x - \sinh^2 x$.

SOLUTION (Your value of x will be different from that shown here.)

```
x = Random[]
0.991288
Cosh[x]² - Sinh[x]²
1.
```

> Hyperbolic functions have properties similar to trigonometric functions: $\cosh^2 x - \sinh^2 x = 1$ for all x.

2.28 Obtain an alternate representation of $\tanh(\ln x)$.

SOLUTION

```
Tanh[Log[x]] //TraditionalForm
```

$$\frac{x^2 - 1}{x^2 + 1}$$

2.29 Approximately how many radians are there in one degree?
Approximately how many degrees are there in one radian?

SOLUTION

```
N[Degree]
0.0174533
N[1/Degree]
57.2958
```

> **Degree** is a *Mathematica* constant which represents the number of radians in one degree. **1/Degree** represents the number of degrees in one radian.

2.30 How much is $\infty + 100{,}000$?

SOLUTION

```
∞ + 100 000
∞
```

2.31 What is the square root of the complex number $3+4i$?

SOLUTION

```
√3 + 4 i    or   Sqrt[3 + 4 I]
2 + i
```

2.32 The number of permutations of n objects taken k at a time is $P(n,k) = \dfrac{n!}{(n-k)!}$. How many permutations of 20 objects taken 10 at a time are there?

SOLUTION

```
n = 20;
k = 10;
n!/(n-k)!   or   Factorial[n]/Factorial[n-k]
670 442 572 800
```

2.33 Between what two consecutive integers does the natural logarithm of 100,000 lie?

SOLUTION

```
Floor[Log[100 000]]
11
Ceiling[Log[100 000]]
12
```

ln 100,000 lies between 11 and 12.

2.34 What is the quotient and remainder if 62,173,467 is divided by 9,542?

SOLUTION

```
Quotient[62 173 467, 9542]
```

6515

```
Mod[62 173 467, 9542]
```

7337

2.35 Find the greatest common divisor and least common multiple of 1,001 and 1,331.

SOLUTION

```
GCD[1001, 1331]
```

11

```
LCM[1001, 1331]
```

121 121

2.36 How long does it take your computer to find the prime factorization of 10! ?

SOLUTION

```
FactorInteger[10!] //Timing
```

$\{0.016, \{\{2, 8\}, \{3, 4\}, \{5, 2\}, \{7, 1\}\}\}$

The factorization is $2^8 3^4 5^2 7^1$; times will vary depending on the speed of your CPU.

2.37 Find an algebraic expression for $\cos\left(\sin^{-1}\left(\frac{x^2}{x^2+1}\right)\right)$.

SOLUTION

$$\mathrm{Cos}\left[\mathrm{ArcSin}\left[\frac{x^2}{x^2+1}\right]\right]$$

$$\sqrt{1 - \frac{x^4}{(1+x^2)^2}}$$

2.38 Is 15,485,863 prime?

SOLUTION

```
PrimeQ[15 485 863]
```

True

2.3 Basic Arithmetic Operations

As we have seen, basic arithmetic operations such as addition are performed by inserting an operation symbol between two numbers. Thus, the sum of 3 and 5 is obtained by typing 3 + 5. However, in more advanced applications it is sometimes useful to represent these operations as functions. Towards this end, *Mathematica* includes the following:

- **Plus[a, b, ...]** computes the sum of a, b, ... **Plus[a, b]** is equivalent to **a + b**.
- **Times[a, b, ...]** computes the product of a, b, ... **Times[a, b]** is equivalent to **a * b**.
- **Subtract[a, b]** computes the difference of a and b. Only two arguments are permitted. **Subtract[a, b]** is equivalent to **a – b**.
- **Divide[a, b]** computes the quotient of a and b. Only two arguments are permitted. **Divide[a, b]** is equivalent to **a/b**.
- **Minus[a]** produces the additive inverse (negative) of a. **Minus[a]** is equivalent to **–a**.
- **Power[a, b]** computes a^b, **Power[a, b, c]** produces a^{b^c}, etc.

EXAMPLE 30

```
Plus[2, 3, 4]
9
Times[2, 3, 4]
24
Power[2, 3, 4]
2 417 851 639 229 258 349 412 352
```

In order to see the way in which *Mathematica* handles functions internally, the command **FullForm** is quite useful.

- **FullForm[*expression*]** exhibits the internal form of *expression*.

EXAMPLE 31

```
FullForm[a + b + c]
Plus[a, b, c]
FullForm[a – b]
Plus[a, Times[-1, b]]
FullForm[(a * b)^c]
Power[Times[a, b], c]
```

FullForm may be used for any *Mathematica* function, not only arithmetic operators.

EXAMPLE 32

```
FullForm[Sin[x^3]^2]
Power[Sin[Power[x,3]],2]
```

In addition to the standard operational symbols discussed previously, there are a few additional commands that are useful in special situations. (Note: In order for the following to work, x and y must have *numerical* values.)

- **Increment[x]** or **x ++** increases the value of x by 1 but returns the *old* value of x.
- **Decrement[x]** or **x --** decreases the value of x by 1 but returns the *old* value of x.
- **PreIncrement[x]** or **++ x** increases the value of x by 1 and returns the *new* value of x.
- **PreDecrement[x]** or **-- x** decreases the value of x by 1 and returns the *new* value of x.
- **AddTo[x,y]** or **x += y** adds y to x and returns the *new* value of x.
- **SubtractFrom[x,y]** or **x -= y** subtracts y from x and returns the *new* value of x.
- **TimesBy[x,y]** or **x *= y** multiplies x by y and returns the *new* value of x.
- **DivideBy[x,y]** or **x /= y** divides x by y and returns the *new* value of x.

The next two examples illustrate the various addition commands. The commands for subtraction, multiplication, and division are similar.

EXAMPLE 33

```
x = 3;

x ++

3            ←The old value of x is returned.

x

4            ←Tthe actual value of x is 4.
```

```
x = 3;

++ x

4            ←The new value of x is returned.

x

4            ←The actual value of x is 4.
```

x ++ is equivalent to the sequence
x   x = x + 1;

++x is equivalent to the statement
x = x + 1

EXAMPLE 34

x = 3; y = 4;   x = 3; y = 4;

x + y   x += y

7   ← The sum is returned.   7   ← The sum is returned.

x   x

3   ← x remains unchanged.   7   ← The new value of x is 7.

y   y

4   ← y remains unchanged.   4   ← y remains unchanged.

> x + = y is equivalent to the statement
> x = x + y

SOLVED PROBLEMS

2.39 How does *Mathematica* evaluate the expression $a + bc / d$?

SOLUTION

FullForm[a + b * c/d]
Plus[a,Times[b, c, Power[d, -1]]]

2.40 How is the function **Minus[x]** treated internally in *Mathematica*?

SOLUTION

FullForm[Minus[x]]
Times[-1, x]

2.4 Strings

A string is an (ordered) sequence of characters. Strings have no numerical value and are often used as labels for tables, graphs, and other displays.

In *Mathematica*, a string is enclosed within quotation marks. Thus "abcde" is a string of five characters. Do not confuse "abcde" with abcde, as the latter is *not* a string.

Mathematica comes equipped with a number of string manipulation commands.

- **StringLength[*string*]** returns the number of characters in *string*.
- **StringJoin[*string1*, *string2*, ...]** or *string1* <> *string2* <> ... concatenates two or more strings to form a new string whose length is equal to the sum of the individual string lengths.
- **StringReverse[*string*]** reverses the characters in *string*.

StringDrop eliminates characters from a string. There are five forms of this command.

- **StringDrop[*string*, n]** returns *string* with its first n characters dropped.
- **StringDrop[*string*, -n]** returns *string* with its last n characters dropped.
- **StringDrop[*string*, {n}]** returns *string* with its nth character dropped.
- **StringDrop[*string*, {-n}]** returns *string* with the nth character from the end dropped.
- **StringDrop[*string*, {m, n}]** returns *string* with characters m through n dropped.

StringTake returns characters from a string. Its format is similar to **StringDrop**.

- **StringTake[*string*, n]** returns the first n characters of *string*.
- **StringTake[*string*, -n]** returns the last n characters of *string*.
- **StringTake[*string*, {n}]** returns the nth character of *string*.
- **StringTake[*string*, {-n}]** returns the nth character from the end of *string*.
- **StringTake[*string*, {m, n}]** returns characters m through n of *string*.

EXAMPLE 35   In this example we define **string** = "abcdefg". The output is shown to the right of the command. (Please observe the difference between the *Mathematica* symbol **String** and the user-defined symbol **string**.)

string = "abcdefg"	abcdefg
string <> "hijklmnop"	abcdefghijklmnop
StringLength[string]	7
StringReverse[string]	gfedcba
StringDrop[string, 2]	cdefg
StringDrop[string, -2]	abcde
StringDrop[string, {2}]	acdefg
StringDrop[string, {-2}]	abcdeg
StringDrop[string, {2, 5}]	afg
StringTake[string, 2]	ab
StringTake[string, -2]	fg
StringTake[string, {2}]	b
StringTake[string, {-2}]	f
StringTake[string, {2, 5}]	bcde

StringInsert allows you to insert characters within existing strings.

- **StringInsert** [*string1*, *string2*, **n**] yields a string with *string2* inserted starting at position n in *string1*.
- **StringInsert** [*string1*, *string2*, **-n**] yields a string with *string2* inserted starting at the nth position from the end of *string1*.
- **StringInsert** [*string1*, *string2*, {n1, n2, ...}] inserts a copy of *string2* at each of the positions n1, n2, . . . of *string1*.

StringReplace allows you to replace part of a string with another string.

- **StringReplace** [*string*, *string1* → *newstring1*] replaces *string1* by *newstring1* whenever it appears in *string*.
- **StringReplace** [*string*, {*string1* → *newstring1*, *string2* → *newstring2*, ...}] replaces *string1* by *newstring1*, *string2* by *newstring2*, . . . whenever they appear in *string*.
- **StringPosition** [*string*, *substring*] returns a list of the start and end positions of all occurrences of *substring* within *string*. (Lists are discussed in detail in Chapter 3.)

EXAMPLE 36

```
string1 = "abcdefg";
string2 = "123";
StringInsert[string1, string2, 3]
ab123cdefg
StringInsert[string1, string2, -3]
abcde123fg
StringInsert[string1, string2, {1, 3, 5, 7}]
123ab123cd123ef123g
StringReplace[string1, "ab" → "AB"]
ABcdefg
StringReplace[string1, {"ab" → "AB", "fg" → "FG"}]
ABcdeFG
```

EXAMPLE 37

```
string = "abcxabcxxabcxxxabc";
StringLength[string]
18
StringPosition[string, "abc"]
{{1, 3}, {5, 7}, {10, 12}, {16, 18}}
```

2.5 Assignment and Replacement

All programming languages must have the ability to make assignments in order to transfer the result of a calculation to a symbol which can be recalled for later use. *Mathematica* offers two types of assignment and there is often confusion as to which one to use in a given situation.

- **lhs = rhs** is an *immediate* assignment in which rhs is evaluated at the time the assignment is made.
- **lhs := rhs** is a *delayed* assignment in which rhs is evaluated each time the value of lhs is called.

In many situations both assignments produce identical results. There are, however, a few instances where one must be careful. The following examples use ideas that are discussed in later chapters. They are self-explanatory, however, and will be easily understood.

EXAMPLE 38 When defining functions recursively, := *must* be used. For example,

```
f[0] = 1;
f[n_] := n f[n-1]
```

produces *n* factorial. Since *Mathematica* cannot compute f[n] until the value of n is specified, the delayed assignment, :=, must be used. Using = causes recursion errors.

```
f[5]
120
f[10]
3628800
```

EXAMPLE 39 When defining piecewise functions, one *must* use :=. For example,

```
g[x_] := x² /; x ≥ 0          ← /; is a conditional. Assignment will be made only if x ≥ 0.
g[x_] := -x² /; x < 0
g[3]
9
g[-3]
-9
```

Using = would cause trouble, as *Mathematica* cannot determine which branch should be taken until a value of x is supplied.

EXAMPLE 40 You may think that the := assignment is more general and can be safely used in any given situation. This is true to a certain extent, but there are times when one should use =. As an extreme, but reasonable, example, let us define

$$F[x\_] := \int_0^x t \; \text{Exp}[t] \; \text{Sin}[t] \; dt$$

Each time a value of F is computed, *Mathematica* performs several "integration by parts" evaluations. Now imagine that many different values of F are needed, for example in the instruction

`Plot[F[x],{x,0,4}]`. This plots `F[x]` from 0 to 4 using many points. Every time the value of `F` is computed, the integral is evaluated—from scratch—applying integration by parts each time. The result is a lengthy delay in displaying the graph. Using = causes the graph to be displayed more quickly.

$$F[x\_] := \int_0^x t \ \text{Exp}[t] \ \text{Sin}[t] \ dt$$
`Plot[F[x], {x,0,1}]//Timing`

$$F[x\_] = \int_0^x t \ \text{Exp}[t] \ \text{Sin}[t] \ dt;$$
`Plot[F[x], {x,0,1}]//Timing`

{16.203, }

{0.016, }

Note the significant difference in time required to plot this function.

Often, you will want to evaluate an expression without assigning a value to a symbol. This can be done with the **ReplaceAll (/.)** replacement operator.

- *expression /. rule* applies a rule or list of rules to each subpart of *expression*.

EXAMPLE 41   Suppose we want to evaluate $x^2 + 5x + 6$ when $x = 3$, but do not want to assign a value to x.

```
Clear[x]
```
$$x^2 + 5x + 6 \ /. \ x \rightarrow 3$$
```
30
?x
```

```
Global`x
```

(x is left undefined)

`/.` can also be used to replace an expression by another expression. Several replacements can be made at the same time if braces are used.

EXAMPLE 42

$$\sqrt{2x+3} + (2x+3)^2 \ /. \ 2x+3 \rightarrow 3y+5$$
$$\sqrt{3y+5} + (3y+5)^2$$

EXAMPLE 43

$$x^2 + \sqrt{y} \ /. \ \{y \rightarrow x, \ x \rightarrow y\}$$
$$\sqrt{x} + y^2$$

SOLVED PROBLEMS

2.41 The *Mathematica* command **Expand[*expression*]**, which is discussed in Chapter 7, expands *expression* algebraically. Define two symbols, **a** and **b**, as **Expand[(x + 1)^3]**, using = and :=, respectively. Then let **x = u + v** and compute **a** and **b**.

SOLUTION

```
a = Expand[(x + 1)^3]
```

$1 + 3x + 3x^2 + x^3$ ← Expansion occurs immediately.

```
b := Expand[(x + 1)^3]
```
← Expansion does not occur until b is called.

```
x = u + v;
a
```

$1 + 3(u + v) + 3(u + v)^2 + (u + v)^3$ ← u+v replaces x *after* expansion.

```
b
```

$1 + 3u + 3u^2 + u^3 + 3v + 6uv + 3u^2 v + 3v^2 + 3uv^2 + v^3$ ← u+v replaces x *before* expansion.

2.42 The command **Together**, which is discussed in Chapter 7, combines the sum or difference of two or more fractions into one fraction. Define two symbols, y and z, as Together[a+b] using, respectively, = and := . Then let a = 1/x and b = 1/(x+1) and compute y and z.

SOLUTION

```
y = Together[a + b]
```

a + b   ← At this point a and b are not fractions so **Together** does nothing.

```
z := Together[a + b]
a = 1/x;
b = 1/(x + 1);
y
```

$\dfrac{1}{x} + \dfrac{1}{1+x}$ ← Since **Together** was executed prior to the introduction
                                                     of the fractions, the result is the sum of a and b.

```
z
```

$\dfrac{1+2x}{x(1+x)}$ ← **Together** is executed *after* the fractions are introduced so
                                              the fractions are combined into one.

2.43 The *Mathematica* command **Factor[*expression*]** attempts to factor the algebraic expression, *expression*. Type **a = Factor[poly]** and **b := Factor[poly]**. Then let **poly = x² + 2x + 1**. Compute **a** and **b** and explain the difference in output.

SOLUTION

```
a = Factor[poly];
b := Factor[poly];
poly = x^2 + 2x + 1;
a
```

$1 + 2x + x^2$

```
b
```

$(1 + x)^2$

Since a is computed before poly is defined, its value is the factored form of the symbol poly, which is just poly. Then poly is replaced by $x^2 + 2x + 1$. On the other hand, b is not evaluated until called in the next to last line, so *Mathematica* factors the polynomial.

2.44 Replace *x* with $x^2 + 2x + 3$ in the expression $x^2 + 5x + 6$.

SOLUTION

```
x^2 + 5x + 6 /. x → x^2 + 2x + 3
```

$6 + 5(3 + 2x + x^2) + (3 + 2x + x^2)^2$

2.45 Replace y with $x + 1$ and z with $x + 2$ in the expression $(x + y + z)^2$.

SOLUTION

```
(x + y + z)^2 /. {y → x + 1, z → x + 2}
(3 + 3 x)^2
```

2.6 Logical Relations

Do not confuse = with ==, a "logical" equality. **lhs == rhs** is True if and only if lhs and rhs have the same value; otherwise it is False. Logical equalities are used extensively in connection with equation solving (Chapter 6).

Other logical relations are available. The following list summarizes them.

- **Equal[x, y]** or **x == y** is True if and only if x and y have the same value.
- **Unequal[x, y]** or **x != y** or **x ≠ y** is True if and only if x and y have different values.
- **Less[x, y]** or **x < y** is True if and only if x is numerically less than y.
- **Greater[x, y]** or **x > y** is True if and only if x is numerically greater than y.
- **LessEqual[x, y]** or **x <= y** or **x ≤ y** is True if and only if x is numerically less than y or equal to y.
- **GreaterEqual[x, y]** or **x >= y** or **x ≥ y** is True if and only if x is numerically greater than y or equal to y.

Note that **Equal** and **Unequal** can be used for comparing both numerical and certain non-numerical quantities, while **Less**, **Greater**, **LessEqual**, and **GreaterEqual** are strictly numerical comparisons.

EXAMPLE 44

1 == 2	1 != 2	1 <= 2	a + a == 2a
False	True	True	True

2 == 2	2 != 2	2 <= 2	a < a
True	False	True	a < a

No conclusion can be drawn since a is undefined.

Mathematica also includes the following logical operations:

- **And[p, q]** or **p && q** or **p ∧ q** is True if both p and q are True; False otherwise.
- **Or[p, q]** or **p || q** or **p ∨ q** is True if p or q (or both) are True; False otherwise.
- **Xor[p, q]** is True if p or q (but not both) are True; False otherwise.
- **Not[p]** or **!p** or **¬p** is True if p is False and False if p is True.
- **Implies[p, q]** or **p ⇒ q** is False if p is True and q is False; True otherwise.

Note: ⇒ can be obtained with the key sequence [ESC], [=], [>], [ESC].

Logical expressions can be compared using **LogicalExpand**.

- **LogicalExpand[*expression*]** applies the distributive laws for logical operations to *expression* and puts it into disjunctive normal form.

EXAMPLE 45 Use *Mathematica* to verify the distributive law: p∧(q∨r) = (p∧q)∨(p∧r).

```
lhs = p && (q || r);
rhs = (p && q) || (p && r);
lhs == rhs
(p && (q || r)) == (p && q || p && r)
LogicalExpand[lhs] == LogicalExpand[rhs]
True
```

SOLVED PROBLEMS

2.46 Use *Mathematica* to verify De Morgan's laws:

$$\neg(p \wedge q) = \neg p \vee \neg q \quad \text{and} \quad \neg(p \vee q) = \neg p \wedge \neg q$$

SOLUTION

```
LogicalExpand[!(p && q)] == LogicalExpand[!p || !q]
True
LogicalExpand[!(p || q)] == LogicalExpand[!p && !q]
True
```

2.47 Show that $((p \wedge q) \vee (p \wedge \neg q)) \vee ((\neg p \wedge q) \vee (\neg p \wedge \neg q))$ is a tautology.

SOLUTION

```
LogicalExpand[((p && q) || (p && !q)) || ((!p && q) || (!p && !q))]
True
```

2.7 Sums and Products

Sums and products are of fundamental importance in mathematics, and *Mathematica* makes their computation simple. Unlike other computer languages, initialization is automatic and the syntax is easy to apply, particularly if the Basic Math Input palette is used. Any symbol may be used as the index of summation. (i is used in the following description.) Negative increments are permitted wherever increment is used.

- **Sum[a[i],{i,imax}]** or $\sum_{i=1}^{imax} a[i]$ evaluates the sum $\sum_{i=1}^{imax} a_i$

- **Sum[a[i],{i,imin,imax}]** or $\sum_{i=imin}^{imax} a[i]$ evaluates the sum $\sum_{i=imin}^{imax} a_i$

- **Sum[a[i],{i,imin,imax, increment}]** evaluates the sum $\sum_{i=imin}^{imax} a_i$ in steps of increment. Summation continues as long as i ≤ imax.

EXAMPLE 46 To compute the sum of the squares of the first 20 consecutive integers, we can type

```
Sum[i^2, {i,1,20}] or
```
$\sum_{i=1}^{20} i^2$

```
2870
```

> Note: Even though *Mathematica* allows the form `Sum[i^2,{i,20}]`, the use of the initial index, 1, is recommended for clarity.

EXAMPLE 47 Compute the sum $\frac{1}{15} + \frac{1}{17} + \frac{1}{19} + \cdots + \frac{1}{51}$.

```
Sum[1/i, {i,15,51,2}]
```

$$\frac{63\,501\,391\,475\,806\,044\,193}{96\,845\,140\,757\,687\,397\,075}$$

- **NSum** has the same syntax as **Sum** and works in a similar manner to yield numerical approximations.

EXAMPLE 48 Approximate the sum $\frac{1}{15} + \frac{1}{17} + \frac{1}{19} + \cdots + \frac{1}{51}$.

```
NSum[1/i, {i, 15, 51, 2}]
0.6557
```

The limits of a sum can be infinite. *Mathematica* uses sophisticated techniques to evaluate infinite summations.

EXAMPLE 49   Compute $\frac{1}{1}+\frac{1}{4}+\frac{1}{9}+\frac{1}{16}+\cdots$

`Sum[1/i^2,{i, 1, Infinity}]` or $\displaystyle\sum_{i=1}^{\infty}\frac{1}{i^2}$

$\dfrac{\pi^2}{6}$

Double sums can be computed using the following syntax or, more conveniently, by clicking twice on the Σ symbol in the **Basic Math Input** palette. The syntax extends in a natural way to triple sums, quadruple sums, and so forth.

- `Sum[a[i,j],{i,imax},{j,jmax}]` or $\displaystyle\sum_{i=1}^{imax}\sum_{j=1}^{jmax} a[i, j]$ evaluates the sum $\displaystyle\sum_{i=1}^{imax}\sum_{j=1}^{jmax} a_{i,j}$

- `Sum[a[i,j],{i,imin,imax},{j,jmin,jmax}]` or $\displaystyle\sum_{i=imin}^{imax}\sum_{j=jmin}^{jmax} a[i, j]$ evaluates the sum $\displaystyle\sum_{i=imin}^{imax}\sum_{j=jmin}^{jmax} a_{i,j}$

- `Sum[a[i,j],{i,imin,imax,i_increment},{j,jmin,jmax, j_increment}]` evaluates the sum $\displaystyle\sum_{i=imin}^{imax}\sum_{j=jmin}^{jmax} a_{i,j}$ in steps of `i_increment` and `j_increment`.

- `NSum`, with identical syntax, returns numerical approximations to each of the sums described in `Sum`.

EXAMPLE 50   Compute the value of

$$\left(\frac{1}{1}+\frac{1}{2}+\frac{1}{3}+\frac{1}{4}\right)+\left(\frac{2}{1}+\frac{2}{2}+\frac{2}{3}+\frac{2}{4}\right)+\left(\frac{3}{1}+\frac{3}{2}+\frac{3}{3}+\frac{3}{4}\right)$$

`Sum[i/j,{i,1,3},{j,1,4}]` or $\displaystyle\sum_{i=1}^{3}\sum_{j=1}^{4}\frac{i}{j}$

$\dfrac{25}{2}$

Just as **Sum** computes sums, the *Mathematica* function **Product** computes products. Its syntax is much the same as **Sum**.

- `Product[a[i],{i,imax}]` or $\displaystyle\prod_{i=1}^{imax} a[i]$ evaluates the product $\displaystyle\prod_{i=1}^{imax} a_i$

- `Product[a[i],{i,imin, imax}]` or $\displaystyle\prod_{i=imin}^{imax} a[i]$ evaluates the product $\displaystyle\prod_{i=imin}^{imax} a_i$

- `Product[a[i],{i,imin,imax,increment}]` evaluates the product $\displaystyle\prod_{i=imin}^{imax} a_i$ in steps of `increment`.

- `NProduct`, with identical syntax, returns numerical approximations to each of the products described in `Product`.

Multiple products are also easily computed. The syntax for a double product is listed in the following, but the concept extends to triple products and higher.

- `Product[a[i,j],{i,imax},{j,jmax}]` or $\displaystyle\prod_{i=1}^{imax}\prod_{j=1}^{jmax} a[i, j]$ evaluates the product $\displaystyle\prod_{i=1}^{imax}\prod_{j=1}^{jmax} a_{i,j}$

- `Product[a[i,j],{i,imin,imax},{j,jmin,jmax}]` or $\displaystyle\prod_{i=imin}^{imax}\prod_{j=jmin}^{jmax} a[i, j]$ evaluates the product $\displaystyle\prod_{i=imin}^{imax}\prod_{j=jmin}^{jmax} a_{i,j}$

- `Product[a[i,j],{i,imin,imax,i_increment},{j,jmin,jmax, j_increment}]` evaluates the product $\displaystyle\prod_{i=imin}^{imax}\prod_{j=jmin}^{jmax} a_{i,j}$ in steps of `i_increment` and `j_increment`.

EXAMPLE 51 Compute the product of the consecutive integers 4 through 9.

Product[i, {i, 4, 9}] or $\prod_{i=4}^{9} i$

60 480

EXAMPLE 52 The binomial coefficient $C(n,k) = \dfrac{n!}{k!(n-k)!}$ can be expressed as $\left(\dfrac{n}{k}\right)\left(\dfrac{n-1}{k-1}\right)\left(\dfrac{n-2}{k-2}\right)\cdots\left(\dfrac{n-k+1}{1}\right)$ for more efficient computation. Use this representation to compute $C(10, 4)$.

n = 10;

k = 4;

Product[(n - i)/(k - i), {i, 0, k - 1}] or $\prod_{i=0}^{k-1}\dfrac{n-i}{k-i}$

210

SOLVED PROBLEMS

2.48 Compute the sum of the first 25 prime numbers.

SOLUTION

Sum[Prime[k], {k, 1, 25}] or $\sum_{k=1}^{25}\text{Prime}[k]$

1060

2.49 Compute the square root of the sum of the squares of the integers 15 through 30, inclusive.

SOLUTION

Sqrt[Sum[k^2, {k, 15, 30}]] or $\sqrt{\sum_{k=15}^{30} k^2}$

$6\sqrt{10}$

2.50 Compute the infinite sum $1 + \dfrac{1}{2} + \dfrac{1}{4} + \dfrac{1}{8} + \dfrac{1}{16} + \cdots$

SOLUTION

Sum[1/2^i, {i, 0, Infinity}] or $\sum_{i=0}^{\infty}\dfrac{1}{2^i}$

2

2.51 Compute the sum $\dfrac{1}{2} + \dfrac{2}{3} + \dfrac{3}{4} + \cdots + \dfrac{99}{100}$

SOLUTION

$\sum_{i=1}^{99}\dfrac{i}{i+1}$

$\dfrac{264\ 414\ 864\ 639\ 329\ 557\ 497\ 913\ 717\ 698\ 145\ 082\ 779\ 489}{2\ 788\ 815\ 009\ 188\ 499\ 086\ 581\ 352\ 357\ 412\ 492\ 142\ 272}$

2.52 Obtain a general formula for the sum of squares of the consecutive integers 1 through n.

SOLUTION

Sum[k^2, {k, 1, n}] or $\sum_{k=1}^{n} k^2$

$\dfrac{1}{6}(n)(1+n)(1+2n)$ ←*Mathematica* has "memorized" these standard formulas.

2.53 Compute the product of the first 20 Fibonacci numbers.

SOLUTION

$\prod_{i=1}^{20}\text{Fibonacci}[i]$ or Product[Fibonacci[i], {i, 1, 20}]

9 692 987 370 815 489 224 102 512 784 450 560 000

2.54 Compute the product of the natural logarithms of the integers 2 through 20. Obtain an approximation to 20 significant digits.

SOLUTION

```
N[Product[Log[i], {i, 2, 20}], 20]
```

$1.3632878207490815857 \times 10^6$

2.55 Compute the sum $1+\left(1+\frac{1}{2}\right)+\left(1+\frac{1}{2}+\frac{1}{3}\right)+ \cdots +\left(1+\frac{1}{2}+\frac{1}{3}+ \cdots +\frac{1}{20}\right)$

SOLUTION

```
Sum[1/j, {i, 1, 20}, {j, 1, i}]
```
or $\displaystyle\sum_{i=1}^{20}\sum_{j=1}^{i}\frac{1}{j}$

$\dfrac{41\,054\,655}{739\,024}$

2.56 Compute a numerical approximation of $\left(1+\frac{1}{2}\right)\left(1+\frac{1}{2}+\frac{1}{3}\right) \cdots \left(1+\frac{1}{2}+\frac{1}{3}+ \cdots +\frac{1}{10}\right)$

SOLUTION

```
NProduct[Sum[1/j, {j, 1, i}], {i, 2, 10}]
```
or $\displaystyle\prod_{i=2}^{10}\sum_{j=1}^{i}\frac{1}{j}$ //N

1871.44

2.8   Loops

Often you may need to repeat an operation or sequence of operations several times. Although *Mathematica* offers the ability to compute sums and products conveniently using the **Sum** and **Product** commands, there are times when your work may require the use of looping techniques. *Mathematica* offers three basic looping functions: **Do**, **While**, and **For**.

- **Do** [*expression*, **{k}**] evaluates *expression* precisely k times.
- **Do** [*expression*, **{i, imax}**] evaluates *expression* imax times with the value of i changing from 1 to imax in increments of 1.
- **Do** [*expression*, **{i, imin, imax}**] evaluates *expression* with the value of i changing from imin to imax in increments of 1.
- **Do** [*expression*, **{i, imin, imax, increment}**] evaluates *expression* with the value of i changing from imin to imax in increments of increment.
- **Do** [*expression*, **{i, imin, imax}**, **{j, jmin, jmax}**] evaluates *expression* with the value of i changing from imin to imax and j changing from jmin to jmax in increments of 1. The variable i changes by 1 for each cycle of j. This is known as a nested Do loop.
- **Do** [*expression*, **{i, imin, imax, i_increment}**,
            {j, jmin, jmax, j_increment}, **...**]
  forms a nested Do loop allowing for incrimination values other than 1.

The last two forms of the command may be extended to three or more variables.

EXAMPLE 53

```
Do[Print["This line will be repeated 5 times."], {5}]
```
```
This line will be repeated 5 times.
This line will be repeated 5 times.
This line will be repeated 5 times.
This line will be repeated 5 times.
This line will be repeated 5 times.
```

EXAMPLE 54   This example computes the sum of consecutive odd integers from 5 to 25. (Of course, the **Sum** command is more convenient.)

```
mysum = 0;
Do[mysum = mysum + k, {k, 5, 25,2}]
mysum
165
```
← Initialization of mysum. This step is important. It is not needed if the command **Sum** is used.

EXAMPLE 55   This example computes the sum of all fractions whose numerators and denominators are positive integers not exceeding 5.

```
fracsum = 0;
Do[fracsum = fracsum + i/j, {i,1,5}, {j,1,5}]
fracsum
```

$$\frac{137}{4}$$

- **While**[*condition*, *expression*] evaluates *condition*, then *expression*, repetitively, until *condition* is False.

If *expression* consists of multiple statements, they are separated by semicolons.

EXAMPLE 56

```
n = 1; While[n < 6, Print[n]; n ++]
```
```
1
2
3
4
5
```

| n = n+1 may be used in place of n++ |
| See page 37. |

- **For**[*initialization*, *test*, *increment*, *expression*] executes *initialization*, then repeatedly evaluates *expression*, *increment*, and *test* until *test* becomes False.

After *initialization*, the order of evaluation is *test*, *expression*, and then *increment*. The **For** loop terminates as soon as *test* gives False. If *initialization, test, increment,* or *expression* consists of multiple statements, they are separated by semicolons.

EXAMPLE 57

```
For[i = 1, i ≤ 5, i ++, Print[i]]
```
```
1
2
3
4
5
```

Although it is not a loop, the **If** instruction is often used in conjunction with other loop commands.

- **If**[*condition*, *true*, *false*] evaluates *condition* and executes *true* if *condition* is True and executes *false* if *condition* is False.
- **If**[*condition*, *true*] evaluates *condition* and executes *true* if *condition* is True. If *condition* is False no action is taken and Null is returned.
- **If**[*condition*, , *false*] evaluates *condition* and executes *false* if *condition* is False. If *condition* is True no action is taken and Null is returned. (Note the double comma.)
- **If**[*condition*, *true*, *false*, *neither*] evaluates *condition* and executes *true* if *condition* is True, executes *false* if *condition* is False, and executes *neither* if *condition* is neither True nor False.

EXAMPLE 58

```
If[2 == 2,Print["TRUE"],Print["FALSE"],Print["NEITHER"]]
```
```
TRUE
```
```
If[2 == 3, Print["TRUE"],Print["FALSE"],Print["NEITHER"]]
```
```
FALSE
```
```
If[7, Print["TRUE"],Print["FALSE"],Print["NEITHER"]]
```
```
NEITHER
```

| 7 is neither True nor False. |

The next example, which separates primes from non-primes, illustrates how the **If** instruction can be used in a **Do** loop.

EXAMPLE 59

```
Do[If[PrimeQ[k], Print[k], Print["  ", k]], {k, 1, 20}]
```
```
         1
  2
  3
           4
  5
           6
  7
           8
           9
          10
 11
          12
 13
          14
          15
          16
 17
          18
 19
          20
```

SOLVED PROBLEMS

2.57 Compute 10! using a **Do** loop.

SOLUTION

```
factorial = 1;
n = 10;
Do[factorial = factorial*k, {k, n}]
factorial
3 628 800
```

2.58 Compute 10! using a **While** loop.

SOLUTION

```
factorial = 1;
n = 10;
While[n > 0, factorial = factorial * n; n --]
factorial
3 628 800
```

2.59 Compute 10! using a **For** loop.

SOLUTION

```
For[factorial = 1; n = 1, n ≤ 10,n++, factorial = n * factorial]
factorial
3 628 800
```

2.60 Print all numbers from 1 to 20 which are *not* multiples of 2, 3, or 5.

SOLUTION

```
Do[If[Mod[k,2]≠0 && Mod[k,3]≠0 && Mod[k,5]≠0, Print[k]],{k,1,20}]
```
or
```
Do[If[Mod[k,2]==0 || Mod[k,3]==0 || Mod[k,5]==0,, Print[k]], {k, 1, 20}]
1
7
11
13
17
19
```

2.61 For each number k from 1 to 10, print half the number if k is even and twice the number if k is odd.

SOLUTION

```
Do[If[EvenQ[k], Print[k/2], Print[2 k]], {k, 1, 10}]
2
1
6
2
10
3
14
4
18
5
```

2.9 Introduction to Graphing

The graph of a function offers tremendous insight into the function's behavior and can be of great value in the solution of problems in mathematics. *Mathematica* offers some very powerful graphics commands that are remarkably easy to implement. Although there is a vast array of options available for customization of output, in this section we deal only with the most rudimentary forms using *Mathematica*'s defaults. A more detailed discussion of graphics commands appears in Chapters 4 and 5.

The **Plot** command plots a two-dimensional graph of a function.

- **Plot[f[x], {x, xmin, xmax}]** plots a two-dimensional graph of the function $f(x)$ on the interval $xmin \leq x \leq xmax$.
- **Plot[{f[x],g[x]}, {x, xmin, xmax}]** plots two functions on one set of axes. This extends in a natural way to three or more functions.

EXAMPLE 60 Plot the graph of $y = x^2$ on the interval $-5 \leq x \leq 5$.

```
Plot[x², {x, -5, 5}]
```

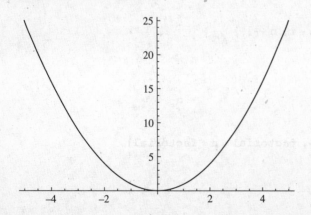

EXAMPLE 61   Plot the functions $y = x^2$ and $y = 2x + 10$, $-5 \leq x \leq 5$, on the same set of axes.

```
Plot[{x², 2 x + 10}, {x, -5, 5}]
```

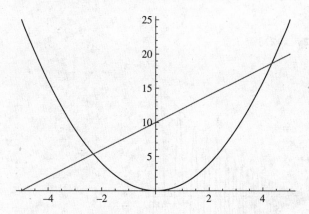

SOLVED PROBLEMS

2.62   Sketch the graphs of $y = x^2$, $y = x^3$, and $y = x^4$, $0 \leq x \leq 1$, on the same set of axes.

SOLUTION

```
Plot[{x², x³, x⁴}, {x, 0, 1}]
```

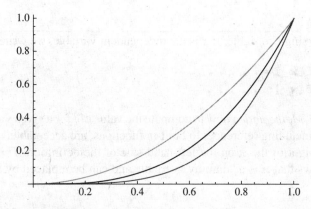

2.63   Sketch the graphs of the functions $y = -x$, $y = x$, and $y = x \sin x$ on the interval $-6\pi \leq x \leq 6\pi$ on one set of axes.

SOLUTION

```
Plot[{x, -x, x Sin[x]}, {x, -6π, 6π}]
```

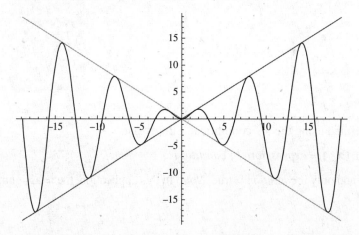

2.64 Sketch the graphs of the functions $y = -x^2$, $y = x^2$, and $y = x^2 \sin\left(\frac{1}{x}\right)$ on the interval $[-.02, .02]$ on one set of axes.

SOLUTION

```
Plot[{-x², x², x² Sin[1 / x]},{x, -.02, .02}]
```

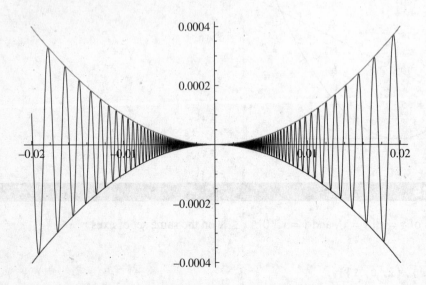

2.10   User-Defined Functions

Suppose we want to define a function, f, of a single variable. If x is the independent variable, we write

$$f[x\_] = \ldots \ldots$$

or

$$f[x\_] := \ldots \ldots$$

where the right-hand side of the definition tells *Mathematica* how to compute the value of f for a given value of x. All legitimate *Mathematica* operations, including references to built-in functions, are acceptable.

Note the underscore immediately to the right of the **x** on the left-hand side of the definition. *This is crucial.* It is the only way *Mathematica* knows that **x** is a "dummy" variable and can be replaced by any expression, numerical or symbolic.

EXAMPLE 62

```
f[x_] = x² + x³;
f[2]
12
f[2 x]
4 x² + 8 x³
f[Exp[x]]
e²ˣ + e³ˣ
f[λ]
λ² + λ³
```

A "piecewise" function can be defined using the **/;** conditional. Simply put,

$$f[x\_] := expression \;/; \; condition$$

assigns **f[x]** the value *expression* if and only if *condition* is true. *Note:* In this application, the **:=** assignment *must* be used.

EXAMPLE 63 We define the function $f(x) = \begin{cases} x^2 & \text{if } x \leq 2 \\ 8 - 2x & \text{if } x > 2 \end{cases}$

```
f[x_]:= x² /; x ≤ 2
f[x_]:= 8 - 2x /; x > 2
f[-4]
16
f[4]
0
Plot[f[x], {x, 0, 4}]
```

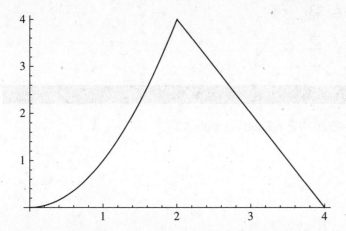

Functions are sometimes defined *recursively*. One or several values of the function are specified and later values are defined in terms of their predecessors.

EXAMPLE 64 The Fibonacci sequence can be defined recursively by defining $f(1) = 1$, $f(2) = 1$, and $f(n) = f(n - 2) + f(n - 1)$ for $n \geq 3$. We will compute the 35th Fibonacci number using this definition.

```
f[1] = 1;
f[2] = 1;
f[n_] := f[n - 2] + f[n - 1]
f[35]
9 227 465
```

> Note the use of := here. This is important. Experiment and see what happens if = is used.

You may have noticed a long pause in the calculation of this number. To see this more precisely, we will time the operation. (Your times may be slightly different, depending upon your computer.)

```
f[35]//Timing
{49.422,9227465}
```

Intermediate calculations have not been stored. Each computation of f[n] necessitates the computation of f[n-2] and f[n-1], each of which causes all values of f down to f[3] to be computed. Since each intermediate value of f is computed recursively based upon the values of f[1] and f[2], the result is that it takes an extremely large number of iterations to compute f[35]. To eliminate this problem, we can store each value of f in memory as it is computed. The values can then be recalled almost instantaneously.

EXAMPLE 65

```
f[1] = 1;
f[2] = 1;
f[n_] := f[n] = f[n - 2] + f[n - 1]
f[35]//Timing
{0., 9227465}
```

← This causes *Mathematica* to store each f[n] value.
Type ?f after computing f[35] to confirm this.

Functions of two or more variables can be defined in an analogous manner. The syntax is self-explanatory.

EXAMPLE 66

```
f[x_, y_] = x² + y³;
f[2, 3]
31
f[3, 2]
17
```

EXAMPLE 67

```
g[x_, y_, z_] = x + y * z;
g[2, 3, 4]
14
```

SOLVED PROBLEMS

2.65 Define $f(x)$ to be the polynomial $x^5 + 3x^4 - 7x^2 + 2$ and compute $f(2)$.

SOLUTION

```
f[x_] = x⁵ + 3x⁴ - 7x² + 2
2 - 7x² + 3x⁴ + x⁵
f[2]
54
```

2.66 Let $f(x) = \begin{cases} -x & \text{if } x \le 0 \\ x^2 & \text{if } 0 < x \le 3 \\ 18 - 3x & \text{if } x > 3 \end{cases}$

Sketch the graph of $f(x)$ for $-6 \le x \le 6$.

SOLUTION

```
f[x_] := -x /; x ≤ 0
f[x_] := x² /; 0 < x ≤ 3
f[x_] := 18 - 3x /; x > 3
Plot[f[x], {x, -6, 6}]
```

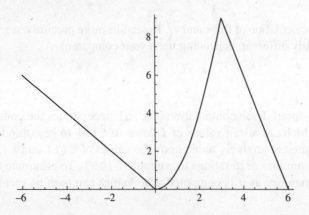

2.67 If $f(x)$ is defined on an interval $[a, b]$, the periodic extension of f with period $T = b - a$ is the function F such that

$$F(x) = \begin{cases} f(x) & \text{if } a \le x \le b \\ f(x - T) & \text{otherwise} \end{cases}$$

Let $f(x) = x^2$ if $-1 \le x \le 1$. Plot the periodic extension of f with period 2 from $x = 0$ to $x = 10$.

SOLUTION

```
f[x_]=x²;
F[x_]:=f[x]/;-1≤x≤1
F[x_]:=F[x-2]/;x>1
Plot[F[x],{x, 0, 10}]
```

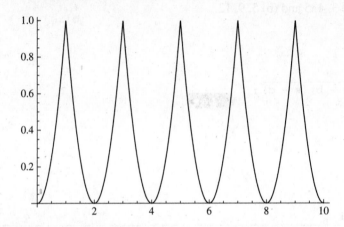

2.68 Define the function $f(n)$:
$$\begin{cases} f(1) = 1 \\ f(2) = 2 \\ f(3) = 3 \\ f(n) = f(n-3) + f(n-2) + f(n-1) & \text{if } n \ge 4 \end{cases}$$

Compute $f(20)$.

SOLUTION

```
Clear[f]
f[1]=1;
f[2]=2;
f[3]=3;
f[n_]:=f[n]=f[n-3]+f[n-2]+f[n-1];
f[20]
101902
```

2.69 Define a function that represents the distance from the point (x, y) to $(3, 4)$ and compute the value of the function at the point $(5, -2)$.

SOLUTION

```
f[x_, y_]= √((x - 3)² + (y - 4)²);
f[5,-2]
2√10
```

2.70 Define a function that represents the distance between the points (x_1, y_1) and (x_2, y_2) and use it to compute the distance from $(2, 3)$ to $(8, 11)$.

SOLUTION

```
d[x1_, y1_, x2_, y2_] = √((x2 - x1)² + (y2 - y1)²);
d[2, 3, 8, 11]
10
```

2.71 The area enclosed by a triangle whose sides have length a, b, and c is given by Heron's formula:

$$K = \sqrt{s(s-a)(s-b)(s-c)}$$

where $s = \dfrac{a+b+c}{2}$. Express the area of a triangle as a function of a, b, and c and compute the area of the triangle whose sides are (a) 3, 4, 5 and (b) 5, 9, 12

SOLUTION

```
s = a+b+c/2 ;
k[a_, b_, c_] = √(s (s - a) (s - b) (s - c)) ;
k[3, 4, 5]
6
k[5, 9, 12]
4√26
```

2.11 Operations on Functions

If f and g are two functions with the same domain, D, we define their sum, difference, product, and quotient pointwise, that is,

$$(f + g)(x) = f(x) + g(x) \qquad \text{for all } x \text{ in D}$$
$$(f - g)(x) = f(x) - g(x) \qquad \text{for all } x \text{ in D}$$
$$(fg)(x) = f(x)\, g(x) \qquad \text{for all } x \text{ in D}$$
$$(f/g)(x) = f(x)/g(x) \qquad \text{for all } x \text{ in D for which } g(x) \neq 0$$

If x is a number in the domain of g such that $g(x)$ is in the domain of f, we define the composite function $f \circ g$:

$$(f \circ g)(x) = f(g(x))$$

The function $g \circ f$ can be defined in a similar manner. The following example illustrates how to construct these functions.

EXAMPLE 68

```
f[x_] = √x ;
g[x_] = x² + 2 x + 3;
h1[x_] = f[x] + g[x]
3 + √x + 2 x + x²
h2[x_] = f[x] – g[x]
-3 + √x - 2 x - x²
h3[x_] = f[x] g[x]
√x (3 + 2 x + x²)
```

```
h4[x_] = f[x]/g[x]
```

$$\frac{\sqrt{x}}{3 + 2x + x^2}$$

```
h5[x_] = f[g[x]]
```

$$\sqrt{3 + 2x + x^2}$$

```
h6[x_] = g[f[x]]
```

$$3 + 2\sqrt{x} + x$$

The composition of two or more functions can be accomplished with the **Composition** command. Note that **Composition** is a *functional* operation and as such, its arguments are functions, f, not f[x].

- **Composition[f1, f2, f3, ...]** constructs the composition f1∘f2∘f3...

EXAMPLE 69

```
f[x_] = √x ;
g[x_] = x² + 2x + 3;
h1 = Composition[f, g];
h1[x]
```

$$\sqrt{3 + 2x + x^2}$$

```
h2 = Composition[g, f];
h2[x]
```

$$3 + 2\sqrt{x} + x$$

If we wish to compute the composition of a function with itself we could, of course, use **Composition[f, f]**, **Composition[f, f, f]**, and so forth. A more convenient tool is **Nest** or **NestList**.

- **Nest[f, *expression*, n]** applies f to *expression* successively n times.
- **NestList[f, *expression*, n]** applies f to *expression* successively n times and returns a list of all the intermediate calculations from 0 to n. (Lists are discussed in detail in Chapter 3.)

EXAMPLE 70

```
f[x_] = x²;
Nest[f, x, 5]
```
$$x^{32}$$

```
NestList[f, x, 5]
```
$$\{x, x^2, x^4, x^8, x^{16}, x^{32}\}$$

```
Nest[f, 2x + 3, 5]
```
$$(3 + 2x)^{32}$$

```
NestList[f, 2x + 3, 5]
```
$$\{3 + 2x, (3 + 2x)^2, (3 + 2x)^4, (3 + 2x)^8, (3 + 2x)^{16}, (3 + 2x)^{32}\}$$

EXAMPLE 71 The function **Framed** [*symbol*] draws a frame around *symbol*. We can use **Nest** and **NestList** to show the effect of repetitive framing.

`Nest[Framed,x,10]`

`NestList[Framed,x,10]`

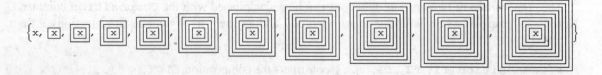

SOLVED PROBLEMS

2.72　If $f(x) = \sin x + 2 \cos x$ and $g(x) = 2 \sin x - 3 \cos x$, construct $(f + g)(x)$, $(f - g)(x)$, $(f g)(x)$, and $(f/g)(x)$ and evaluate them at $\pi/2$.

SOLUTION

```
f[x_] = Sin[x] + 2 Cos[x];
g[x_] = 2 Sin[x] - 3 Cos[x];
h1[x_] = f[x] + g[x]
```

$-$Cos[x] + 3 Sin[x]

```
h2[x_] = f[x] - g[x]
```

5 Cos[x] $-$ Sin[x]

```
h3[x_] = f[x] g[x]
```

(2 Cos[x] + Sin[x]) ($-$3 Cos[x] + 2 Sin[x])

```
h4[x_] = f[x]/g[x]
```

$$\frac{2\,\text{Cos}[x] + \text{Sin}[x]}{-3\,\text{Cos}[x] + 2\,\text{Sin}[x]}$$

```
h1[π/2]
```

3

```
h2[π/2]
```

-1

```
h3[π/2]
```

2

```
h4[π/2]
```

$\dfrac{1}{2}$

2.73　Let $f(x) = \sqrt{1 + x}$. Compute $(fofofofof)(x)$.

SOLUTION

```
f[x_] = √1+x ;
Nest[f, x, 5]
```

$$\sqrt{1 + \sqrt{1 + \sqrt{1 + \sqrt{1 + \sqrt{1 + x}}}}}$$

2.74 Let $f(x) = \dfrac{1}{1+x}$. Let $f^n(x) = (\underbrace{f \circ f \circ ... \circ f}_{n})(x)$. Evaluate $f(x)$, $f^2(x)$, $f^3(x)$, $f^4(x)$, and $f^5(x)$. Then evaluate $f(1), f^2(1), f^3(1), \ldots, f^{20}(1)$. What do you observe? Convert to a decimal form and approximate $\lim\limits_{n \to \infty} f^n(1)$.

SOLUTION

```
f[x_] = 1/(1 + x)
```

```
NestList[f, x, 5]
```

$$\left\{ x, \frac{1}{1+x}, \frac{1}{1+\frac{1}{1+x}}, \frac{1}{1+\frac{1}{1+\frac{1}{1+x}}}, \frac{1}{1+\frac{1}{1+\frac{1}{1+\frac{1}{1+x}}}}, \frac{1}{1+\frac{1}{1+\frac{1}{1+\frac{1}{1+\frac{1}{1+x}}}}} \right\}$$

```
NestList[f, 1, 20]
```

$$\left\{ 1, \frac{1}{2}, \frac{2}{3}, \frac{3}{5}, \frac{5}{8}, \frac{8}{13}, \frac{13}{21}, \frac{21}{34}, \frac{34}{55}, \frac{55}{89}, \frac{89}{144}, \frac{144}{233}, \frac{233}{377}, \frac{377}{610}, \frac{610}{987}, \frac{987}{1597}, \frac{1597}{2584}, \frac{2584}{4181}, \frac{4181}{6765}, \frac{6765}{10\,946}, \frac{10\,946}{17\,711} \right\}$$

The numerators (and denominators) appear to be terms of the Fibonacci sequence.

```
NestList[f, 1, 20] //N
```

$\{1., 0.5, 0.666667, 0.6, 0.625, 0.615385, 0.619048, 0.617647, 0.618182, 0.617978, 0.618056, 0.618026, 0.618037, 0.618033, 0.618034, 0.618034, 0.618034, 0.618034, 0.618034, 0.618034, 0.618034\}$

The numbers appear to be approaching a limit of approximately 0.618034.

2.75 If x is an approximation to $\sqrt{a}$, it can be shown that $\dfrac{1}{2}\left(x + \dfrac{a}{x}\right)$ is a better approximation. (This is a special case of Newton's method.) Use **NestList** to observe the first 10 approximations obtained in computing $\sqrt{3}$, starting with $x = 100$.

SOLUTION

```
a = 3;
```

```
f[x_] = 1/2 (x + a/x);
```

```
NestList[f, 100, 10] //N
```

$\{100., 50.015, 25.0375, 12.5787, 6.40858, 3.43835, 2.15543, 1.77363, 1.73254, 1.73205, 1.73205\}$

CHAPTER 3

Lists

3.1 Introduction

Lists are general objects that contain collections of other objects. In reading this chapter you will see that lists are used for a variety of applications. Therefore, *Mathematica* offers an extensive collection of list manipulation commands.

The objects within a list are contained within curly brackets, { }. Alternatively, the **List** command may be used to define a list.

- **List [*elements*]** represents a list of objects. *elements* represents the members of the list separated by commas. **List [*elements*]** is equivalent to **{*elements*}**.

{1, 2, 3, 4} is a list of numbers. **List[1, 2, 3, 4]** represents the same list.

EXAMPLE 1

```
List[a, b, c, d]
{a, b, c, d}                          ← List [a, b, c, d] is equivalent to {a, b, c, d}.
```

Lists can be given symbolic names so they can be easily referenced. Any operation performed on a list will be performed on *each* element of the list.

EXAMPLE 2

```
list = {1, 2, 3, 4, 5, 6, 7, 8, 9, 10}
{1, 2, 3, 4, 5, 6, 7, 8, 9, 10}
1/list
```
$$\left\{1, \frac{1}{2}, \frac{1}{3}, \frac{1}{4}, \frac{1}{5}, \frac{1}{6}, \frac{1}{7}, \frac{1}{8}, \frac{1}{9}, \frac{1}{10}\right\}$$

> When executing these instructions, care must be taken to use a lowercase **l** in **list** to avoid conflict with the *Mathematica* command **List**.

```
list²
{1, 4, 9, 16, 25, 36, 49, 64, 81, 100}
√list
```
$$\{1, \sqrt{2}, \sqrt{3}, 2, \sqrt{5}, \sqrt{6}, \sqrt{7}, 2\sqrt{2}, 3, \sqrt{10}\}$$

If two or more lists contain the same number of elements, new lists can be created using standard operations.

EXAMPLE 3

```
list1 = {1, 2, 3, 4, 5};
list2 = {2, 3, 2, 3, 2};
list1 + list2
{3, 5, 5, 7, 7}
```

```
list1 * list2
{2, 6, 6, 12, 10}
list1/list2
```

$$\left\{\frac{1}{2}, \frac{2}{3}, \frac{3}{2}, \frac{4}{3}, \frac{5}{2}\right\}$$

```
list1^list2
{1, 8, 9, 64, 25}
```

The following list commands are simple but extremely useful:

- **Total** [*list*] gives the sum of the elements of *list*.
- **Accumulate** [*list*] returns a list having the same length as *list* containing the successive partial sums of *list*.
- **Max** [*list*] returns the largest number in *list*.
- **Min** [*list*] returns the smallest number in *list*.

EXAMPLE 4

```
list = {1, 2, 3, 4, 5}
Total[list]
15
Accumulate[list]
{1, 3, 6, 10, 15}
Max[list]
5
Min[list]
1
```

SOLVED PROBLEMS

3.1 Construct a list of the factorials of the integers 1 through 10.

SOLUTION

```
list = {1, 2, 3, 4, 5, 6, 7, 8, 9, 10};
list!
{1, 2, 6, 24, 120, 720, 5040, 40 320, 362 880, 3 628 800}
```

3.2 Construct a list of the first ten positive integer powers of 2.

SOLUTION

```
list = {1, 2, 3, 4, 5, 6, 7, 8, 9, 10};
2^list
{2, 4, 8, 16, 32, 64, 128, 256, 512, 1024}
```

3.3 Construct a list whose elements are the sum of the squares of the first five positive integers added to their respective cubes.

SOLUTION

```
list = (1, 2, 3, 4, 5}
list^2 + list^3   or   list^2 + list^3
{2, 12, 36, 80, 150}
```

3.4 Define list1 = {1, 3, 5, 7, 9} and list2 = {2, 4, 6, 8, 10}. Construct a list whose five elements are the products of the entries of the two lists.

SOLUTION

```
list1 = {1, 3, 5, 7, 9};
list2 = {2, 4, 6, 8, 10};
list1 * list2
{2, 12, 30, 56, 90}
```

3.2 Generating Lists

The most common lists are lists of equally spaced numbers. The **Range** command allows convenient construction. The values of m, n, and d in the following description need not be integer valued. Negative values are acceptable as well.

- **Range[n]** generates a list of the first n consecutive integers.
- **Range[m, n]** generates a list of numbers from m to n in unit increments.
- **Range[m, n, d]** generates a list of numbers from m through n in increments of d.

EXAMPLE 5

```
Range[10]
{1, 2, 3, 4, 5, 6, 7, 8, 9, 10}
Range[5, 10]
{5, 6, 7, 8, 9, 10}
Range[25, 5, -2]
{25, 23, 21, 19, 17, 15, 13, 11, 9, 7, 5}
Range[1/3, 1, 1/12]
```
$$\left\{\frac{1}{3}, \frac{5}{12}, \frac{1}{2}, \frac{7}{12}, \frac{2}{3}, \frac{3}{4}, \frac{5}{6}, \frac{11}{12}, 1\right\}$$
```
Range[1, 2, .1]
{1, 1.1, 1.2, 1.3, 1.4, 1.5, 1.6, 1.7, 1.8, 1.9, 2.}
```

Lists with more complicated structures can be constructed with the **Table** command. There are several different forms.

- **Table[*expression*, {n}]** generates a list containing n copies of the object *expression*.
- **Table[*expression*, {k, n}]** generates a list of the values of *expression* as k varies from 1 to n.
- **Table[*expression*, {k, m, n}]** generates a list of the values of *expression* as k varies from m to n.
- **Table[*expression*, {k, m, n, d}]** generates a list of the values of *expression* as k varies from m to n in steps of d.

EXAMPLE 6

```
Table["Mathematica", {10}]
{Mathematica, Mathematica, Mathematica, Mathematica, Mathematica,
  Mathematica, Mathematica, Mathematica, Mathematica, Mathematica}
Table[k², { k, 10}]
{1, 4, 9, 16, 25, 36, 49, 64, 81, 100}
Table[1/k, {k, 5, 13}]
```
$$\left\{\frac{1}{5}, \frac{1}{6}, \frac{1}{7}, \frac{1}{8}, \frac{1}{9}, \frac{1}{10}, \frac{1}{11}, \frac{1}{12}, \frac{1}{13}\right\}$$
```
Table[√k , {k, 5, 13, 2}]
```
$$\left\{\sqrt{5}, \sqrt{7}, 3, \sqrt{11}, \sqrt{13}\right\}$$

The command **Array** is useful for defining sequences.

- **Array[f, n]** generates a list consisting of n values, f[1], f[2], ..., f[n].
- **Array[f, n, r]** generates a list consisting of n values, f[i], starting with f[r], i.e., f[r], f[r+1], ..., f[r+n-1].

EXAMPLE 7

```
Clear[f]
Array[f,7]
{f[1],f[2],f[3],f[4],f[5],f[6],f[7]}
Array[f,7,3]
{f[3],f[4],f[5],f[6],f[7],f[8],f[9]}
```

EXAMPLE 8

```
f[x_] = x² + x + 1;
Array[f, 7]
{3, 7, 13, 21, 31, 43, 57}
Array[f,7,3]
{13, 21, 31, 43, 57, 73, 91}
Array[f, 7, 0]
{1, 3, 7, 13, 21, 31, 43}          ← The first element is f[0].
Array[f, 7, -2]                     ← Negative values are allowed in the third position only.
{3, 1, 1, 3, 7, 13, 21}
```

Nested lists are lists that contain lists. For example,

```
{{1, 2, 3, 4}, {2, 3, 4, 5}, {3, 4, 5, 6}}
```

is a nested list of depth two, consisting of three lists, each of which is a list of four integers.

Nested lists can be generated using the **Table** and **Array** commands. All indices have unit increments.

- **Table[*expression*, {m}, {n}]** generates a two-dimensional list, each element of which is the object *expression*.
- **Table[*expression*, {i, m_i, n_i}, {j, m_j, n_j}]** generates a nested list whose values are *expression*, computed as j goes from m_j to n_j and as i goes from m_i to n_i. The index j varies most rapidly.
- **Array[f, {m, n}]** generates a nested list consisting of an array of m elements, each of which is an array of n elements, whose values are f[i, j] as j goes from 1 to n and i goes from 1 to m. Here f is a function of two variables. The second index varies most rapidly.
- **Array[f, {m, n}, {r, s}]** generates a nested list consisting of an array of m elements, each of which is an array of n elements. The first element of the first sublist is f[r, s].

Each of the previous descriptions extends in a natural way to lists of greater depth.

EXAMPLE 9

```
Table["Mathematica", {3}, {4}]
{{Mathematica, Mathematica, Mathematica, Mathematica},
 {Mathematica, Mathematica, Mathematica, Mathematica},
 {Mathematica, Mathematica, Mathematica, Mathematica}}
```

EXAMPLE 10

```
Table[i + j, {i, 1, 3}, {j, 1, 5}]
{{2, 3, 4, 5, 6}, {3, 4, 5, 6, 7}, {4, 5, 6, 7, 8}}
```
```
    i=1              i=2              i=3
  j=1,2,3,4,5      j=1,2,3,4,5      j=1,2,3,4,5
```

```
Table[i + j, {i, 1, 5}, {j, 1, 3}]
```

$$\underbrace{\{\{2, 3, 4\}}_{\substack{i=1\\j=1,2,3}}, \underbrace{\{3, 4, 5\}}_{\substack{i=2\\j=1,2,3}}, \underbrace{\{4, 5, 6\}}_{\substack{i=3\\j=1,2,3}}, \underbrace{\{5, 6, 7\}}_{\substack{i=4\\j=1,2,3}}, \underbrace{\{6, 7, 8\}\}}_{\substack{i=5\\j=1,2,3}}$$

EXAMPLE 11

```
Clear[f]
Array[f,{3,4}]
{{f[1,1], f[1,2], f[1,3], f[1,4]},
 {f[2,1], f[2,2], f[2,3], f[2,4]},
 {f[3,1], f[3,2], f[3,3], f[3,4]}}
```

EXAMPLE 12

```
f[x_, y_] = x² + 3 y
```
$$f[x\_, y\_] = x^2 + 3\,y$$
```
Array[f, {3, 4}]
{{4, 7, 10, 13}, {7, 10, 13, 16}, {12, 15, 18, 21}}
Array[f, {4, 3}]
{{4, 7, 10}, {7, 10, 13}, {12, 15, 18}, {19, 22, 25}}
Array[f, {4, 3}, {0, 0}]
{{0, 3, 6}, {1, 4, 7}, {4, 7, 10}, {9, 12, 15}}
```

Often it will be convenient to construct lists of letters and other characters.

- **Characters** [*string*] produces a list of characters in *string*.
- **CharacterRange** ["*char1*", "*char2*"] produces a list of characters from *char1* to *char2*, based upon their standard ASCII values (assuming an American English alphabet).

EXAMPLE 13

```
Characters["Mathematica"]
{M, a, t, h, e, m, a, t, i, c, a}
```

EXAMPLE 14

```
CharacterRange["a","e"]
{a, b, c, d, e}
CharacterRange[" "," ~"]
{ , !, \, #, $, %, &, ', (,), *, +, , , -, ., ., /, 0, 1, 2, 3, 4, 5, 6, 7, 8, 9, :, , ;, <, =, >, ?, @,
A, B, C, D, E, F, G, H, I, J, K, L, M, N, O, P, Q, R, S, T, U, V, W, X, Y, Z, [, \\, ], ^, _, `, a, b,
c, d, e, f, g, h, i, j, k, l, m, n, o, p, q, r, s, t, u, v, w, x, y, z, {, |, }, ~}
```

Even though the output of **Characters** and **CharacterRange** appears to be individual characters, in actuality they are strings of length 1. By *Mathematica*'s convention, quotation marks are not printed.

EXAMPLE 15

```
digits = CharacterRange["0"," 9"]
{0, 1, 2, 3, 4, 5, 6, 7, 8, 9}        ← These are not numbers but strings of characters of length 1.
FullForm[digits]
List["0", "1", "2", "3", "4", "5", "6", "7", "8", "9"]
```

SOLVED PROBLEMS

3.5 Construct a list of the positive multiples of 7 that do not exceed 100.

SOLUTION 1

```
Range[7, 100, 7]
{7, 14, 21, 28, 35, 42, 49, 56, 63, 70, 77, 84, 91, 98}
```

SOLUTION 2

```
Table[7 k, {k, 1, 14}]
{7, 14, 21, 28, 35, 42, 49, 56, 63, 70, 77, 84, 91, 98}
```

3.6 Construct a list of the first ten prime numbers.

SOLUTION 1

```
Table[Prime[k], {k, 1, 10}]
{2, 3, 5, 7, 11, 13, 17, 19, 23, 29}
```

SOLUTION 2

```
Array[Prime, 10]
{2, 3, 5, 7, 11, 13, 17, 19, 23, 29}
```

SOLUTION 3

```
Prime[Range[10]]
{2, 3, 5, 7, 11, 13, 17, 19, 23, 29}
```

3.7 Construct a list of the reciprocals of the first ten even integers.

SOLUTION 1

```
Table[1/k, {k, 2, 20, 2}]
```
$$\left\{ \frac{1}{2}, \frac{1}{4}, \frac{1}{6}, \frac{1}{8}, \frac{1}{10}, \frac{1}{12}, \frac{1}{14}, \frac{1}{16}, \frac{1}{18}, \frac{1}{20} \right\}$$

SOLUTION 2

$$\frac{1}{\text{Range}[2, 20, 2]}$$
$$\left\{ \frac{1}{2}, \frac{1}{4}, \frac{1}{6}, \frac{1}{8}, \frac{1}{10}, \frac{1}{12}, \frac{1}{14}, \frac{1}{16}, \frac{1}{18}, \frac{1}{20} \right\}$$

3.8 Construct a list of five objects, each of which is a list consisting of six integers. The first list is to contain the first six multiples of 2, the second, multiples of 3, the third, multiples of 4, and so forth.

SOLUTION 1

```
Table[i * j, {i, 2, 6}, {j, 1, 6}]
{{2, 4, 6, 8, 10, 12}, {3, 6, 9, 12, 15, 18}, {4, 8, 12, 16, 20, 24},
  {5, 10, 15, 20, 25, 30}, {6, 12, 18, 24, 30, 36}}
```

SOLUTION 2

```
                              Times[x, y] = x * y
Array[Times, {5, 6}, {2, 1}]
{{2, 4, 6, 8, 10, 12}, {3, 6, 9, 12, 15, 18}, {4, 8, 12, 16, 20, 24},
  {5, 10, 15, 20, 25, 30}, {6, 12, 18, 24, 30, 36}}
```

3.9 Let $p(x) = x^2 - 8x + 10$. Construct a list of values of $p(x)$ for $x = 1, 2, 3, \ldots, 10$.

SOLUTION 1

```
p[x_] = x^2 - 8 x + 10;
Array[p, 10]
{3, -2, -5, -6, -5, -2, 3, 10, 19, 30}
```

SOLUTION 2

```
p[x_] = x^2 - 8 x + 10;
p[Range[10]]
{3, -2, -5, -6, -5, -2, 3, 10, 19, 30}
```

3.10 Approximate the sum of the square roots of the first 100 positive integers.

SOLUTION

```
Total[Sqrt[Range[100]]] //N
671.463
```

3.3   List Manipulation

- **Length** [*list*] returns the length of *list*, i.e., the number of elements in *list*.
- **First** [*list*] returns the element of *list* in the first position.
- **Last** [*list*] returns the element of *list* in the last position.

EXAMPLE 16

```
list = {a, b, c, d, e, f, g};
Length[list]
7
First[list]
a
Last[list]
g
```

The function **Part** returns individual elements of a list.

- **Part[list, k]** or **list[[k]]** returns the kth element of **list**.
- **Part[list, -k]** or **list[[-k]]** returns the kth element from the end of **list**.

Note: **Part[list, 1]** and **Part[list, -1]** are equivalent to **First[list]** and **Last[list]**, respectively.

EXAMPLE 17

```
list = {a, b, c, d, e, f, g};
Part[list, 1]   or   list[[1]]
a
Part[list, 3]   or   list[[3]]
c
Part[list, -3]   or   list[[-3]]
e
Part[list, -1]   or   list[[-1]]
g
```

Lists may be nested. The elements of a list may themselves be lists.

EXAMPLE 18

```
list = {{a, b, c, d}, {e, f, g, h}, {i, j, k, l}};
First[list]
{a, b, c, d}
Last[list]
{i, j, k, l}
list[[2]]
{e, f, g, h}
```

Since **list[[2]]** is itself a list, its third entry, for example, can be obtained as **list[[2]][[3]]** (the third entry of the second list). For convenience, this can be represented as **list[[2, 3]]** or **Part[list, 2, 3]**. **Part[Part[list, 2], 3]** can also be used, but is somewhat clumsy.

- **Part[list, m, n]** or **list[[m, n]]** returns the nth entry of the mth element of **list**, provided **list** has depth at least 2.

This command extends to lists of depth greater than 2 in a natural way provided the **Part** specification does not exceed the depth of the list.

EXAMPLE 19

```
list = {{a, b, c, d}, {e, f, g, h}, {i, j, k, l}};
list[[2]][[3]]
g
list[[2, 3]]
g
Part[list, 2, 3]
g
Part[Part[list, 2], 3]
g
```

Lists can be modified several different ways. If **list** is any list of objects,

- **Rest**[*list*] returns *list* with its *first* element deleted.
- **Take**[*list*, n] returns a list consisting of the first n elements of *list*.
- **Take**[*list*, {n}] returns a list consisting of the nth element of *list*.
- **Take**[*list*, −n] returns a list consisting of the last n elements of *list*.
- **Take**[*list*, {−n}] returns a list consisting of the nth element from the end of *list*.
- **Take**[*list*, {m, n}] returns a list consisting of the elements of *list* in positions m through n inclusive.
- **Take**[*list*, {m, n, k}] returns a list consisting of the elements of *list* in positions m through n in increments of k.

EXAMPLE 20

```
list = {a, b, c, d, e, f, g};
Rest[list]
{b, c, d, e, f, g}
Take[list, 3]
{a, b, c}
Take[list, -3]
{e, f, g}
Take[list, {3}]
{c}
```

```
Take[list, {-3}]
{e}
Take[list, {2, 5}]
{b, c, d, e}
Take[list, {1, 5, 2}]
{a, c, e}
```

Elements can be deleted from a list by using the **Delete** command.

- **Delete**[*list*, n] deletes the element in the nth position of *list*.
- **Delete**[*list*, −n] deletes the element in the nth position from the end of *list*.
- **Delete**[*list*, {{p_1}, {p_2}, ...}] deletes the elements in positions p_1, p_2, ...

EXAMPLE 21

```
list = {a, b, c, d, e, f, g};
Delete[list, 3]
{a, b, d, e, f, g}
Delete[list, -3]
{a, b, c, d, f, g}
Delete[list,{{2}, {5}, {6}}]
{a, c, d, g}
```

Delete can also be used for lists of greater depth.

- **Delete**[list, {p, q}] deletes the element in position q of part p.
- **Delete**[list, {{p_1, q_1}, {p_2, q_2}, . . .] deletes the elements in position q_1 of part p_1, position q_2 of part p_2, ...

This command extends in a natural way to lists of greater depth.

EXAMPLE 22

```
list = {{1, 2, 3}, {4, 5}, {6, 7, 8, 9}};
Delete[list, 2]
{{1, 2, 3}, {6, 7, 8, 9}}
Delete[list, {3, 2}]
{{1, 2, 3}, {4, 5}, {6, 8, 9}}        ← The second element of the third sublist is deleted.
Delete[list, {{1, 2}, {3, 3}}]
{{1, 3}, {4, 5}, {6, 7, 9}}           ← The second element of the first sublist and the third element of
                                        the third sublist are deleted.
```

The function **Drop** is similar to **Delete** and allows a little more flexibility.

- **Drop**[*list*, n] returns *list* with its first n objects deleted.
- **Drop**[*list*, −n] returns *list* with its last n objects deleted.
- **Drop**[*list*, {n}] returns *list* with its nth object deleted.
- **Drop**[*list*, {−n}] returns *list* with the nth object from the end deleted.
- **Drop**[*list*, {m, n}] returns *list* with objects m through n deleted.
- **Drop**[*list*, {m, n, k}] returns *list* with objects m through n in increments of k deleted.

Note: **Drop**[*list*, {n}] is equivalent to **Delete**[*list*, n] and **Drop**[*list*, {−n}] is equivalent to **Delete**[*list*, −n].

EXAMPLE 23

```
list = {a, b, c, d, e, f, g};
Drop[list, 2]
```

```
{c, d, e, f, g}
Drop[list, -2]
{a, b, c, d, e}
Drop[list, {2}]
{a, c, d, e, f, g}
Drop[list, {-2}]
{a, b, c, d, e, g}
Drop[list, {2, 4}]
{a, e, f, g}
Drop[list, {1,7,2}]
{b, d, f}
```

There are a variety of list functions that allow elements to be inserted into a list.

- **Append**[*list*, **x**] returns *list* with x inserted to the right of its last element.
- **Prepend**[*list*, **x**] returns *list* with x inserted to the left of its first element.
- **Insert**[*list*, **x**, **n**] returns *list* with x inserted in position n.
- **Insert**[*list*, **x**, **-n**] returns *list* with x inserted in the nth position from the end.

If *list* has a depth of 2, the following form can be used to insert elements:

- **Insert**[*list*, **x**, **{m, n}**] returns *list* with x inserted in the nth position of the mth entry in the outer level.

This command extends in a natural way to lists of greater depth.

EXAMPLE 24

```
list = {1, 2, 3, 4, 5, 6, 7, 8, 9, 10};
Append[list, x]
{1, 2, 3, 4, 5, 6, 7, 8, 9, 10, x}
Prepend[list, x]
{x, 1, 2, 3, 4, 5, 6, 7, 8, 9, 10}
Insert[list, x, 4]
{1, 2, 3, x, 4, 5, 6, 7, 8, 9, 10}
Insert[list, x, -4]
{1, 2, 3, 4, 5, 6, 7, x, 8, 9, 10}
```

EXAMPLE 25

```
list = {{1, 2, 3}, {4, 5}, {6, 7, 8, 9}};
Insert[list, x, {3, 2}]
{{1, 2, 3}, {4, 5}, {6, x, 7, 8, 9}}
```

Objects in a list can be replaced by other objects using **ReplacePart**.

- **ReplacePart**[*list*, **x**, **n**] replaces the object in the nth position of *list* by x.
- **ReplacePart**[*list*, **x**, **-n**] replaces the object in the nth position from the end by x.

ReplacePart can also be invoked using the following syntax, which allows a bit more flexibility:

- **ReplacePart**[*list*, **i** → *new*] replaces the ith part of *list* with *new*.
- **ReplacePart**[*list*, **{i_1 → new_1, i_2 → new_2, . . . , i_n → new_n}**] replaces parts i_1, i_2, . . . , i_n with *new*$_1$, *new*$_2$, . . ., *new*$_n$, respectively.
- **ReplacePart**[*list*, **{{i_1}, {i_2}, . . . , {i_n}}** → *new*] replaces all elements in positions i_1, i_2, . . . , i_n with *new*.

If *list* has a depth of 2, the following form can be used to replace elements:

- **ReplacePart** [*list*, {i, j} → *new*] replaces the element in position j of the ith outer level entry with *new*.
- **ReplacePart** [*list*, {i$_1$, j$_1$} → *new*$_1$, {i$_2$, j$_2$} → *new*$_2$, ..., {i$_n$, j$_n$} → *new*$_n$] replaces the entries in positions j$_k$ of entry i$_k$ in the outer level with *new*$_k$.
- **ReplacePart** [*list*, {{i$_1$, j$_1$}, {i$_2$, j$_2$}, ..., {i$_n$, j$_n$}} → *new*] replaces all entries in positions j$_k$ of entry i$_k$ in the outer level with *new*.

This command extends in a natural way to lists of greater depth.

EXAMPLE 26

```
list = Range[10]
{1, 2, 3, 4, 5, 6, 7, 8, 9, 10}
ReplacePart[list, x, 7]
{1, 2, 3, 4, 5, 6, x, 8, 9, 10}
ReplacePart[list, x, -7]
{1, 2, 3, x, 5, 6, 7, 8, 9, 10}
ReplacePart[list, 2 → x]
{1, x, 3, 4, 5, 6, 7, 8, 9, 10}
ReplacePart[list, {2 → x, 4 → y, 7 → z}]
{1, x, 3, y, 5, 6, z, 8, 9, 10}
ReplacePart[list, {{3}, {5}, {7}} → x]
{1, 2, x, 4, x, 6, x, 8, 9, 10}
```

EXAMPLE 27

```
list = {{a, b, c}, {d, e}, {f, g, h, i, j}};
ReplacePart[list, {3, 2} → x]
{{a, b, c}, {d, e}, {f, x, h, i, j}}
ReplacePart[list, {{1, 3} → x, {3, 2} → y}]
{{a, b, x}, {d, e}, {f, y, h, i, j}}
ReplacePart[list, {{1, 2}, {2, 1}, {3, 4}} → x]
{{a, x, c}, {x, e}, {f, g, h, x, j}}
```

Lists can be rearranged using **Sort** and **Reverse**.

- **Sort** [*list*] sorts *list* in increasing order. Real numbers are ordered according to their numerical value. Letters are arranged lexicographically, with capital letters coming after lowercase letters.
- **Reverse** [*list*] reverses the order of the elements of *list*.

EXAMPLE 28

```
list = {1, 5, -3, 0, 2.5};
Sort[list]
{-3, 0, 1, 2.5, 5}
```

EXAMPLE 29

```
list = {z, x, Y, w, X, y, Z, W};
Sort[list]
{w, W, x, X, y, Y, z, Z}
```

EXAMPLE 30

```
list = {a, b, c, d, e, f, g};
```

```
Reverse[list]
{g, f, e, d, c, b, a}
```

Cycling of lists is made possible by use of the functions **RotateLeft** and **RotateRight**.

- **RotateLeft** [*list*] cycles each element of *list* one position to the left. The leftmost element is moved to the extreme right of the list.
- **RotateLeft** [*list*, n] cycles the elements of *list* precisely n positions to the left. The leftmost n elements are moved to the extreme right of the list in their same relative positions. If n is negative, rotation occurs to the right.
- **RotateRight** [*list*] cycles each element of *list* one position to the right. The rightmost element is moved to the extreme left of the list.
- **RotateRight** [*list*, n] cycles the elements of *list* precisely n positions to the right. The rightmost n elements are moved to the extreme left of the list in their same relative positions. If n is negative, rotation occurs to the left.

EXAMPLE 31

```
list = {1, 2, 3, 4, 5, 6, 7, 8, 9, 10};
RotateLeft[list]
{2, 3, 4, 5, 6, 7, 8, 9, 10, 1}
RotateLeft[list, 3]
{4, 5, 6, 7, 8, 9, 10, 1, 2, 3}
RotateLeft[list, -3]
{8, 9, 10, 1, 2, 3, 4, 5, 6, 7}
RotateRight[list]
{10, 1, 2, 3, 4, 5, 6, 7, 8, 9}
RotateRight[list, 3]
{8, 9, 10, 1, 2, 3, 4, 5, 6, 7}
RotateRight[list, -3]
{4, 5, 6, 7, 8, 9, 10, 1, 2, 3}
```

Lists can be concatenated using **Join**.

- **Join** [*list1*, *list2*] combines the two lists *list1* and *list2* into one list consisting of the elements from *list1* and *list2*.

Join makes no attempt to eliminate repetitive elements. However, repetition can be conveniently eliminated with the **Union** command (see Section 3.4).

Join can be generalized in a natural way to combine more than two lists.

EXAMPLE 32

```
list1 = {1, 2, 3, 4, 5};
list2 = {3, 4, 5, 6, 7};
Join[list1, list2]
{1, 2, 3, 4, 5, 3, 4, 5, 6, 7}
```

Nested lists, which are very common, can have a complicated structure. There are a few *Mathematica* commands that can help you understand and manipulate them.

- **Depth** [*list*] returns *one more* than the number of levels in the list structure. Raw objects, i.e., objects that are not lists, have a depth of 1.
- **Level** [*list*, {*levelspec*}] returns a list consisting of those objects that are at level *levelspec* of *list*.
- **Level** [*list*, *levelspec*] returns a list consisting of those objects that are at or below level *levelspec* of *list*.

EXAMPLE 33

```
Depth[x]
1                       ← x is not a list.
Depth[{x}]
2
Depth[{{x}}]
3
```

EXAMPLE 34

```
list = {1, {2, {3, 4, 5}}};
Depth[list]
4
Level[list, {1}]
{1, {2, {3, 4, 5}}}
Level[list, {2}]
{2, {3, 4, 5}}
Level[list, {3}]
{3, 4, 5}
Level[list, 3]
{1, 2, 3, 4, 5, {3, 4, 5}, {2, {3, 4, 5}}}
```

> $4 - 1 = 3$. This tells us that `list` contains lists within lists within itself. Note that **Depth** always returns one more than the actual number of levels in the list. This is for technical reasons dealing with the structure of *Mathematica* commands. For now, just remember that the number of levels is always 1 less than **Depth**.

- **Flatten** [*list*] converts a nested list to a simple list containing the innermost objects of *list*.
- **Flatten** [*list*, n] flattens a nested list n times, each time removing the outermost level. The depth of each level is reduced by n or to a minimum level of 1.
- **FlattenAt** [*list*, n] flattens the sublist which is at the nth position of the list by one level. If n is negative, *Mathematica* counts backward, starting at the end of the list.

EXAMPLE 35

```
list = {1, {2, 3}, {4, 5, {6}}, {7, {8, {9, 10}}}}
Flatten[list]
{1, 2, 3, 4, 5, 6, 7, 8, 9, 10}
Flatten[list, 1]
{1, 2, 3, 4, 5, {6}, 7, {8, {9, 10}}}
Flatten[list, 2]
{1, 2, 3, 4, 5, 6, 7, 8, {9, 10}}
FlattenAt[list, 3]
{1, {2, 3}, 4, 5, {6}, {7, {8, {9, 10}}}}    ← Only the *third* sublist of *list* is flattened one level.
FlattenAt[list, -3]
{1, 2, 3, {4, 5, {6}}, {7, {8, {9, 10}}}}
```

Flatten converts a nested list into a simpler list. **Partition** takes simple lists and converts them into nested lists in a very organized and convenient way.

- **Partition** [*list*, k] converts *list* into sublists of length k. If *list* contains k n + m elements, where m < k, **Partition** will create n sublists and the remaining m elements will be dropped.
- **Partition** [*list*, k, d] partitions *list* into sublists of length k, offsetting each sublist from the previous sublist by d elements. In other words, each sublist (other than the first) begins with the d+1st element of the previous sublist.

Note that **Partition** [*list*, k] is equivalent to **Partition** [*list*, k, k].

Partition is a very convenient command for generating tables and matrices. Only the simplest forms of the command have been described. The reader, if interested, is urged to investigate other forms in *Mathematica*'s Documentation Center.

EXAMPLE 36

```
list = Range[12]
{1, 2, 3, 4, 5, 6, 7, 8, 9, 10, 11, 12}
Partition[list, 4]
{{1, 2, 3, 4}, {5, 6, 7, 8}, {9, 10, 11, 12}}
Partition[list, 5]
{{1, 2, 3, 4, 5}, {6, 7, 8, 9, 10}}
Partition[list, 5, 1]
{{1, 2, 3, 4, 5}, {2, 3, 4, 5, 6}, {3, 4, 5, 6, 7}, {4, 5, 6, 7, 8}, {5, 6, 7, 8, 9},
 {6, 7, 8, 9, 10}, {7, 8, 9, 10, 11}, {8, 9, 10, 11, 12}}
Partition[list, 5, 2]
{{1, 2, 3, 4, 5}, {3, 4, 5, 6, 7}, {5, 6, 7, 8, 9}, {7, 8, 9, 10, 11}}
Partition[list, 5, 3]
{{1, 2, 3, 4, 5}, {4, 5, 6, 7, 8}, {7, 8, 9, 10, 11}}
```

SOLVED PROBLEMS

3.11 The *Mathematica* function **IntegerDigits** returns a list containing the digits of an integer. How many digits are there in 100! and what is the 50th digit from the left and from the right?

SOLUTION

```
list = IntegerDigits[100!]
{9, 3, 3, 2, 6, 2, 1, 5, 4, 4, 3, 9, 4, 4, 1, 5, 2, 6, 8, 1, 6, 9, 9, 2, 3, 8, 8, 5, 6,
 2, 6, 6, 7, 0, 0, 4, 9, 0, 7, 1, 5, 9, 6, 8, 2, 6, 4, 3, 8, 1, 6, 2, 1, 4, 6, 8, 5, 9,
 2, 9, 6, 3, 8, 9, 5, 2, 1, 7, 5, 9, 9, 9, 9, 3, 2, 2, 9, 9, 1, 5, 6, 0, 8, 9, 4, 1, 4,
 6, 3, 9, 7, 6, 1, 5, 6, 5, 1, 8, 2, 8, 6, 2, 5, 3, 6, 9, 7, 9, 2, 0, 8, 2, 7, 2, 2, 3,
 7, 5, 8, 2, 5, 1, 1, 8, 5, 2, 1, 0, 9, 1, 6, 8, 6, 4, 0, 0, 0, 0, 0, 0, 0, 0, 0, 0, 0,
 0, 0, 0, 0, 0, 0, 0, 0, 0, 0, 0, 0, 0}
Length[list]
158
Part[list, 50]  or  list[[50]]
1
Part[list, -50]  or  list[[-50]]
2
```

3.12 Compute the sum of the digits of the 100th Fibonacci number.

SOLUTION

We use **IntegerDigits** (see previous problem).

```
list = IntegerDigits[Fibonacci[100]]
{3, 5, 4, 2, 2, 4, 8, 4, 8, 1, 7, 9, 2, 6, 1, 9, 1, 5, 0, 7, 5}
```

$$\text{Sum[list[[k]], \{k, 1, Length[list]\}]} \quad \text{or} \quad \sum_{k=1}^{Length[list]} \text{list[[k]]}$$

93

3.13 The command **Table[i*j, {i, 3, 10}, {j, 2, 7}]** generates a nested list of numbers. Add the fourth number in the fifth sublist to the third number in the sixth sublist.

SOLUTION

```
list = Table[i*j, {i, 3, 10}, {j, 2, 7}]
```

```
{{6, 9, 12, 15, 18, 21}, {8, 12, 16, 20, 24, 28}, {10, 15, 20, 25, 30, 35},
 {12, 18, 24, 30, 36, 42},{14, 21, 28, 35, 42, 49}, {16, 24, 32, 40, 48, 56},
 {18, 27, 36, 45, 54, 63}, {20, 30, 40, 50, 60, 70}}
list[[5, 4]] + list[[6, 3]]
67                                              ←35 + 32 = 67
```

3.14 The *Mathematica* function **RealDigits** returns a list containing a list of the digits of an approximate real number followed by the number of digits that are to the left of the decimal point. Compute a 15 significant digit approximation of π and determine the next to the last decimal digit.

SOLUTION

```
approx = N[Pi,15]
3.14159265358979
list = RealDigits[approx]
{{3, 1, 4, 1, 5, 9, 2, 6, 5, 3, 5, 8, 9, 7, 9}, 1}
list[[1, -2]]
7
```

3.15 Construct a list consisting of the consecutive integers from 1 to 10 followed by the consecutive integers from 20 to 30.

SOLUTION 1

```
Drop[Range[30], {11, 19}]
{1, 2, 3, 4, 5, 6, 7, 8, 9, 10, 20, 21, 22, 23, 24, 25, 26, 27, 28, 29, 30}
```

SOLUTION 2

```
Join[Range[1, 10], Range[20, 30]]
{1, 2, 3, 4, 5, 6, 7, 8, 9, 10, 20, 21, 22, 23, 24, 25, 26, 27, 28, 29, 30}
```

3.16 Construct a list consisting of the consecutive integers 1 to 10, followed by 99, followed by 11 to 20.

SOLUTION

```
Insert[Range[20], 99, 11]
{1, 2, 3, 4, 5, 6, 7, 8, 9, 10, 99, 11, 12, 13, 14, 15, 16, 17, 18, 19, 20}
```

3.17 Construct a list of the integers 1 to 20 in descending order.

SOLUTION 1

```
Range[20, 1, -1]
{20, 19, 18, 17, 16, 15, 14, 13, 12, 11, 10, 9, 8, 7, 6, 5, 4, 3, 2, 1}
```

SOLUTION 2

```
Range[20] //Reverse              ← This is equivalent to Reverse[Range[20]].
{20, 19, 18, 17, 16, 15, 14, 13, 12, 11, 10, 9, 8, 7, 6, 5, 4, 3, 2, 1}
```

3.18 Sort the letters of the word MISSISSIPPI alphabetically.

SOLUTION

```
list = Characters["MISSISSIPPI"]
{M,I,S,S,I,S,S,I,P,P,I}
Sort[list]
{I, I, I, I, M, P, P, S, S, S, S}
```

3.19 Construct a list of numbers from 0 to 2π in increments of $\pi/6$.

SOLUTION

```
Range[0, 2π, π/6]
```

$$\left\{0, \frac{\pi}{6}, \frac{\pi}{3}, \frac{\pi}{2}, \frac{2\pi}{3}, \frac{5\pi}{6}, \pi, \frac{7\pi}{6}, \frac{4\pi}{3}, \frac{3\pi}{2}, \frac{5\pi}{3}, \frac{11\pi}{6}, 2\pi\right\}$$

3.20 Flavius Joseph was a Jewish historian of the first century. He wrote about a group of ten Jews in a cave who, rather than surrender to the Romans, chose to commit suicide, one by one. They formed a circle and every other one was killed. Who was the lone survivor?

SOLUTION

We number the people 1 through 10 and define a list consisting of these ten integers.

```
list = Range[10]
{1, 2, 3, 4, 5, 6, 7, 8, 9, 10}
```

The first person to go is number 2. We eliminate him by rotating the list one position to the left and dropping his number from the list.

```
list = Rest[RotateLeft[list]]
{3, 4, 5, 6, 7, 8, 9, 10, 1}
```

The new list begins with 3 and omits the number 2. To determine the survivor, we repeat the process until only one number remains.

```
list = Rest[RotateLeft[list]]
{5, 6, 7, 8, 9, 10, 1, 3}

list = Rest[RotateLeft[list]]
{7, 8, 9, 10, 1, 3, 5}

list = Rest[RotateLeft[list]]
{9, 10, 1, 3, 5, 7}

list = Rest[RotateLeft[list]]
{1, 3, 5, 7, 9}

list = Rest[RotateLeft[list]]
{5, 7, 9, 1}

list = Rest[RotateLeft[list]]
{9, 1, 5}

list = Rest[RotateLeft[list]]
{5, 9}

list = Rest[RotateLeft[list]]
{5}
```

Number 5 is the survivor.

Although it is interesting to see how the list progresses from step to step, the above technique would not be appropriate for a long list. A more efficient procedure would involve a simple **While** loop.

```
list = Range[10];
While[Length[list] > 1, list = Rest[RotateLeft[list]]]
list
{5}
```

3.21 Determine which elements are in the highest level of the list

```
{a, {b, c}, {{d, e}, {f, g}, {{h, i}}, {j, {k, l, m}}}}
```

SOLUTION

```
list = {a, {b, c}, {{d, e}, {f, g}, {{h, i}}, {j, {k, l, m}}}};
Depth[list]
5
```

```
Level[list, {4}]          ← Remember to subtract 1 to determine the highest level.
{h, i, k, l, m}
```

3.22 Reduce the depth of the list

```
{a, {b, c}, {{d, e}, {f, g}, {{h, i}}, {j, {k, l, m}}}}
```

by 1 level; by 2 levels.

SOLUTION

```
list = {a, {b, c}, {{d, e}, {f, g}, {{h, i}}, {j, {k, l, m}}}};
Flatten[list, 1]
{a, b, c, {d, e}, {f, g}, {{h, i}}, {j, {k, l, m}}}
Flatten[list, 2]
{a, b, c, d, e, f, g, {h, i}, j, {k, l, m}}
```

3.23 Take the list of characters A through X and construct a list with six sublists, each containing four distinct letters.

SOLUTION

```
list = CharacterRange["A", "X"]
{A, B, C, D, E, F, G, H, I, J, K, L, M, N, O, P, Q, R, S, T, U, V, W, X}
Partition[list, 4]
{{A, B, C, D}, {E, F, G, H}, {I, J, K, L}, {M, N, O, P}, {Q, R, S, T}, {U, V, W, X}}
```

3.4 Set Theory

Sets are represented as lists in *Mathematica*. Sets are manipulated using the basic list functions **Union**, **Intersection**, and **Complement**.

- **Union** [*list1*, *list2*] combines *list1* and *list2* into one sorted list, eliminating any duplicate elements. Although only two lists are presented in this description, any number of lists may be used. As a special case, **Union** [*list*] will eliminate duplicate elements in *list*.
- **Intersection** [*list1*, *list2*] returns a sorted list of elements common to *list1* and *list2*. If *list1* and *list2* are disjoint, i.e., they have no common elements, the command returns the empty list, { }.
- **Complement** [*universe*, *list*] returns a sorted list consisting of those elements of *universe* that are not in *list*. In this context, *universe* represents the universal set.
- **Complement** [*universe*, *list1*, *list2*] returns a sorted list consisting of those elements of *universe* that are not in *list1* or *list2*. This command extends in a natural way to more than two sets.

EXAMPLE 37

```
list = {a, b, c, a, c, c, c, b, b};
Union[list]
{a, b, c}
```

EXAMPLE 38

```
universe = {1, 2, 3, 4, 5, 6, 7, 8, 9, 10};
list1 = {1, 3, 5, 7};
list2 = {5, 7, 8, 10};
Union[list1, list2]
{1, 3, 5, 7, 8, 10}
```

```
Intersection[list1, list2]
{5, 7}

Complement[universe, list1]
{2, 4, 6, 8, 9, 10}

Complement[universe, list1, list2]
{2, 4, 6, 9}
```

Using the Basic Math Input palette, the symbols ∪ and ∩ may be used to represent union and intersection, respectively.

- *list1* ∪ *list2* is equivalent to **Union**[*list1*, *list2*].
- *list1* ∩ *list2* is equivalent to **Intersection**[*list1*, *list2*].

EXAMPLE 39

```
list1 = {1, 2, 3, 4, 5};
list2 = {3, 4, 5, 6, 7};
list1 ∪ list2
{1, 2, 3, 4, 5, 6, 7}
list1 ∩ list2
{3, 4, 5}
```

A subset of A is any set, each of whose elements are members of A. The empty set is a subset of every set. Including the empty set, a set of *n* elements has 2^n subsets. The set of all subsets of A is called the *power set* of A.

- **Subsets**[*list*] returns a list containing all subsets of *list*, including the empty set, i.e., the power set of *list*.

There are a number of useful set commands available in the package **Combinatorica`**. Among them are **CartesianProduct** and **KSubsets**.

By definition, the Cartesian product of two sets, A and B, is the set of ordered pairs of elements, the first taken from A and the second from B.

- **CartesianProduct**[*list1*, *list2*] returns the Cartesian product of *list1* and *list2*.
- **KSubsets**[*list*, **k**] returns a list containing all subsets of *list* of size k.

EXAMPLE 40

```
« Combinatorica`              ← This loads the package. See Chapter 1.
list1 = {a, b, c, d};
list2 = {x, y, z};
CartesianProduct[list1, list2]
{{a, x}, {a, y}, {a, z}, {b, x}, {b, y}, {b, z}, {c, x}, {c, y},
 {c, z}, {d, x}, {d, y}, {d, z}}
```

EXAMPLE 41

```
list = {a, b, c, d};
Subsets[list]
{{}, {a}, {b}, {c}, {d}, {a, b}, {a, c}, {a, d}, {b, c}, {b, d},
 {c, d}, {a, b, c}, {a, b, d}, {a, c, d}, {b, c, d}, {a, b, c, d}}
« Combinatorica`              ← Omit if you have already loaded the package.
KSubsets[list, 3]
{{a, b, c}, {a, b, d}, {a, c, d}, {b, c, d}}
```

SOLVED PROBLEMS

3.24 Which distinct letters are contained in the word MISSISSIPPI? (Compare with Problem 3.18)

SOLUTION

```
Union[Characters["MISSISSIPPI"]]
{I, M, P, S}
```

3.25 Find the union and intersection of the sets {a, b, c, d, e, f, g}, {c, d, e, f, g, h, i}, and {e, f, g, h, i, j, k}.

SOLUTION

```
set1 = {a, b, c, d, e, f, g};
set2 = {c, d, e, f, g, h, i};
set3 = {e, f, g, h, i, j, k};
Union[set1, set2, set3]  or  set1 ∪ set2 ∪ set3
{a, b, c, d, e, f, g, h, i, j, k}
Intersection[set1, set2, set3]  or  set1 ∩ set2 ∩ set3
{e, f, g}
```

3.26 Find all the elements of the set {a, b, c, d, e, f, g} that are *not* in {a, c, d, e}.

SOLUTION

```
universe = {a, b, c, d, e, f, g};
set = {a, c, d, e};
Complement[universe, set]
{b, f, g}
```

3.27 The 20th prime is 71. Find all the numbers not exceeding 71 that are *not* prime.

SOLUTION

```
universe = Range[71];
primes = Prime[Range[20]]
{2, 3, 5, 7, 11, 13, 17, 19, 23, 29, 31, 37, 41, 43, 47, 53, 59, 61, 67, 71}
Complement[universe, primes]
{1, 4, 6, 8, 9, 10, 12, 14, 15, 16, 18, 20, 21, 22, 24, 25, 26, 27, 28, 30, 32, 33,
   34, 35, 36, 38, 39, 40, 42, 44, 45, 46, 48, 49, 50, 51, 52, 54, 55, 56, 57, 58, 60,
   62, 63, 64, 65, 66, 68, 69, 70}
```

3.28 Construct a list consisting of the consonants of the alphabet.

SOLUTION

```
letters = CharacterRange["a", "z"];
vowels = Characters["aeiou"];
consonants = Complement[letters, vowels]
{b, c, d, f, g, h, j, k, l, m, n, p, q, r, s, t, v, w, x, y, z}
```

3.29 Find all the numbers less than 1000 that are *both* prime and Fibonacci.

SOLUTION

```
k = 1; list1 = {};
While[Fibonacci[k] ≤ 1000, list1 = Append[list1, Fibonacci[k]]; k++]
```

```
k = 1; list2 = {};
While[Prime[k] ≤1000, list2 = Append[list2, Prime[k]]; k++]
list1 ∩ list2
{2, 3, 5, 13, 89, 233}
```

3.30 Create a list that contains all the subsets of {a, b, c, d, e}. How many subsets are there?

SOLUTION

```
letters = {a, b, c, d, e};
Subsets[letters]
{{}, {a}, {b}, {c}, {d}, {e}, {a, b}, {a, c}, {a, d}, {a, e}, {b, c}, {b, d}, {b, e},
  {c, d}, {c, e}, {d, e}, {a, b, c}, {a, b, d}, {a, b, e}, {a, c, d}, {a, c, e},
  {a, d, e}, {b, c, d}, {b, c, e}, {b, d, e}, {c, d, e}, {a, b, c, d}, {a, b, c, e},
  {a, b, d, e}, {a, c, d, e}, {b, c, d, e}, {a, b, c, d, e}}
Length[%]
32
```

3.31 Create a list of all the subsets of {a, b, c, d, e} that contain precisely three elements. How many are there?

SOLUTION

```
≪ Combinatorica`
letters={a, b, c, d, e};
KSubsets[letters, 3]
{{a, b, c}, {a, b, d}, {a, b, e}, {a, c, d}, {a, c, e}, {a, d, e}, {b, c, d}, {b, c, e},
  {b, d, e}, {c, d, e}}
Length[%]
10
```

3.5 Tables and Matrices

Mathematica represents tables and matrices as nested lists. Internally, there is no difference in the way they are stored, but they are represented differently using the functions **MatrixForm** and **TableForm**. It is often more convenient to use **//MatrixForm** or **//TableForm** to the right of the matrix or table name.

- **MatrixForm**[*list*] prints double nested lists as a rectangular array enclosed within parentheses. The innermost lists are printed as rows. Single nested lists are printed as columns enclosed within parentheses.
- **TableForm**[*list*] prints *list* the same way as **MatrixForm** except the surrounding parentheses are omitted.

Matrices and tables can be entered directly as nested lists. A matrix or table having *m* rows and *n* columns would be a nested list of *m* sublists, each containing *n* entries.

EXAMPLE 42

```
list = {{1, 2, 3, 4}, {5, 6, 7, 8}, {9, 10, 11, 12}};
MatrixForm[list]  or  list //MatrixForm
```

$$\begin{pmatrix} 1 & 2 & 3 & 4 \\ 5 & 6 & 7 & 8 \\ 9 & 10 & 11 & 12 \end{pmatrix}$$

```
TableForm[list] or list //TableForm
```

```
1  2   3   4
5  6   7   8
9  10  11  12
```

Matrices and tables can also be conveniently entered by going to Insert ⇒ Table/Matrix ⇒ New.

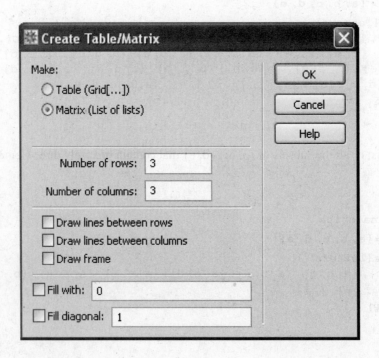

Clicking OK yields an empty grid—use the [TAB] key to cycle from entry to entry.

$$\begin{pmatrix} \square & \square & \square \\ \square & \square & \square \\ \square & \square & \square \end{pmatrix}$$

EXAMPLE 43

```
list = {{1, 2, 3, 4}, {5, 6, 7, 8}, {9, 10, 11, 12}};
{{1, 2, 3, 4}, {5, 6, 7, 8}, {9, 10, 11, 12}}
```

```
MatrixForm[list]  or  list //MatrixForm
```

$$\begin{pmatrix} 1 & 2 & 3 & 4 \\ 5 & 6 & 7 & 8 \\ 9 & 10 & 11 & 12 \end{pmatrix}$$

Two special matrix-generating commands are worth remembering because of their frequency in applications. These generate a nested list. Use **MatrixForm** to get a matrix.

- **IdentityMatrix[n]** produces an $n \times n$ matrix with 1s on the main diagonal and 0s elsewhere.
- **DiagonalMatrix[*list*]** creates a diagonal matrix whose diagonal entries are the elements of *list*.

EXAMPLE 44

```
IdentityMatrix[3] //MatrixForm
```

$$\begin{pmatrix} 1 & 0 & 0 \\ 0 & 1 & 0 \\ 0 & 0 & 1 \end{pmatrix}$$

```
DiagonalMatrix[{1, 2, 3}] //MatrixForm
```

$$\begin{pmatrix} 1 & 0 & 0 \\ 0 & 2 & 0 \\ 0 & 0 & 3 \end{pmatrix}$$

Once defined, matrices can be combined using the operations of addition, subtraction, scalar, and matrix multiplication. The operation of matrix multiplication is represented by a period (.). Matrices are discussed in greater detail in Chapter 12.

EXAMPLE 45

$$A = \begin{pmatrix} 1 & 2 & 3 \\ 4 & 5 & 6 \\ 7 & 8 & 9 \end{pmatrix}$$

> The matrix is input using Insert ⇒ Table/Matrix ⇒ New. *Mathematica* outputs the matrix as a nested list of numbers.

```
{{1, 2, 3}, {4, 5, 6}, {7, 8, 9}}
```

$$B = \begin{pmatrix} 2 & 1 & 5 \\ 4 & 7 & 2 \\ 1 & 3 & 2 \end{pmatrix}$$

```
{{2, 1, 5}, {4, 7, 2}, {1, 3, 2}}
A + B //MatrixForm
```

$$\begin{pmatrix} 3 & 3 & 8 \\ 8 & 12 & 8 \\ 8 & 11 & 11 \end{pmatrix}$$

```
A − B //MatrixForm
```

$$\begin{pmatrix} -1 & 1 & -2 \\ 0 & -2 & 4 \\ 6 & 5 & 7 \end{pmatrix}$$

```
3 A //MatrixForm
```

$$\begin{pmatrix} 3 & 6 & 9 \\ 12 & 15 & 18 \\ 21 & 24 & 27 \end{pmatrix}$$

```
A.B //MatrixForm
```

$$\begin{pmatrix} 13 & 24 & 15 \\ 34 & 57 & 42 \\ 55 & 90 & 69 \end{pmatrix}$$

It is useful to remember that if list is a simple list of numbers, list.list yields the sum of their squares. The result is printed as a single number without braces.

EXAMPLE 46

```
list = {1, 2, 3, 4, 5};
list.list
55
```

Tables are also stored as nested lists, but are represented as tables with **TableForm**. Although this command allows representation of tables of any dimension, we will discuss only one- and two-dimensional tables in this book.

EXAMPLE 47

```
list = {{12, 7, 10}, {105, 205, 7}, {3, 30, 300}};
list//TableForm
```

12	7	10
105	205	7
3	30	300

- **TableForm[*list, options*]** allows the use of various formatting options in determining the appearance of a table.

From Example 47 we can observe that the numbers in a table are, by default, left justified. This can sometimes make the table confusing to read. Justification can be controlled with the **TableAlignments** option.

- **TableAlignments → Left** justifies the columns to the left (default).
- **TableAlignments → Right** justifies the columns to the right.
- **TableAlignments → Center** centers the columns.

EXAMPLE 48

```
list = {{12,7,10},{105,205,7},{3,30,300}};
TableForm[list, TableAlignments → Right]
```

12	7	10
105	205	7
3	30	300

```
TableForm[list, TableAlignments → Center]
```

12	7	10
105	205	7
3	30	300

Row and column headings can be inserted by using the option **TableHeadings** within the **TableForm** command. The default is **TableHeadings → None**.

- **TableHeadings → Automatic** produces consecutive integer labels for both rows and columns.

Each row and column of a table can be labeled separately using strings (characters enclosed within double quotes) or *Mathematica* expressions. The general form of this option is

- **TableHeadings → {*rowlist, columnlist*}**

where *rowlist* is a list of row labels and *columnlist* is a list of column labels. If you desire to have row labels but not column labels, or column labels but not row labels, simply replace *rowlist* or *columnlist* by **None**.

EXAMPLE 49

```
list = {{a, b, c}, {d, e, f},{g, h, i}};
TableForm[list, TableHeadings → Automatic]
```

	1	2	3
1	a	b	c
2	d	e	f
3	g	h	i

EXAMPLE 50

```
list = {{a, b, c}, {d, e, f},{g, h, i}};
TableForm[list, TableHeadings→{{"Row1","Row2","Row3"},
                                {"Column1","Column2","Column3"}}]
```

	Column1	Column2	Column3
Row1	a	b	c
Row2	d	e	f
Row3	g	h	i

EXAMPLE 51

```
list = {{a, b, c}, {d, e, f}, {g, h, i}};
TableForm[list, TableHeadings → {{"Row1","Row2","Row3"},
                                {"Column1","Column2","Column3"}},
            TableAlignments → Center]
```

	Column1	Column2	Column3
Row1	a	b	c
Row2	d	e	f
Row3	g	h	i

EXAMPLE 52

```
TableForm[list, TableHeadings → {None, {"Column1","Column2","Column3"}},
            TableAlignments → Center]
```

Column1	Column2	Column3
a	b	c
d	e	f
g	h	i

```
TableForm[list,TableHeadings → {{"Row1","Row2","Row3"}, None},
            TableAlignments → Center]
```

Row1	a	b	c
Row2	d	e	f
Row3	g	h	i

TableDirections is an option that determines how the entries of the table should be placed. If *list* represents a two-dimensional nested list, then

- **TableDirections → Column** prints the table with the first element of each inner list in the first column, the second element of each inner list in the second column, and so forth. (This is the default.)
- **TableDirections → Row** interchanges the positions of the columns with the rows.

EXAMPLE 53

```
list = Array[a, {3, 4}]
```

```
{{a[1, 1], a[1, 2], a[1, 3], a[1, 4]}, {a[2, 1], a[2, 2], a[2, 3], a[2, 4]},
 {a[3, 1], a[3, 2], a[3, 3], a[3, 4]}}
```

```
TableForm[list, TableDirections → Column]
a[1, 1]  a[1, 2]  a[1, 3]  a[1, 4]
a[2, 1]  a[2, 2]  a[2, 3]  a[2, 4]
a[3, 1]  a[3, 2]  a[3, 3]  a[3, 4]
```

> The elements a[1, 1], a[2, 1], and a[3, 1] form the first column.

```
TableForm[list, TableDirections → Row]
a[1, 1]  a[2, 1]  a[3, 1]
a[1, 2]  a[2, 2]  a[3, 2]
a[1, 3]  a[2, 3]  a[3, 3]
a[1, 4]  a[2, 4]  a[3, 4]
```

> The elements a[1, 1], a[2, 1], and a[3, 1] form the first row.

By default, *Mathematica* prints real numbers to a specified number of significant digits. So numbers that vary in magnitude will appear to have different formats. The command **PaddedForm** allows the output of a calculation to be precisely formatted.

- **PaddedForm[*expression*, n]** prints the value of *expression* leaving space for a total of n digits. This form of the command can be used for integers or real number approximations. *Note*: The decimal point is not counted as a position.
- **PaddedForm[*expression*, {n, f}]** prints the value of *expression* leaving space for a total of n digits, f of which are to the right of the decimal point. The fractional portion of the number is rounded if any digits are deleted.

EXAMPLE 54

```
a = 123.456789;
PaddedForm[a, 12]
   123.456789          ← 3 spaces to the left of the number.
PaddedForm[a, 20]
           123.456789  ← 11 spaces to the left of the number.
PaddedForm[a, {20, 3}]
              123.457  ← 14 spaces to the left of the number, the third decimal is rounded to 7.
```

EXAMPLE 55   The following prints a table of values of a polynomial $p(x)$ along with its corresponding value of x. First we will print the table using the standard **TableForm** command.

```
p[x_] = x^5 - 3 x^4 + 2 x^3 - 7 x + 12;
list = Table[{x, p[x]}, {x, -3, 3, .5}];
TableForm[list]
-3.    -507.
-2.5   -216.594
-2.    -70.
-1.5   -7.03125
-1.    13.
-0.5   15.0313
0.     12.
0.5    8.59375
1.     5.
1.5    0.65625
2.     -2.
2.5    6.21875
3.     45.
```

Now we use **PaddedForm** to pad the entire table.

```
PaddedForm[TableForm[list],{10, 6}]
```
```
    -3.000000    -507.000000
    -2.500000    -216.594000
    -2.000000     -70.000000
    -1.500000      -7.031000
    -1.000000      13.000000
    -0.500000      15.031000
     0.000000      12.000000
     0.500000       8.594000
     1.000000       5.000000
     1.500000       0.656000
     2.000000      -2.000000
     2.500000       6.219000
     3.000000      45.000000
```

If we wish to format the individual columns differently for a more customized appearance, we can pad the individual entries of the list, rather than the whole table.

```
list = Table[{PaddedForm[x, {5, 1}], PaddedForm[p[x], {10, 3}]}, {x, -3, 3, .5}];
TableForm[list]
```
```
    -3.0     -507.000
    -2.5     -216.594
    -2.0      -70.000
    -1.5       -7.031
    -1.0       13.000
    -0.5       15.031
     0.0       12.000
     0.5        8.594
     1.0        5.000
     1.5        0.656
     2.0       -2.000
     2.5        6.219
     3.0       45.000
```

Spacing between rows and columns can be controlled with **TableSpacing**. This specifies the number of spaces to put between entries in each direction.

- **TableSpacing** → *{rowspaces, columnspaces}*

rowspaces specifies the number of blank lines between successive rows of the table; *columnspaces* specifies the number of blank characters between successive columns.

EXAMPLE 56

```
list = {{a, b, c}, {d, e, f}, {g, h, i}};
TableForm[list, TableSpacing → {0, 0}]
```
```
abc                          ← No spacing between rows or columns.
def
ghi
```

```
TableForm[list,TableSpacing → {1, 3}]
```

a b c   ← 1 line between rows, 3 spaces between columns.

d e f

g h i

```
TableForm[list,TableSpacing → {3, 1}]
```

a b c   ← 3 lines between rows, 1 space between columns.

d e f

g h i

Lists can be expressed as single columns with **ColumnForm**.

- **ColumnForm[*list*]** presents *list* as a single column of objects.
- **ColumnForm[*list*, *horizontal*]** specifies the horizontal alignment of each row. Acceptable values of *horizontal* are **Left** (default), **Center**, and **Right**.
- **ColumnForm[*list*, *horizontal*, *vertical*]** allows vertical alignment of the column. Acceptable values of *vertical* are **Above**, **Center**, and **Below** (default).

EXAMPLE 57

```
list = {a, bb, ccc}
ColumnForm[list]
```
a
bb
ccc

```
ColumnForm[list, Right]
```
  a
 bb
ccc

A list may have rows and columns can be controlled with row-length arrays. These extend the features of *Screen* to nonrectangular arrays of rows in each direction.

3.32 Construct a 3 × 3 matrix whose entries are consecutive integers, increasing as we go to the right and down.

SOLUTION

```
list = Table[3i + j, {i, 0, 2}, {j, 1, 3}]
{{1, 2, 3}, {4, 5, 6}, {7, 8, 9}}
list//MatrixForm
```

$$\begin{pmatrix} 1 & 2 & 3 \\ 4 & 5 & 6 \\ 7 & 8 & 9 \end{pmatrix}$$

3.33 The Hilbert matrix is a square matrix whose element in position (i, j) is $\frac{1}{i+j-1}$. Construct the Hilbert matrix of order 5.

SOLUTION

```
a[i_, j_] = 1/(i + j - 1);
hilbert = Array[a,{5, 5}];
hilbert//MatrixForm
```

$$\begin{pmatrix} 1 & \frac{1}{2} & \frac{1}{3} & \frac{1}{4} & \frac{1}{5} \\ \frac{1}{2} & \frac{1}{3} & \frac{1}{4} & \frac{1}{5} & \frac{1}{6} \\ \frac{1}{3} & \frac{1}{4} & \frac{1}{5} & \frac{1}{6} & \frac{1}{7} \\ \frac{1}{4} & \frac{1}{5} & \frac{1}{6} & \frac{1}{7} & \frac{1}{8} \\ \frac{1}{5} & \frac{1}{6} & \frac{1}{7} & \frac{1}{8} & \frac{1}{9} \end{pmatrix}$$

3.34 Construct the 5×5 identity matrix.

SOLUTION

```
IdentityMatrix[5]//MatrixForm
```

$$\begin{pmatrix} 1 & 0 & 0 & 0 & 0 \\ 0 & 1 & 0 & 0 & 0 \\ 0 & 0 & 1 & 0 & 0 \\ 0 & 0 & 0 & 1 & 0 \\ 0 & 0 & 0 & 0 & 1 \end{pmatrix}$$

3.35 Construct a 5×5 matrix having the first five primes as diagonal entries and 0s elsewhere.

SOLUTION

```
diag = Table[Prime[k],{k, 1, 5}]
{2, 3, 5, 7, 11}
DiagonalMatrix[diag] //MatrixForm
```

$$\begin{pmatrix} 2 & 0 & 0 & 0 & 0 \\ 0 & 3 & 0 & 0 & 0 \\ 0 & 0 & 5 & 0 & 0 \\ 0 & 0 & 0 & 7 & 0 \\ 0 & 0 & 0 & 0 & 11 \end{pmatrix}$$

3.36 Construct a table having three columns. The first column lists the consecutive integers 1 through 10 and the second and third columns are their squares and cubes. Label the three columns *integers*, *squares*, and *cubes*.

SOLUTION

```
list = Table[{k, k², k³},{k, 1, 10}];
TableForm[list, TableHeadings → {None, {"integers","squares", "cubes"}},
              TableAlignments → Right]
```

integers	squares	cubes
1	1	1
2	4	8
3	9	27
4	16	64
5	25	125
6	36	216
7	49	343
8	64	512
9	81	729
10	100	1000

3.37 If c represents the temperature in degrees Celsius, its corresponding Fahrenheit temperature is $f = \frac{9}{5}c + 32°$. Construct a labeled table showing, horizontally, the Fahrenheit equivalents of Celsius temperatures from 1° to 10° in increments of 1°.

SOLUTION

$f = \frac{9}{5}c + 32$

```
list = Table[{c, PaddedForm[N[f], {3, 1}]}, {c, 1, 10}]
TableForm[list, TableDirections → Row,
        TableHeadings → {None, {"Celsius", "Fahrenheit"}},
        TableAlignments → Center]
```

Celsius	1	2	3	4	5	6	7	8	9	10
Fahrenheit	33.8	35.6	37.4	39.2	41.0	42.8	44.6	46.4	48.2	50.0

3.38 Construct a table showing the radian equivalents of angles from 0° to 30° in increments of 5°.

SOLUTION

> **Degree** is a *Mathematica* constant (see Chapter 2).

```
list = Table[{deg, N[deg Degree]}, {deg, 0, 30, 5}];
TableForm[list, TableDirections → Row,
        TableHeadings → {None, {"Degrees","Radians"}},
        TableAlignments → Center]
```

Degrees	0	5	10	15	20	25	30
Radians	0	0.0872665	0.174533	0.261799	0.349066	0.436332	0.523599

3.39 If *p* dollars is invested for *t* years in a bank account paying an annual interest rate of *r* compounded *n* times a year, the amount of money after *k* periods is $p\left(1 + \frac{r}{n}\right)^k$ dollars. If \$1,000 is invested in an account paying 6% compounded quarterly, make a table showing how much money has accumulated during a three-year period.

SOLUTION

```
p = 1000; r = .06; n = 4; t = 3;
a[k_] = p (1 + r/n)^k;
list = Table[{k, a[k]}, {k, 1, n*t}]
TableForm[list, TableHeadings → {None, {"period", "amount"}}]
```

period	amount
1	1015.
2	1030.22
3	1045.68
4	1061.36
5	1077.28
6	1093.44
7	1109.84
8	1126.49
9	1143.39
10	1160.54
11	1177.95
12	1195.62

3.40 If p dollars is invested in a bank account paying a rate of r compounded n times a year, the amount of money after t years is $p\left(1+\dfrac{r}{n}\right)^{nt}$ dollars. If interest is compounded continuously, the amount after t years is pe^{rt}. If $1,000 is invested in an account paying 6% annually, make a table showing how much money is in the account at the end of each year for 10 years if interest is compounded quarterly, monthly, daily, and continuously.

SOLUTION

```
p = 1000;
r = .06;
a = p (1 + r / 4)^4 t ;
b = p (1 + r / 12)^12 t ;
c = p (1 + r / 365)^365 t ;
d = p Exp[r t];

tt = PaddedForm[t, 2];
aa = PaddedForm[a, {7, 2}];
bb = PaddedForm[b, {7, 2}];
cc = PaddedForm[c, {7, 2}];
dd = PaddedForm[d, {7, 2}];
list = Table[{tt, aa, bb, cc, dd},{t, 1, 10}];
TableForm[list, TableHeadings →
    {None, {"year"," quarterly"," monthly","  daily","continuously"}}]
```

year	quarterly	monthly	daily	continuously
1	1061.36	1061.68	1061.83	1061.84
2	1126.49	1127.16	1127.49	1127.50
3	1195.62	1196.68	1197.20	1197.22
4	1268.99	1270.49	1271.22	1271.25
5	1346.86	1348.85	1349.83	1349.86
6	1429.50	1432.04	1433.29	1433.33
7	1517.22	1520.37	1521.91	1521.96
8	1610.32	1614.14	1616.01	1616.07
9	1709.14	1713.70	1715.93	1716.01
10	1814.02	1819.40	1822.03	1822.12

3.41 The payment on a monthly mortgage of a dollars is $\dfrac{a \times \frac{r}{12}}{1-(1+\frac{r}{12})^{-12n}}$ where n is the number of years the money is borrowed and r is the annual rate of interest. Construct a table showing the monthly payments on a 30-year mortgage of \$250,000 at rates of 6% to 8% in increments of 25%.

SOLUTION

```
a = 250 000;
n = 30;
```

$$\text{payment} = \frac{a\frac{r}{12}}{1-\left(1+\frac{r}{12}\right)^{-12n}}$$

```
list = Table[{PaddedForm[r, {4, 4}],
        PaddedForm[payment, {6, 2}]}, {r, .06, .08, .0025}];
TableForm[list, TableHeadings → {None, {" rate", " payment"}}]
```

rate	payment
0.0600	1498.88
0.0625	1539.29
0.0650	1580.17
0.0675	1621.50
0.0700	1663.26
0.0725	1705.44
0.0750	1748.04
0.0775	1791.03
0.0800	1834.41

CHAPTER 4

Two-Dimensional Graphics

4.1 Plotting Functions of a Single Variable

Anyone who has ever tried to plot a graph using one of the standard programming languages will appreciate the ease with which graphs can be produced in *Mathematica*. In many instances, only one simple instruction is all that is needed to produce a pictorial representation of a function or a more general relationship between two variables.

Although *Mathematica*'s defaults work well in most instances, there are many options available to control subtleties. We shall describe the more common ones in this section and present a variety of examples that illustrate the ease with which graphs may be constructed.

The basic command for drawing the graph of a function is **Plot**. Although x is used as the independent variable in the following description, any symbol may be used in its place.

- **Plot[f[x], {x, xmin, xmax}]** plots a two-dimensional graph of the function $f(x)$ on the interval $\text{xmin} \le \text{x} \le \text{xmax}$.

EXAMPLE 1   Plot the parabola $f(x) = x^2$ from –3 to 3.

```
Plot[x², {x, -3, 3}]
```

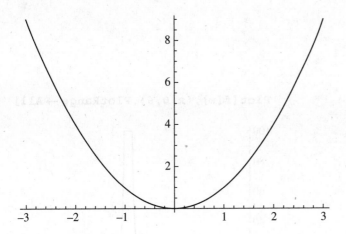

Two functions can be plotted on the same set of axes. Mathematica draws each in a different color.

- **Plot[{f[x], g[x]}, {x, xmin, xmax}]** plots the graphs of $f(x)$ and $g(x)$ from xmin to xmax on the same set of axes. This command can be generalized in a natural way to plot three or more functions.

EXAMPLE 2   Plot $f(x) = x^2$ and $g(x) = 9 - x^2$ from -3 to 3.

```
Plot[{x², 9 - x²}, {x, -3, 3}]
```

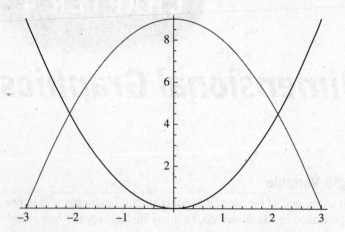

When plotting points over a specified interval, *Mathematica* makes a decision on the range of points to plot in order to produce a pleasing graph. **PlotRange** is an option that allows the user to override *Mathematica*'s default.

- **PlotRange → Automatic** is *Mathematica*'s default. Any points whose vertical coordinates appear to be too large (e.g., outliers) are omitted from the graph.
- **PlotRange → All** forces *Mathematica* to plot all points.
- **PlotRange → {ymin, ymax}** plots only those points whose vertical coordinates fall between ymin and ymax.
- **PlotRange → {{xmin, xmax}, {ymin, ymax}}** plots those points whose horizontal coordinates fall between xmin and xmax and whose vertical coordinates fall between ymin and ymax.

EXAMPLE 3

```
f[x_] := 1/(x - 3)²   /; x < 2.9 || x > 3.1
f[x_] := 100 /; 2.9 ≤ x ≤ 3.1
Plot[f[x], {x, 0, 6}]                    Plot[f[x], {x, 0, 6}, PlotRange → All]
```

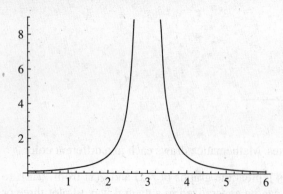
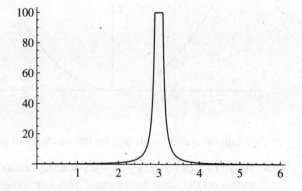

The **Show** command is useful for plotting several graphs simultaneously, particularly when their domains are different intervals.

- **Show[g1, g2, . . .]** plots several graphs on a common set of axes.

EXAMPLE 4 Suppose we wish to plot the graph of $y = x^2 - 9$ on the interval $[-4, 4]$ and the graph of $y = \sin x$ on the interval $[0, 2\pi]$, but wish to plot them on one set of axes. We define two graphics objects, g1 and g2.

```
g1 = Plot[x² - 9, {x, -4, 4}]
```

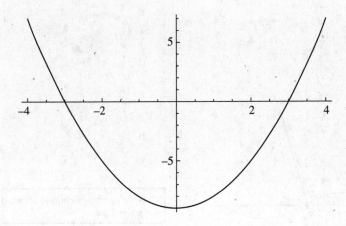

```
g2 = Plot[Sin[x], {x, 0, 2π}]
```

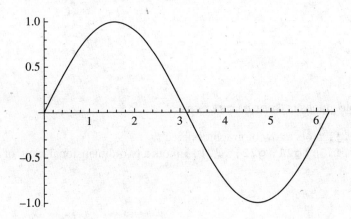

Now we apply the **Show** command. Note how the axes are adjusted to exhibit both graphs:

```
Show[g1, g2, PlotRange → All]
```

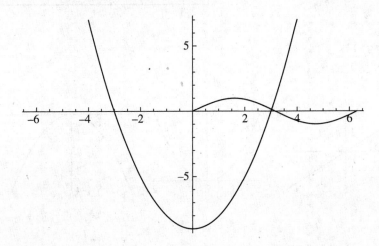

You will notice that in defining g1 and g2, each curve was drawn individually on its own axis. To suppress this output, a semicolon (;) can be placed at the right side of each plot command.

EXAMPLE 5

```
g1 = Plot[x² - 9];
g2 = Plot[Sin[x], {x, 0, 2π}];
Show[g1, g2, PlotRange → All]
```

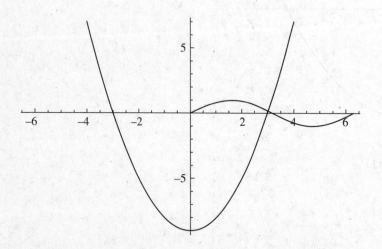

> Only the combined graph is drawn, not **g1** and **g2**.

A useful command for drawing multiple graphs is **GraphicsArray**.

- **GraphicsArray[{g1, g2, ...}]** plots a row of graphics objects.
- **GraphicsArray[{g11, g12, ...}, {g21, g22, ...}}]** plots a two-dimensional array of graphics objects.

EXAMPLE 6

```
g1 = Plot[x, {x, -2, 2}];
g2 = Plot[-x, {x, -2, 2}];
g3 = Plot[x², {x, -2, 2}];
g4 = Plot[-x², {x, -2, 2}];
GraphicsArray[{g1, g2, g3, g4}]
```

> Note the use of the semicolon (;) to suppress intermediate graphics from being plotted.

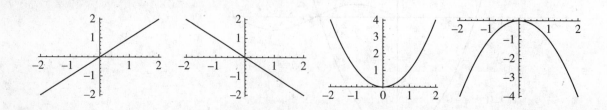

```
GraphicsArray[{{g1, g2}, {g3, g4}}]
```

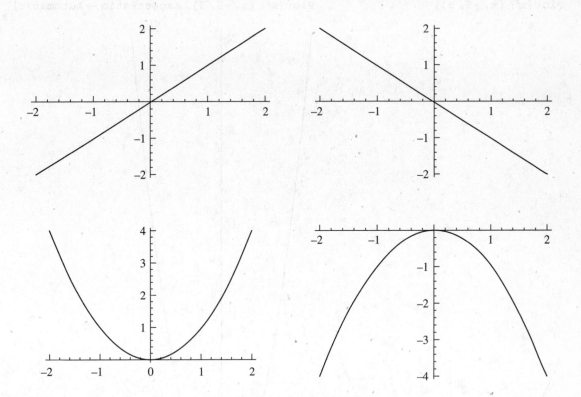

Plot has a variety of options that can be viewed by typing **?? Plot** or **Options[Plot]**. These options may be used individually or in conjunction with one another. Some of the more common options are described in the remainder of this section.

Since *Mathematica* obviously cannot plot an infinite number of points, it selects a finite number of equally spaced points as "sample" points and uses an adaptive algorithm to construct a smooth-looking curve. The initial number of points it will use, **PlotPoints**, is set to 50 by default. If the curve "wiggles" excessively, a larger number might be necessary to obtain a smooth-looking curve.

- **PlotPoints → n** specifies that an initial number of n sample points should be used in the construction of the graph.
- **MaxRecursion → n** specifies that up to n levels of recursion should be made in the adaptive algorithm. Recursive subdivision is done only in those places where more samples seem to be needed in order to achieve results with a certain level of quality.

When you plot a graph, you will notice that the horizontal and vertical axes are usually not the same length. By default, the ratio of vertical axis length to horizontal axis length is **1/GoldenRatio**, where **GoldenRatio** = $(1 + \sqrt{5})/2$. The designers of *Mathematica* felt that this ratio was the most comfortable and pleasing to the eye. It can be changed with the option **AspectRatio**, which determines the height-to-width ratio of the graph.

- **AspectRatio → Automatic** computes the aspect ratio from the actual coordinate values of the plot.
- **AspectRatio → *ratio*** sets the ratio of height to width to the value *ratio*.

EXAMPLE 7

Plot[x², {x, −5, 5}]   Plot[x², {x, −5, 5}, AspectRatio → Automatic]

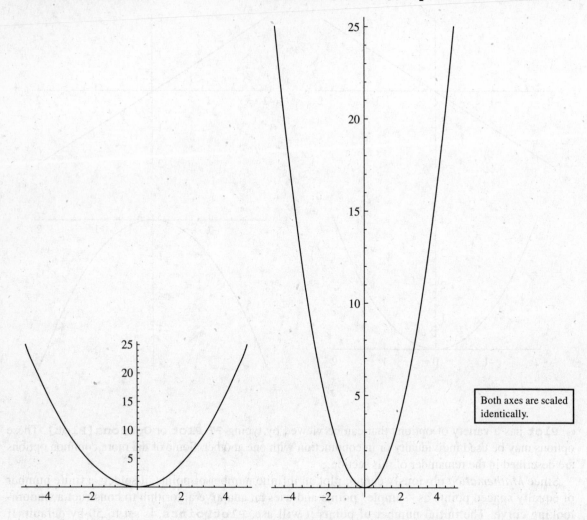

Both axes are scaled
identically.

EXAMPLE 8   The following command should produce a circle of radius 3 centered at the origin. However, because
of unequal axis scaling, the graph appears as an ellipse.

Plot[{−Sqrt[9 − x²], Sqrt[9 − x²]}, {x, −3, 3}]

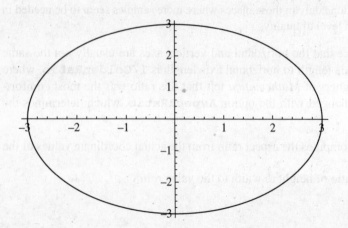

We can make the circle appear round by setting **AspectRatio → Automatic**.

```
Plot[{-Sqrt[9 - x²], Sqrt[9 - x²]}, {x, -3, 3}, AspectRatio → Automatic]
```

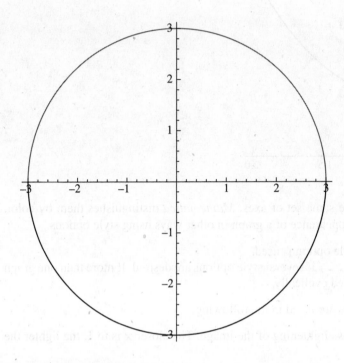

When graphing a function, *Mathematica* makes a calculated decision where to place the origin. If (0, 0) is within the plotting region, the axes will cross at that location. If not, an algorithm decides where the axes should cross. This can sometimes lead to a confusing (and misleading) rendering of the function. The option **AxesOrigin** gives control over the placement of the intersection point.

- **AxesOrigin → Automatic** is the default. If the point (0, 0) is within, or close to, the plotting region, then it is usually chosen as the axis origin.
- **AxesOrigin → {x, y}** forces the intersection of the axes to be the point (x, y).

EXAMPLE 9

```
Plot[5 + x⁴, {x, 1, 2}]
```

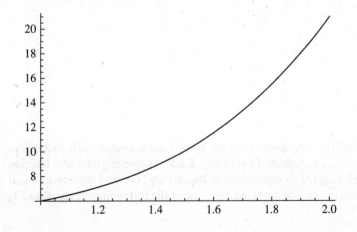

The axes intersect at (1, 6). The graph is drawn completely, however, from x = 1 to x = 2.

```
Plot[5 + x⁴, {x, 1, 2}, AxesOrigin → {0, 0}]
```

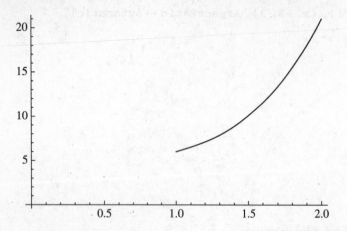

When multiple graphs are drawn on the same set of axes, *Mathematica* distinguishes them by color. **PlotStyle** allows the user to alter the appearance of a graph in other ways using style options.

- **PlotStyle** → *style* if only one style option is used.
- **PlotStyle** → {*style1*, *style2*, . . .} if several style options are desired. If more than one graph is to be modified, the styles are applied cyclically.

Some of the more common style options are listed in the following:

- **GrayLevel[x]** for $0 \le x \le 1$ allows lightening of the image. The closer x is to 1, the lighter the image will appear.

EXAMPLE 10
```
Plot[{Sin[x], Sin[2x], Sin[3x]}, {x, -π, π},
      PlotStyle → {GrayLevel[0.0], GrayLevel[0.5], GrayLevel[0.8]}]
```

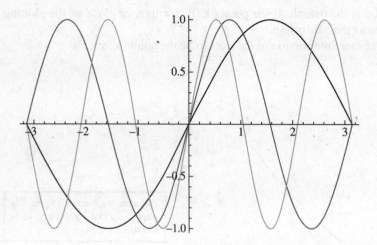

- **Dashing[{r_1, r_2, . . . , r_m}]** specifies that the curves are to be drawn dashed with successive segments and spaces of lengths r_1, r_2, . . . , r_m repeated cyclically. Each r value is given as a fraction of the total width of the graph. **Dashing[r]** is equivalent to **Dashing[{r, r}]** and gives equal size dashes and spaces. For convenience, r can be replaced with one of the following: **Tiny**, **Small**, **Medium**, or **Large**.

- **AbsoluteDashing**[$\{d_1, d_2, \ldots, d_m\}$] specifies that the curve is to be drawn dashed, with successive segments having absolute lengths $d_1, d_2, \ldots, d_m$ repeated cyclically. **AbsoluteDashing**[d] is equivalent to **AbsoluteDashing**[$\{d, d\}$] and gives equal size dashes and spaces. The absolute lengths are measured in units of printer's points, equal to $\dfrac{1}{72}$ of an inch. For convenience, d can be replaced with one of the following: **Tiny**, **Small**, **Medium**, or **Large**.

EXAMPLE 11

```
Plot[{x², 2x², 3x²}, {x, -3, 3},
     PlotStyle → {Dashing[.01], Dashing[.03], Dashing[{.03, .1}]}]
```

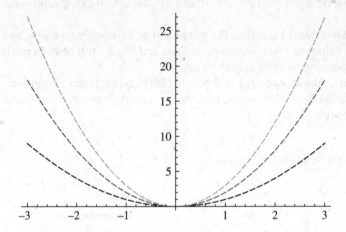

- **Thickness**[r] specifies that the graph is to be drawn with a thickness r. The thickness r is given as a fraction of the total width of the graph. The default value for two-dimensional graphs is 0.004. For convenience, r can be replaced with one of the following: **Tiny**, **Small**, **Medium**, or **Large**. These yield thicknesses independent of the width of the graph.

- **AbsoluteThickness**[d] specifies that the graph is to be drawn with absolute thickness d. The absolute thickness is measured in units of printer's points, equal to $\dfrac{1}{72}$ of an inch.

EXAMPLE 12

```
Plot[{x², 2x², 3x²}, {x, -3, 3},
     PlotStyle → {Thickness[.005], Thickness[.01], Thickness[.02]}]
```

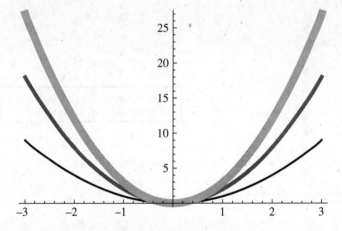

There are several style options that control color.

- **Hue**[*hue*] is a color specification. As *hue* varies from 0 to 1, the corresponding color runs through red, yellow, green, cyan, blue, magenta, and back to red again.

- **Hue [*hue, saturation, brightness*]** specifies colors in terms of hue, saturation, and brightness levels. The values of *saturation* and *brightness* must be between 0 and 1.
- **Hue [*hue, saturation, brightness, opacity*]** specifies colors in terms of hue, saturation, brightness, and opacity levels. The values of *saturation*, *brightness*, and *opacity* must be between 0 and 1. (An opacity of 0 represents perfect transparency.)
- **RGBColor [*red, green, blue*]** specifies the mixture of red, green, and blue to produce a certain color. The values of *red*, *green*, and *blue* must be between 0 and 1. RGBColor[1, 0, 0] produces a pure red display, RGBColor[0, 1, 0] produces green, and RGBColor[0, 0, 1] produces blue.
- **RGBColor [*red, green, blue, opacity*]** is similar to **RGBColor [*red, green, blue*].** The values of *red*, *green*, *blue*, and *opacity* must be between 0 and 1. (An opacity of 0 represents perfect transparency.)
- **CMYKColor [*cyan, magenta, yellow, black*]** specifies the mixture of cyan, magenta, yellow, and black to produce a certain color. The values of *cyan*, *magenta*, *yellow*, and *black* must be between 0 and 1. **CMYKColor** is useful when printing colored graphs on paper.
- **CMYKColor [*cyan, magenta, yellow, black, opacity*]** is similar to **CMYKColor [*cyan, magenta, yellow, black*].** The values of *cyan*, *magenta*, *yellow*, *black*, and *opacity* must be between 0 and 1. (An opacity of 0 represents perfect transparency.)

Certain colors can be mentioned by name. Available choices are:

Red	Green	Blue	Black
White	Gray	Cyan	Magenta
Yellow	Brown	Orange	Pink
Purple	LightRed	LightGreen	LightBlue
LightGray	LightCyan	LightMagenta	LightYellow
LightBrown	LightOrange	LightPink	LightPurple

EXAMPLE 13

Plot [{x^2, 2x^2, 3x^2}, {x, -3, 3}, PlotStyle → {Red, Green, Blue}]

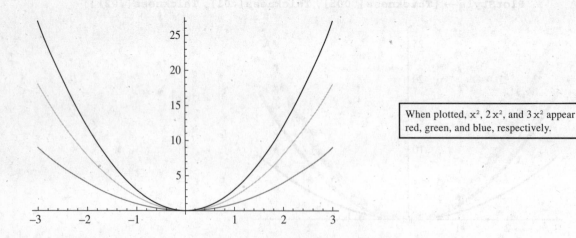

When plotted, x^2, 2x^2, and 3x^2 appear red, green, and blue, respectively.

Mathematica makes it easy to compute the RGB "formula" for custom colors. Simply click on Insert ⇒ Color and select the color of your choice. The exact RGB combination for the color selected will be placed into your *Mathematica* notebook at the cursor position.

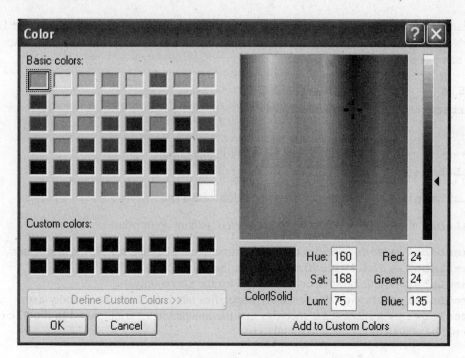

Color selector on a PC.

Color selector on a Macintosh.

The function **ColorData** contains a list of predefined colors. Type **ColorData["Legacy",** "ColorNames"] to see an extensive list of named colors. To see the RGB formula, replace "Names" with the name of the color within quotes.

EXAMPLE 14

    ColorData["Legacy", "AliceBlue"]
    RGBColor[0.941206, 0.972503, 1.]

There are two graphics options that can be used to label graphs.
PlotLabel specifies an overall label for the graph.

- **PlotLabel** → "*description*" labels the graph with a title.

AxesLabel allows one or both axes to be labeled with an appropriate description.

- **AxesLabel** → **None** specifies that neither axis should be labeled. This is *Mathematica*'s default.
- **AxesLabel** → "*label*" specifies a label for the y-axis only.
- **AxesLabel** → {"*label*"} specifies a label for the x-axis only.
- **AxesLabel** → {"*x–label*","*y–label*"} specifies labels for both the x- and y-axes.
- **AxesLabel** → **Automatic** specifies that the independent variable used in the **Plot** command should be printed along the horizontal axis.

EXAMPLE 15

    Plot[Sin[x], {x, 0, 2π}, PlotLabel → "GRAPH OF Y = SIN X",
             AxesLabel → {"Values of x","Values of sin x"}]

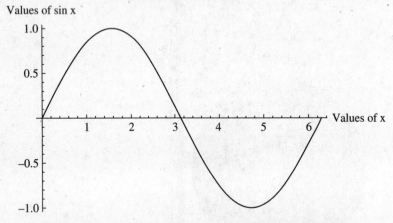

GRAPH OF Y = SIN X

PlotLegend is a useful option that can be used to label the graphs in a legend box. It is contained within the package **PlotLegends`**, which must be loaded prior to its use.

- **PlotLegend[{"*text1*", "*text2*", ...}]** attaches *text1*, *text2*, ... to each description specified in **PlotStyle**.
- **LegendPosition** → {**a, b**} specifies the position for the lower-left corner of the legend box. The center of the graphic is position (0, 0) and the longest side of the graphic runs from –1 to 1.

LegendSize determines the size of the legend box.

- **LegendSize** → *scale* scales the size by a factor of *scale*.
- **LegendSize** → {**a, b**} uses a and b to determine the size of the legend box. The value 1 corresponds to half the length of the longest side of the graphic.

LegendOrientation determines the orientation of the legend box.

- **LegendOrientation → Vertical** (default) prints the descriptions top to bottom.
- **LegendOrientation → Horizontal** prints the descriptions left to right.

LegendShadow determines the positioning of the shadow of the legend box.

- **LegendShadow → Automatic** is the default.
- **LegendShadow → None** produces no shadow. The legend box is transparent.
- **LegendShadow → {*x_offset, y_offset*}** moves the shadow to the right or up for positive values and to the left or down for negative values.

EXAMPLE 16

```
<<PlotLegends`
Plot[{x², 2x², 3x²}, {x, -3, 3}, PlotStyle → {Dashing[{.01}],
    Dashing[{.03}], Dashing[{.03, .08}]}, PlotLegend → {"x²", "2x²", "3x²"}]
```

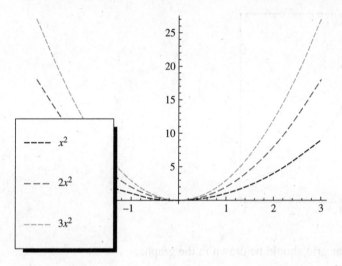

EXAMPLE 17

```
<<PlotLegends`
Plot[{x², 2x², 3x²}, {x, -3, 3},
    PlotStyle → {Dashing[{.01}], Dashing[{.03}], Dashing[{.03, .08}]},
    PlotLegend → {"x²", "2x²", "3x²"}, LegendPosition → {.2, .4},
    LegendSize → .5, LegendOrientation → Horizontal,
    LegendShadow → {-.05,.05}]
```

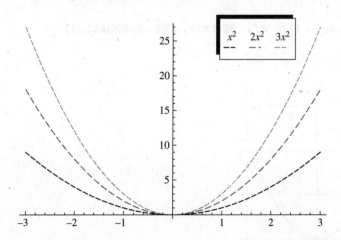

If desired, graphs can be enclosed within a rectangular frame. Additionally, one or both axes of a graph can be suppressed. **Frame** specifies whether a frame should be drawn around the graph.

- **Frame → True** specifies that a rectangular frame is to be drawn around the graph.
- **Frame → False** specifies that no frame is to be drawn (default).

Axes specifies whether the axes should be drawn.

- **Axes → True** specifies that both axes will be drawn (default).
- **Axes → False** draws no axes.
- **Axes → {False, True}** draws a y-axis but no x-axis.
- **Axes → {True, False}** draws an x-axis but no y-axis.

EXAMPLE 18

$$\text{Plot}\left[\frac{1}{x^2+1}, \{x, -3, 3\}, \text{Frame} \to \text{True}, \text{Axes} \to \text{False}\right]$$

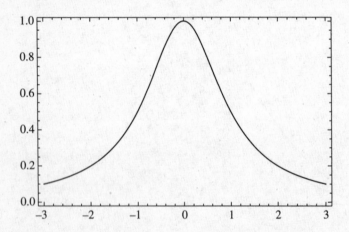

Gridlines specifies that a rectangular grid should be drawn in the graph.

- **GridLines → None** specifies that no grid lines are to be drawn (default).
- **GridLines → Automatic** specifies that the gridline positions are to be chosen by *Mathematica*.
- **GridLines → {xlist, ylist}** specifies that gridline positions are to be drawn at the specified locations. *xlist* and *ylist* are lists of numbers enclosed within { } or may (individually) be specified as **Automatic,** in which case *Mathematica* will choose their location.

EXAMPLE 19 When plotting trigonometric graphs, it is convenient to have vertical grid lines placed at multiples of $\pi/2$.

$$\text{Plot}[\text{Sin}[x], \{x, 0, 2\pi\}, \text{GridLines} \to \{\{0, \pi/2, \pi, 3\pi/2, 2\pi\}, \text{Automatic}\}]$$

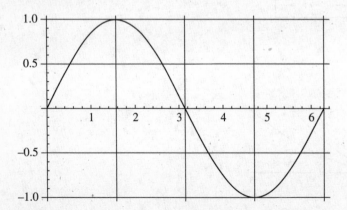

Tick marks and corresponding labeling along the axes can be controlled with the option **Ticks**. **FrameTicks** offers similar options along the edges of a frame when **Frame → True** is set.

- **Ticks → None** specifies that no tick marks are to be drawn. The numerical labeling of the axes is suppressed.
- **Ticks → Automatic** specifies that tick marks will be drawn (default).
- **Ticks → {*xlist, ylist*}** specifies that tick marks will be drawn at the specified locations. *xlist* and *ylist* are lists of numbers enclosed within { } or may be specified as **Automatic**.

EXAMPLE 20   Here are three ways to plot the graph $y = \dfrac{x^2}{x^2+1}$.

$$\texttt{Plot}\left[\frac{\texttt{x}^2}{\texttt{x}^2+\texttt{1}}, \; \{\texttt{x}, \; -3, \; 3\}\right]$$

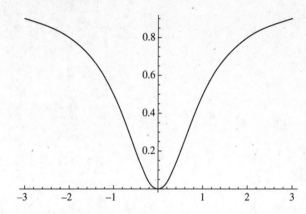

$$\texttt{Plot}\left[\frac{\texttt{x}^2}{\texttt{x}^2+\texttt{1}}, \; \{\texttt{x}, \; -3, \; 3\}, \; \texttt{Ticks} \; \rightarrow \; \texttt{None}\right]$$

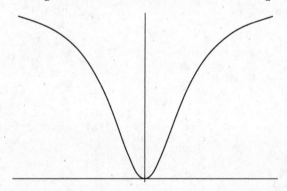

$$\texttt{Plot}\left[\frac{\texttt{x}^2}{\texttt{x}^2+\texttt{1}}, \; \{\texttt{x}, \; -3, \; 3\}, \; \texttt{Ticks} \; \rightarrow \; \{\{-3, \; 3\}, \; \texttt{Automatic}\}\right]$$

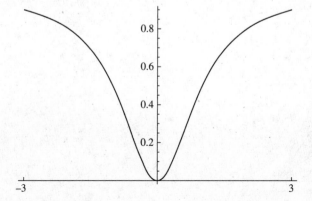

The option **Filling** will plot a shaded graph.

- **Filling** → **Axis** fills from the curve to the x-axis.
- **Filling** → **Top** fills from the curve to the top of the plot.
- **Filling** → **Bottom** fills from the curve to the bottom of the plot.
- **Filling** → y fills from the curve to value y in the vertical direction.
- **Filling** → {m} fills to the mth curve.
- **Filling** → {m → {n}} fills from the mth curve to the nth curve.
- **Filling** → {m → {y, g}} fills from the mth curve to the value y using style option g.
- **Filling** → {m → {{n}, g}} fills from the mth curve to the nth curve using style option g.

EXAMPLE 21

Plot[1 – x², {x, –1, 1}, Filling → Axis]

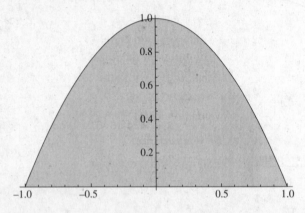

Plot[{1 – x², 2 – 2x²}, {x, –1, 1}, Filling → {1 → {2}}]

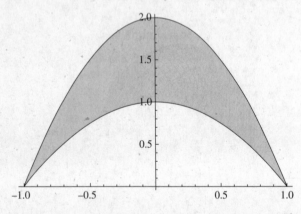

Plot[{1 – x², 2 – 2x², 3 – 3x²}, {x, –1, 1}, Filling → {1 → {2}, 2 → {3}}]

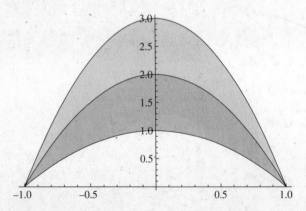

EXAMPLE 22

```
Plot[{1 - x², 2 - 2x², 3 - 3x²}, {x, -1, 1},
     Filling → {1 → {0, Orange}, 1→ {{2}, Green}, 2 → {{3}, Yellow}}]
```

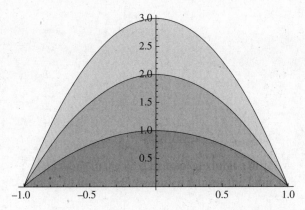

SOLVED PROBLEMS

4.1 Plot the graph of $y = xe^{-x}$ from $x = 0$ to $x = 5$.

SOLUTION

```
Plot[x Exp[-x], {x, 0, 5}]
```

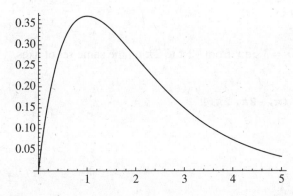

4.2 Plot $f(x) = |1 - |x||$ on the interval $[-3, 3]$.

SOLUTION

```
Plot[Abs[1 - Abs[x]], {x, -3, 3}]
```

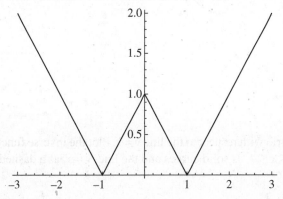

4.3 The standard normal curve used in probability and statistics is defined by the function

$$f(x) = \frac{1}{\sqrt{2\pi}} e^{-\frac{1}{2}x^2}$$

Sketch the graph for $-3 \le x \le 3$.

SOLUTION

```
f[x_] = 1/√(2π) Exp[-1/2 x²];
```

```
Plot[f[x], {x, -3, 3}]
```

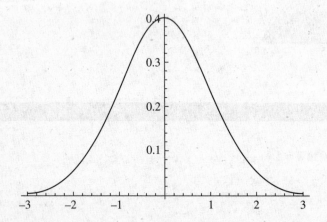

4.4 Plot the graphs $y = \sin x$, $y = 2 \sin x$, and $y = 3 \sin x$ from -2π to 2π on the same set of axes.

SOLUTION

```
Plot[{Sin[x], 2 Sin[x], 3 Sin[x]}, {x, -2π, 2π}]
```

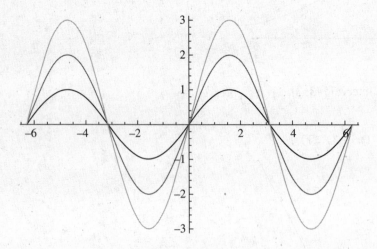

4.5 The graphs of inverse functions are symmetric with respect to the line $y = x$. Plot the inverse functions $f(x) = x^2, 0 \le x \le 2$, and $f^{-1}(x) = \sqrt{x}, 0 \le x \le 4$, as solid curves and the line $y = x$ as a dashed line and observe the symmetry.

SOLUTION

```
g1 = Plot[x², {x, 0, 2}];
g2 = Plot[√x, {x, 0, 4}];
g3 = Plot[x, {x, 0, 4}, PlotStyle → Dashing[{.02, .01}]];
Show[g1, g2, g3, AspectRatio → Automatic, PlotRange → {{0, 4}, Automatic}]
```

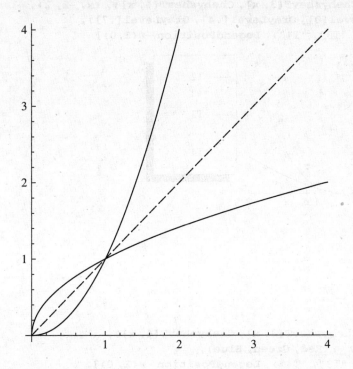

4.6 Sketch the graphs of $y = x^2$, $y = -x^2$, and $y = x^2 \sin 10x$, $-2\pi \le x \le 2\pi$, on a single set of axes enclosed by a frame.

SOLUTION

```
Plot[{x², -x², x²Sin[10x]}, {x, -2π, 2π}, Frame → True]
```

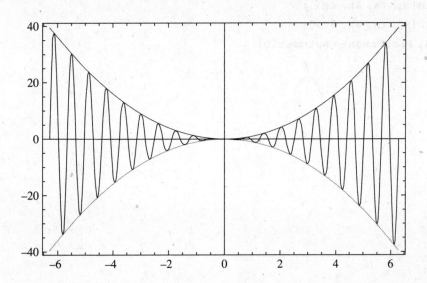

4.7 The family of Chebyshev polynomials is used in approximation theory and numerical analysis. *Mathematica* represents these polynomials as **ChebyshevT[n, x]**. On a single set of axes, using some device to distinguish the curves, plot a labeled graph showing the Chebyshev polynomials of degrees 2, 3, and 4.

SOLUTION 1

```
≪PlotLegends`
Plot[{ChebyshevT[2, x], ChebyshevT[3, x], ChebyshevT[4, x]}, {x, -2, 2},
     PlotStyle → {GrayLevel[0], GrayLevel[.4], GrayLevel[.7]},
     PlotLegend → {"T2", "T3", "T4"}, LegendPosition → {1,0}]
```

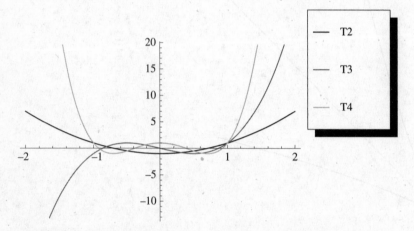

SOLUTION 2

```
≪PlotLegends`
Plot[{ChebyshevT[2, x], ChebyshevT[3, x], ChebyshevT[4, x]},
     {x, -2, 2}, PlotStyle → {Red, Green, Blue},
     PlotLegend → {"T2", "T3", "T4"}, LegendPosition → {1, 0}]
```

> Color graph not shown.

4.8 Sketch the graphs of $y = 1 + \sin x, 0 \leq x \leq 2\pi$, $y = 2 + \sin x, 2\pi \leq x \leq 4\pi$, and $y = 3 + \sin, 4\pi \leq x \leq 6\pi$ on one set of axes.

SOLUTION

```
g1 = Plot[1 + Sin[x], {x, 0, 2π}];
g2 = Plot[2 + Sin[x], {x, 2π, 4π}];
g3 = Plot[3 + Sin[x], {x, 4π, 6π}];
Show[g1, g2, g3, PlotRange → Automatic]
```

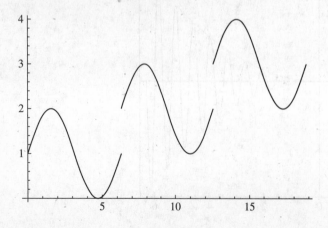

4.2 Additional Graphics Commands

Standard geometric shapes can be constructed with the **Graphics** command and viewed with the **Show** command.

- **Graphics [***primitive***]** creates a two-dimensional graphics object.

The following are a few of the more common graphics primitives available in *Mathematica*:

- **Circle [{x, y}, r]** creates a circle centered at (x, y) having radius r.
- **Disk [{x, y}, r]** creates a disk (filled circle) centered at (x, y) having radius r.
- **Point [{x, y}]** plots a point at coordinate (x, y).
- **Line [{{x1, y1}, {x2, y2}, ...}]** draws lines connecting points (x1, y1), (x2, y2), ...
- **Rectangle [{x1, y1}, {x2, y2}]** creates a filled rectangle having (x1, y1) and (x2, y2) as opposite ends of a diagonal.
- **Polygon [{{x1, y1}, {x2, y2}, ...}]** constructs a filled polygon having points (x1, y1), (x2, y2), ... as vertices.
- **Text [***textstring***, {x, y}]** prints a string of text centered at position (x, y). **TextStyle** allows you to change the default font and size used in the graph's text. **TextStyle → {FontFamily →** *fontname***, FontSize →** *size***}** is a simple, but useful, application.

When viewing graphics objects using **Show**, the default, **Axes → False**, causes the object to be drawn without axes. If desired, **Axes → True** may be included as an option.

EXAMPLE 23

```
g1 = Graphics [Circle [{0, 0}, 1]];
g2 = Graphics [Line [{{-1, -1}, {-1, 1}, {1, 1}, {1, -1}, {-1, -1}}]];
g3 = Graphics [Polygon [{{-1, 0}, {0, 1}, {1, 0}, {0, -1}}]];
g4 = Graphics [Text ["Square in a Circle in a Square", {0, 1.2},
            TextStyle → {FontSize → 20}]];

Show [g1, g2, g3, g4]
```

Square in a Circle in a Square

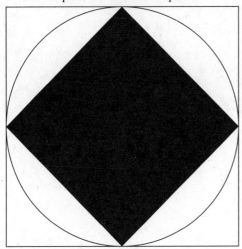

Curves are sometimes defined parametrically, i.e., the *x*- and *y*-coordinates of points are defined as two independent functions of a third variable. Parametric curves, which are usually more complex in their behavior, can be viewed using **ParametricPlot**.

- **ParametricPlot [{x[t], y[t]}, {t, tmin, tmax}]** plots the parametric curve $x = x(t)$, $y = y(t)$ over the interval $tmin \leq t \leq tmax$.

■ `ParametricPlot[{{x1[t], y1[t]}, {x2[t], y2[t]}, ...}, {t, tmin, tmax}]`
plots several sets of parametric equations over $\text{tmin} \leq t \leq \text{tmax}$.

EXAMPLE 24

`ParametricPlot[{t`^{3}`-2 t, t`^{2}`- t}, {t, -2, 2}]`

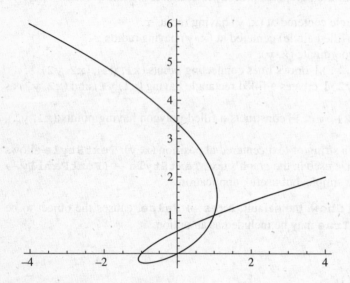

EXAMPLE 25

```
x[t_] = Cos[t] - Cos[100 t] Sin[t];
y[t_] = 2 Sin[t] - Sin[100 t];
ParametricPlot[{x[t], y[t]}, {t, 0, 2π}]
```

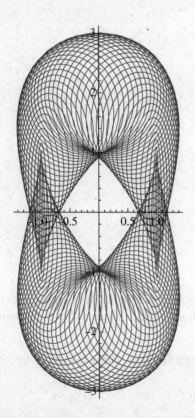

Implicitly defined curves can be plotted with the **ContourPlot** command.

- **ContourPlot[*equation*, {x, xmin, xmax}, {y, ymin, ymax}]** plots *equation* by treating it as a function in three-dimensional space, and generates a contour of the equation cutting through the plane where z equals zero.

equation must be of the form **lhs == rhs**. Note the double equal sign in the middle.

- **ContourPlot[{*equation1*, *equation2*,...}, {x, xmin, xmax}, {y, ymin, ymax}]** plots several implicitly defined curves.

By default, **ContourPlot** sets **Axes → False** and **Frame → True**. Additional options such as **Dashing**, **Graylevel**, **Thickness**, etc. determining the appearance of the graph may be included using **ContourStyle**.

EXAMPLE 26 Plot the equation $x^2 y^2 = (y+1)^2 (4-y^2)$ for $-10 \leq x \leq 10$, $-2 \leq y \leq 2$. (Conchoid of Nicomedes.)

```
ContourPlot[x² y² == (y + 1)²(4 - y²), {x, -10, 10}, {y, -2, 2},
        AspectRatio → Automatic]
```

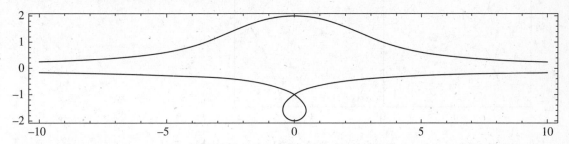

```
ContourPlot[x² y² == (y + 1)²(4 - y²), {x, -10, 10}, {y, -2, 2},
        AspectRatio → Automatic, Axes → True, Frame → False]
```

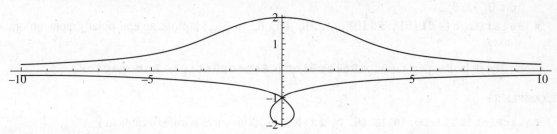

EXAMPLE 27 Plot the equation $x^3 + y^3 = 6xy$ for $-4 \leq x \leq 4$, $-4 \leq y \leq 4$. (Folium of Descartes.)

```
ContourPlot[x³ + y³ == 6 x y, {x, -4, 4}, {y, -4, 4}, Axes → True, Frame → False]
```

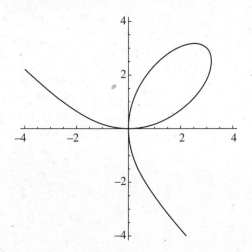

EXAMPLE 28   Plot $\cos(x - y) = y \sin x$ and $\sin(x - y) = y \cos x$, $-2 \leq x \leq 2$, $-2 \leq y \leq 2$ on one set of axes.

```
ContourPlot[{Cos[x - y] == y Sin[x], Sin[x - y] == y Cos[x]}, {x, -2, 2}, {y, -2, 2},
       ContourStyle → {Dashing[.01], Dashing[.03]}, Axes → True]
```

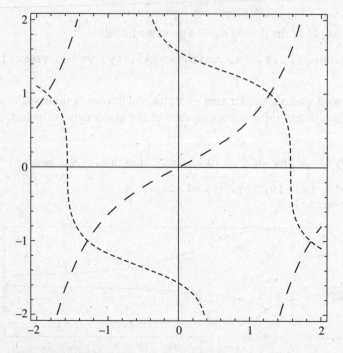

For curves defined in polar coordinates, **PolarPlot** is available.

- **PolarPlot[f[θ], {θ, θ$_{min}$, θ$_{max}$}]** generates a plot of the polar equation $r = f(θ)$ as θ varies from θ$_{min}$ to θ$_{max}$.
- **PolarPlot[{f1[θ], f2[θ], ...}, {θ, θ$_{min}$, θ$_{max}$}]** plots several polar graphs on one set of axes.

Note: The default aspect ratio for **PolarPlot** is **AspectRatio → Automatic**.

EXAMPLE 29

```
PolarPlot[3 (1 - Cos[θ]), {θ, 0, 2π}]
```
        (This curve is called a cardioid.)

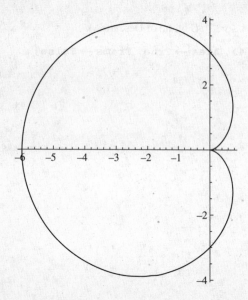

EXAMPLE 30 Plot the three-leaf rose $r = \sin 3\theta$ inside the unit circle $r = 1$.

```
PolarPlot[{1, Sin[3θ]}, {θ, 0, 2π}]
```

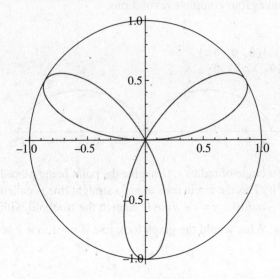

SOLVED PROBLEMS

4.9 Sketch the parabola $y = x^2 - 9$ and a circle of radius 3 centered at the origin.

SOLUTION

```
g1 = Plot[x² - 9, {x, -4, 4}];
g2 = Graphics[Circle[{0, 0}, 3]];
g3 = Graphics[Text["CIRCLE IN A PARABOLA", {0, 6},
                    TextStyle → {FontSize → 16}]];
Show[g1, g2, g3, AspectRatio → Automatic]
```

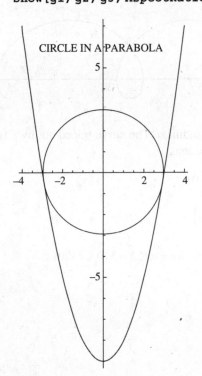

4.10 The curve traced by a point on a circle as the circle rolls along a straight line is called a *cycloid* and has parametric equations $x = r(\theta - \sin\theta)$, $y = r(1 - \cos\theta)$ where r represents the radius of the circle. Plot the cycloid formed as a circle of radius 1 makes four complete revolutions.

SOLUTION

```
ParametricPlot[{θ - Sin[θ], 1 - Cos[θ]}, {θ, 0, 8π},
          Ticks → {Automatic, {0, 1, 2}}]
```

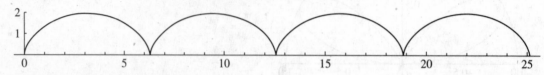

4.11 Let P be a point at a distance a from the center of a circle of radius r. (Imagine the point being placed on a spoke of a bicycle wheel.) The curve traced by P as the circle rolls along a straight line is called a *trochoid*. Its parametric equations are $x = r\theta - a\sin\theta$, $y = r - a\cos\theta$. Sketch the trochoid with $r = 1$, $a = \dfrac{1}{2}$ as the circle makes four revolutions. What would the graph look like if $r = 1$, $a = 2$ so that the point is outside the circle?

SOLUTION

```
r = 1; a = 1/2;
ParametricPlot[{r θ - a Sin[θ], r - a Cos[θ]}, {θ, 0, 8π},
          PlotRange → {Automatic, {0, 2}},
          Ticks → {Automatic, {0, 1, 2}}]
```

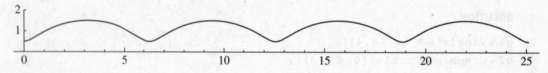

```
r = 1; a = 2;
ParametricPlot[{r θ - a Sin[θ], r - a Cos[θ]}, {θ, 0, 8π}]
```

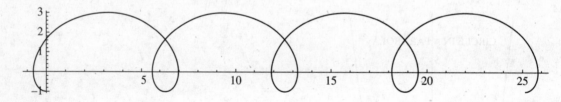

4.12 A circle of radius b rolls on the inside of a larger circle of radius a. The curve traced out by a fixed point initially at $(a, 0)$ is called a hypocycloid and has equations

$$x = (a - b)\cos\theta + b\cos\left(\frac{a - b}{b}\theta\right)$$

$$y = (a - b)\sin\theta - b\sin\left(\frac{a - b}{b}\theta\right)$$

Sketch the hypocycloid for $a = 4$, $b = 1$ ($0 \le x \le 2\pi$) and then again for $a = 8$, $b = 5$ ($0 \le x \le 10\pi$).

SOLUTION

```
x[θ_] := (a - b) Cos[θ] + b Cos[  a - b
                                  ───── θ]
                                    b

y[θ_] := (a - b) Sin[θ] - b Sin[  a - b
                                  ───── θ]
                                    b
```

```
a = 4;
b = 1;
ParametricPlot[{x[θ], y[θ]}, {θ, 0, 2π}]
```

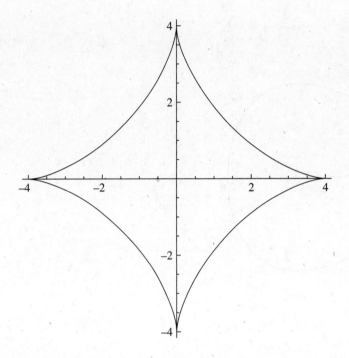

```
a = 8;
b = 5;
ParametricPlot[{x[θ], y[θ]}, {θ, 0, 10π}]
```

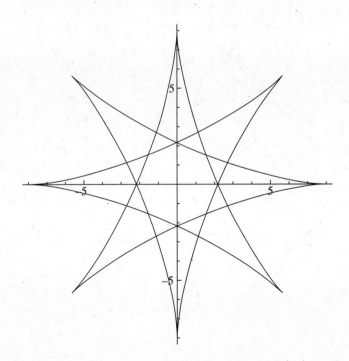

4.13 Sketch the graph defined by the equation $y^2 = x^3(2-x)$, $0 \le x \le 2$, $-2 \le y \le 2$.

SOLUTION

`ContourPlot[y² == x³(2 - x), {x, 0, 2}, {y, -2, 2}, Frame → False, Axes → True]`

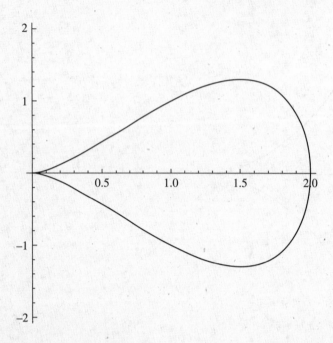

4.14 Sketch the graph of the Tschirnhausen cubic: $y^2 = x^3 + 3x^2$, $-3 \le x \le 3$, $-8 \le y \le 8$.

SOLUTION

`ContourPlot[y² == x³ + 3x², {x, -3, 3}, {y, -8, 8}, Axes → True, Frame → False]`

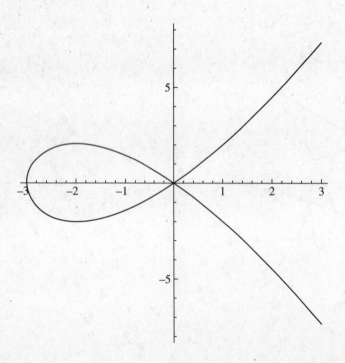

4.15 The polar graph $r = \theta$ is called the Spiral of Archimedes. Sketch the graph for $0 \le \theta \le 10\pi$ and then again for $-10\pi \le \theta \le 10\pi$.

SOLUTION

`PolarPlot[θ, {θ, 0, 10π}]`   `PolarPlot[θ, {θ, -10π, 10π}]`

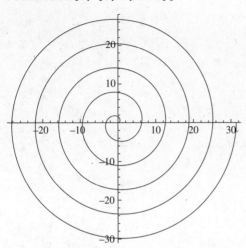 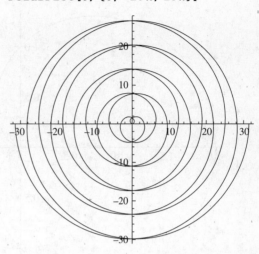

4.16 The equation $r = \sin n\theta$, where n is a positive integer, represents a family of polar curves called roses. Investigate the behavior of this family and form a conjecture about how the number of loops is related to n.

SOLUTION

```
g1 = PolarPlot[Sin[2θ], {θ, 0, 2π}, Ticks → False, PlotLabel → "n = 2"];
g2 = PolarPlot[Sin[3θ], {θ, 0, 2π}, Ticks → False, PlotLabel → "n = 3"];
g3 = PolarPlot[Sin[4θ], {θ, 0, 2π}, Ticks → False, PlotLabel → "n = 4"];
g4 = PolarPlot[Sin[5θ], {θ, 0, 2π}, Ticks → False, PlotLabel → "n = 5"];
GraphicsArray[{{g1, g2}, {g3, g4}}]
```

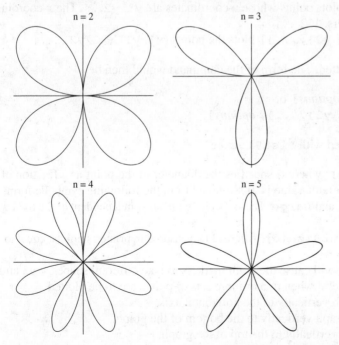

Conclusion: If n is odd, the rose will have n leaves. If n is even, there will be $2n$ leaves.

4.17   Sketch the cardioid $r = 1 - \cos\theta$ and the circle $r = 1$ on the same set of axes.

SOLUTION

```
PolarPlot[{1 - Cos[θ], 1}, {θ, 0, 2π}]
```

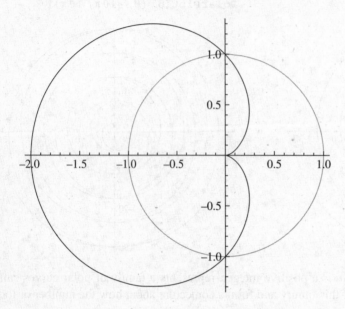

4.3   Special Two-Dimensional Plots

Discrete functions, i.e., functions defined on a discrete set, can be visualized using the special plotting function **ListPlot**.

- **ListPlot[{y1, y2,...}]** plots points whose *y*-coordinates are y1, y2, ... The *x*-coordinates are taken to be the positive integers, 1, 2, ...
- **ListPlot[{{x1, y1}, {x2, y2},...}]** plots the points (x1, y1), (x2, y2), ...

Standard graphics options are permitted. The form of the command would then be

- **ListPlot[{y1, y2, ...},** *options*] or
- **ListPlot[{{x1, y1}, {x2, y2},...},** *options*]

The most useful graphics options used with **ListPlot** are

- **PlotStyle → PointSize[d]**   where d specifies the diameter of the point as a fraction of the overall width of the graph. The default value is .008. In addition, the following symbolic forms can be used: **Tiny**, **Small**, **Medium**, and **Large**. These specify point sizes independent of the total width of the graphic.
- **PlotStyle → AbsolutePointSize[d]**   where d is measured in printer's points, equal to $\frac{1}{72}$ of an inch.
- **PlotMarkers → Automatic** will cause the point markers to take different shapes, e.g., circles, squares, diamonds, etc. This is useful when two or more sets of points are to be plotted.
- **Filling → Axis** fills the graph vertically to the horizontal axis.
- **Filling → Bottom** fills the graph vertically to the bottom of the graph.
- **Filling → Top** fills the graph vertically to the top of the graph.
- **Filling → v** fills the graph vertically to the value v.

EXAMPLE 31   The following plots a list of the squares of the positive integers 1 through 20.

```
squares = Table[k², {k, 1, 20}];
ListPlot[squares]
```

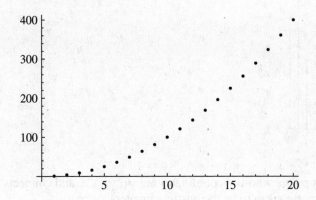

```
ListPlot[squares, PlotStyle → PointSize[.03]]
```

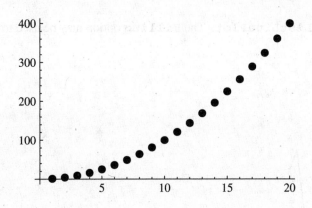

EXAMPLE 32

```
randomintegers = Table[RandomInteger[{1, 20}], {k, 1, 30}];
ListPlot[randomintegers]
```

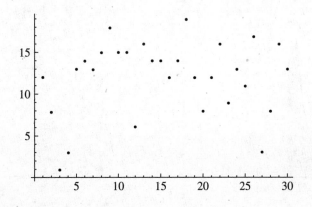

> The **Table** command generates a list of 30 random integers, each between 1 and 20.

```
ListPlot[randomintegers,Filling → Axis]
```

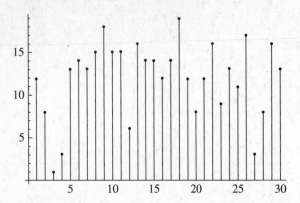

- `ListLinePlot[{y₁, y₂,...}]` plots points whose y-coordinates are $y_1, y_2, \ldots$ and connects them with line segments. The x-coordinates are taken to be the positive integers.
- `ListLinePlot[{{x₁, y₁}, {x₂, y₂},...}]` plots the points $(x_1, y_1), (x_2, y_2), \ldots$ and connects them with line segments.
- `ListLinePlot[list₁, list₂, ...]` plots multiple lines through points defined by $list_1$, $list_2, \ldots$

The options for `ListPlot` may be used for `ListLinePlot`. The `Filling` option may be used to create a filled polygon that describes the data.

EXAMPLE 33   (Continuation of Example 32)

```
ListLinePlot[randomintegers]
```

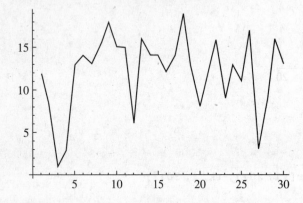

```
ListLinePlot[randomintegers, Filling → Axis]]
```

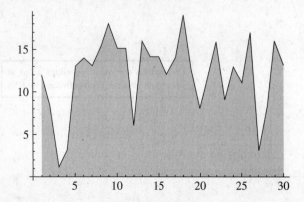

Different types of bar graphs can be drawn with *Mathematica*, using the command **BarChart**.

Note: Starting with version 7, **BarChart** can be found in the *Mathematica* kernel. If you are using version 6, you will find **BarChart** in the package **BarCharts`** which must be loaded prior to use. See the Documentation Center for appropriate usage.

- **BarChart**[*datalist*] draws a simple bar graph. *datalist* is a set of numbers enclosed within braces.
- **BarChart**[{*datalist1*, *datalist2*, ...}] draws a bar graph containing data from multiple data sets. Each data list is a set of numbers enclosed within braces.

EXAMPLE 34

```
dataset1 = {1, 2, 3, 4, 5};
dataset2 = {6, 5, 4, 3, 2};
g1 = BarChart[dataset1];
g2 = BarChart[{dataset1, dataset2}];
GraphicsArray[{g1, g2}]
```

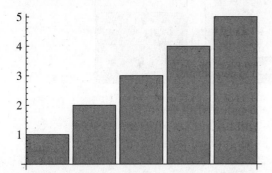 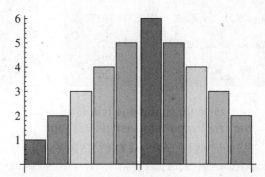

If a customized look is desired, there are a variety of options that can be invoked. The format of the command with options becomes

- **BarChart**[*datalist*, *options*]
- **BarChart**[{*datalist1*, *datalist2*, ...}, *options*]

Some of the more popular options are

- **Chartstyle** → *g* specifies that style option *g* should be used to draw the bars. Examples of style options are **GrayLevel**, **Hue**, **Opacity**, **RGBColor**, and Colors (**Red**, **Blue**, etc.).
- **Chartstyle** → {*g1*, *g2*, ...} specifies that style options *g1*, *g2*, ... should be used cyclically.
- **ChartLayout** → "*layout*" specifies that a layout of type *layout* should be used to draw the graph. Examples of layouts are "**Stacked**", in which case the bars are stacked on top of each other rather than placed side by side, and "**Percentile**", which generates a stacked bar chart with the total height of each bar constant at 100%.

BarSpacing controls the spacing between bars and between groups of bars. The default is **BarSpacing** → **Automatic** which allows *Mathematica* to control the spacing.

- **BarSpacing** → *s* allows a space of *s* between bars within each data set. The value of *s* is measured as a fraction of the width of each bar.
- **BarSpacing** → {*s*, *t*} allows a space of *s* between bars within each data set and a value of *t* determines the space between data sets. The values of *s* and *t* are measured as a fraction of the width of each bar.

In each of the preceding **BarSpacing** commands, the values of *s* and *t* may be replaced by the predefined symbols **None**, **Tiny**, **Small**, **Medium** and **Large**.

- **BarOrigin** → *edge* controls where the bars originate from. The default value of *edge* is **Bottom**. Other acceptable values are **Top**, **Left**, and **Right**.
- **ChartLabels** → {*label1* , *label2* , ...} specifies the labeling for each bar corresponding to each value in the data list.

EXAMPLE 35

```
dataset1 = {1, 2, 3, 4, 5};
dataset2 = {6, 5, 4, 3, 2};
g1 = BarChart[{dataset1, dataset2}, ChartLayout → "Stacked"];
g2 = BarChart[{dataset1, dataset2}, ChartLayout → "Percentile"];
GraphicsArray[{g1, g2}]
```

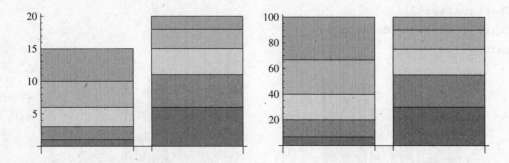

EXAMPLE 36

```
dataset = {6, 3, 4, 1, 5};
BarChart[dataset, ChartLabels → {"Bar1", "Bar2", "Bar3", "Bar4", "Bar5"}]
```

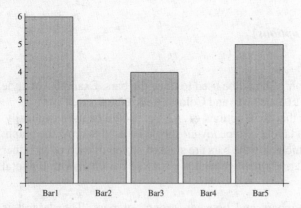

EXAMPLE 37

```
dataset = {6, 3, 4, 1, 5};
g1 = BarChart[dataset];
g2 = BarChart[dataset, BarOrigin → Top];
g3 = BarChart[dataset, BarOrigin → Left];
g4 = BarChart[dataset, BarOrigin → Right];
GraphicsArray[{{g1, g2}, {g3, g4}}]
```

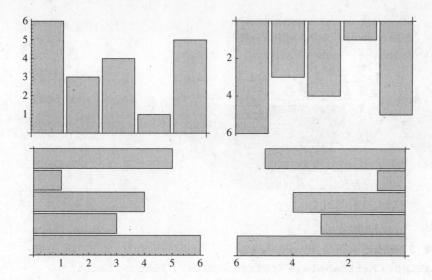

Pie Charts may be constructed using the **PieChart** command.

Note: Starting with version 7, **PieChart** can be found in the *Mathematica* kernel. If you are using version 6, you will find **PieChart** in the package **PieCharts`** which must be loaded prior to use. See the Documentation Center for appropriate usage.

- **PieChart** [*datalist*] draws a simple pie chart. *datalist* is a list of numbers enclosed within braces.
- **PieChart** [{*datalist1*, *datalist2*, ...}] draws a pie chart containing data from multiple data sets. Each data set is a list of numbers enclosed within braces.

Similar to **BarChart**, there are options that can be invoked to enhance the display. The format of the command with options becomes

- **PieChart** [*datalist*, *options*]
- **PieChart** [{*datalist1*, *datalist2*, ...}, *options*]

Some of the available options associated with **PieChart** are

- **Chartstyle** $\to g$ specifies that style option g should be used to draw the bars. Examples of style options are **GrayLevel**, **Hue**, **Opacity**, **RGBColor**, and Colors (**Red**, **Blue**, etc.).
- **Chartstyle** $\to \{g1, g2, \dots\}$ specifies that style options $g1, g2, \dots$ should be used cyclically.

SectorSpacing determines the spacing between concentric sectors for different data sets and the spacing between sectors within a data set.

- **SectorSpacing** $\to s$ determines the spacing between concentric sectors for different data lists. The value of s is measured as a fraction of the radial width of the sectors.
- **SectorSpacing** $\to \{s, t\}$ allows a space of s between sectors corresponding to each data set and a space of t between concentric sectors for different data sets. The values of s and t are measured as a fraction of the radial width of the sectors.

In each of the preceding **SectorSpacing** commands, the values of s and t may be replaced by the predefined symbols **None**, **Tiny**, **Small**, **Medium** and **Large**.

Note: Clicking on any sector of a pie chart will cause it to shift radially outward by an amount s.

EXAMPLE 38

```
dataset = {1.5, 3, 4.5, 9};
g1 = PieChart[dataset];
g2 = PieChart[dataset, SectorSpacing → {Tiny, None}];
GraphicsArray[{g1, g2}]
```

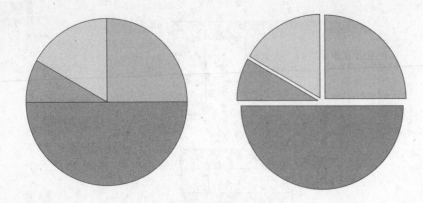

EXAMPLE 39

```
datalist = {1.5, 3, 4.5, 9};
g1 = PieChart[datalist, ChartLabels → {"First Sector", "Second Sector",
              "Third Sector", "Fourth Sector"}]
```

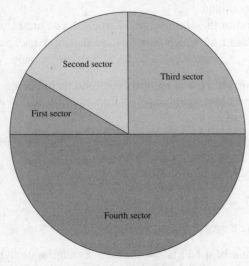

SOLVED PROBLEMS

4.18   Plot the first 50 prime numbers.

SOLUTION

```
primelist = Table[Prime[k], {k, 1, 50}];
ListPlot[primelist]
```

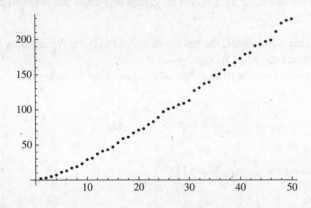

4.19 Plot the points (0, 0), (2, 7), (3, 5), and (4, 11) and connect them with line segments.

SOLUTION

```
list = {{0, 0}, {2, 7}, {3, 5}, {4, 11}};
ListLinePlot[list, PlotMarkers → Automatic]
```

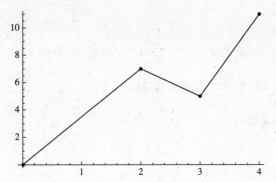

4.20 Plot the set of points corresponding to the first ten primes, the first ten Fibonacci numbers, and the first ten perfect squares. First plot individual points and then plot them connected with line segments.

SOLUTION

```
<<PlotLegends`
list1 = Table[Prime[n], {n, 1, 10}];
list2 = Table[Fibonacci[n], {n, 1, 10}];
list3 = Table[n², {n, 1, 10}];
ListPlot[{list1, list2, list3}, PlotMarkers → Automatic,
        PlotLegend → {"Primes", "Fibonacci", "Squares"},
        LegendPosition → {1,0}]
```

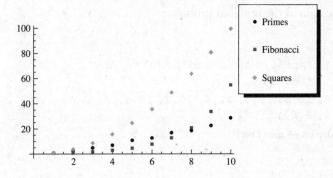

```
ListLinePlot[{list1, list2, list3}, PlotMarkers → Automatic,
        PlotLegend → {"Primes", "Fibonacci", "Squares"},
        LegendPosition → {1,0}]
```

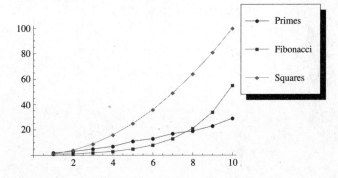

4.21 The monthly sales for XYZ Corp. (in thousands of dollars) were

JAN	FEB	MAR	APR	MAY	JUNE	JULY	AUG	SEPT	OCT	NOV	DEC
13.2	15.7	17.4	12.6	19.7	22.6	20.2	18.3	16.2	15.0	12.1	8.6

Construct a bar graph illustrating this data.

SOLUTION

```
months = {"Jan", "Feb", "Mar", "Apr", "May", "Jun", "Jul", "Aug", "Sep",
          "Oct", "Nov", "Dec"};
salesdata = {13.2, 15.7, 17.4, 12.6, 19.7, 22.6, 20.2, 18.3,
             16.2, 15.0, 12.1, 8.6};
BarChart[salesdata, ChartLabels → months]
```

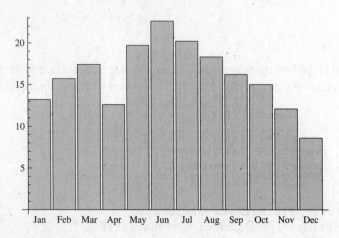

4.22 Construct a pie chart illustrating the data of the previous problem.

SOLUTION

```
months = {"Jan", "Feb", "Mar", "Apr", "May", "Jun", "Jul", "Aug",
          "Sep", "Oct", "Nov", "Dec"};
salesdata = {13.2, 15.7, 17.4, 12.6, 19.7, 22.6, 20.2, 18.3,
             16.2, 15.0, 12.1, 8.6};
PieChart[salesdata, ChartLabels → months]
```

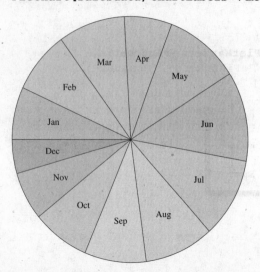

4.4   Animation

Animation effects can be produced quickly and easily through the use of the **Animate** command. This command displays several different graphics images rapidly in succession, producing the illusion of movement. The form of the command is

- **Animate[*expression*, {k, m, n, i}]**

where *expression* is any *Mathematica* command with parameter k which varies from m to n in increments of i (optional; if omitted, i varies continuously from m to n).

The following example gives an interesting animated description of the behavior of the odd powers of x^n as n gets larger.

EXAMPLE 40

Animate[Plot[x^k, {x, −1, 1}, PlotRange → {−1, 1}, Ticks → False], {k, 1, 19, 2}]

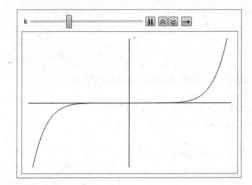

The speed of the animation and the direction are easily controlled by clicking on the ⊻ , ⌃ , and → buttons. The animation can be paused, using the ❚❚ button.

To allow the user more control over the animation, the **Manipulate** command can be used. **Manipulate** works very much the same way as **Animate** except it allows the user to control the parameter directly with a slider.

- **Manipulate[*expression*, {k, m, n, i}]**

EXAMPLE 41

Manipulate[Plot[x^k, {x, −1, 1}, PlotRange → {−1, 1}, Ticks → False], {k, 1, 19, 2}]

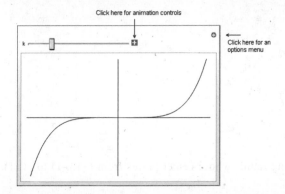

A convenient way of controlling expressions involving integer parameters is by clicking on "radio buttons." This can be accomplished with the option **ControlType → RadioButton**.

EXAMPLE 42

```
Manipulate[Plot[xᵏ, {x, -1, 1}, PlotRange → {-1, 1}, Ticks → False],
          {k, 1, 19, 2}, ControlType → RadioButton]
```

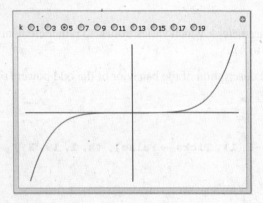

expression may involve two or more parameters. In this case the form of the command is

- `Animate[expression, {k1, m1, n1, i1}, {k2, m2, n2, i2},...]`
- `Manipulate[expression, {k1, m1, n1, i1}, {k2, m2, n2, i2},...]`

Each parameter can be controlled independently (speed, direction, pause).

EXAMPLE 43

```
Animate[Plot[a Sin[b x], {x, 0, 2π}, PlotRange → {-10, 10}],
        {a, 0, 10}, {b, 0, 10}]
```

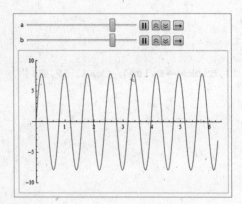

EXAMPLE 44   This animation shows a circle of varying radius whose center varies from (-1, -1) to (1, 1). Pause each variable (x, y, r) to see the effect.

```
Animate[Graphics[Circle[{Sin[x], Cos[y]}, r], Axes → True,
        PlotRange → {{-2, 2}, {-2, 2}}], {x, 0, 2π}, {y, 0, 2π}, {r, 0, 1}]
```

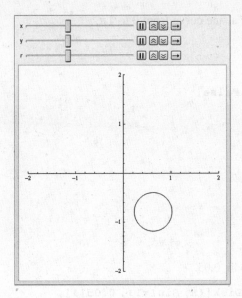

Animate and **Manipulate** are not limited to the presentation of graphics. We will use these commands in other contexts in later chapters.

SOLVED PROBLEMS

4.23 Construct an animation of the Spiral of Archimedes, $r = \theta$ as θ varies from 8π to 10π.

SOLUTION

```
Animate[PolarPlot[θ, {θ, 0, 8π + ϕ}, Ticks → False,
        PlotRange → {{-10π, 10π}, {-10π, 10π}}], {ϕ, 0, 2π}]
```

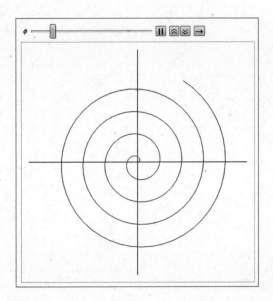

4.24 Use **Manipulate** to simulate a point "rolling" along a sine curve from 0 to 2π.

SOLUTION

First we construct the sine curve.

```
sincurve = Plot[Sin[x], {x, 0, 2π}, Ticks → False]
```

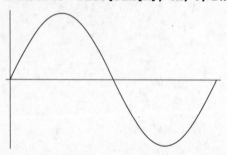

Now we animate the sequence of points as red disks of radii 0.05.

```
Manipulate[Show[sincurve, Graphics[{Red, Disk[{x, Sin[x]}, 0.05]}],
        PlotRange → {{0, 2π}, {-1, 1}},
        AspectRatio → Automatic], {x, 0, 2π}].
```

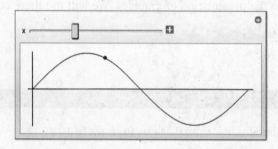

Move the slider to control the movement of the disk.

CHAPTER 5

Three-Dimensional Graphics

5.1 Plotting Functions of Two Variables

A function of two variables may be viewed as a surface in three-dimensional space. The simplest command for plotting a surface is **Plot3D**.

- **Plot3D[f[x, y], {x, xmin, xmax}, {y, ymin, ymax}]** plots a three-dimensional graph of the function f[x, y] above the rectangle $xmin \leq x \leq xmax$, $ymin \leq y \leq ymax$.
- **Plot3D[{f₁[x, y], f₂[x, y],...}, {x, xmin, xmax}, {y, ymin, ymax}]** plots several surfaces on one set of axes.

Mathematica's default axis orientation is as shown in the figure to the right. This is somewhat different from what appears in many calculus textbooks.

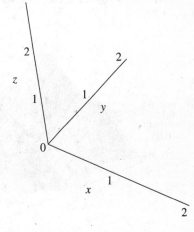

EXAMPLE 1

```
Plot3D[Sin[x - y], {x, -π, π}, {y, -π, π}]
```

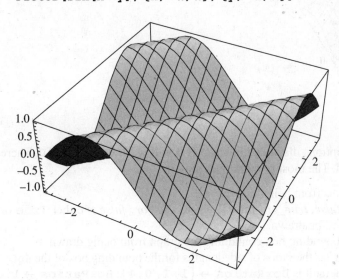

The option **PlotPoints** specifies the number of points to be used in each direction to produce the graph. Unlike two-dimensional graphics, the default for a three-dimensional plot is **PlotPoints → 15**. This often leads to graphs with ragged surfaces. Increasing **PlotPoints** will alleviate this condition.

- **Plotpoints → n** specifies that n initial sample points should be used in each direction. Additional points are selected by adaptive algorithms.
- **PlotPoints → {nx, ny}** specifies that nx and ny initial sample points are to be used along the x-axis and y-axis, respectively.

The next example shows how an increase in the value of **PlotPoints** affects the "smoothness" of the resulting graph.

EXAMPLE 2

```
f[x_, y_] = x²y² Exp[-(x² + y²)];
Plot3D[f[x, y], {x, -2, 2}, {y, -2, 2}]
```

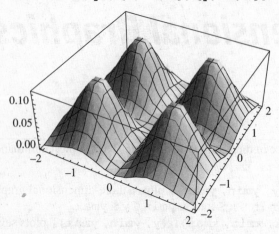

```
Plot3D[f[x, y], {x, -2, 2}, {y, -2, 2}, PlotPoints → 40]
```

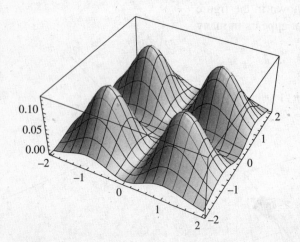

Most of the two-dimensional graphics options discussed in Chapter 4 will work with **Plot3D**. There are a few extra options that are new as well. The most popular ones are:

- **Axes → False** will suppress the axes from being drawn.
- **Axes → {*true_or_false, true_or_false, true_or_false*}**, where *true_or_false* is either **True** or **False**, will determine which axes will be drawn.
- **Boxed → False** will suppress the bounding box containing the graph from being drawn.
- **BoxRatios → {sx, sy, sz}** specifies the ratios of side lengths for the bounding box of the three-dimensional picture. *Mathematica*'s default is **BoxRatios → {1, 1, 0.4}**. **BoxRatios → 1** is equivalent to **BoxRatios → {1, 1, 1}**.
- **Ticks → False** will eliminate tick marks and corresponding labeling along the axes. **Ticks → {*true_or_false, true_or_false, true_or_false*}**, where *true_or_false* is either **True** or **False**, will control ticks on individual axes.

FaceGrids is an option that draws grid lines on the faces of the bounding box.

- **FaceGrids → All** draws grid lines on all six faces of the bounding box.
- **FaceGrids → None** (default) draws no grid lines.

- **FaceGrids** $\rightarrow$ $\{\{x_1, y_1, z_1\}, \{x_2, y_2, z_2\}, \ldots, \{x_6, y_6, z_6\}\}$ allows gridlines to be drawn on individual faces. Two of the three numbers in each sublist must be 0 and the third ± 1 to indicate which of the six possible faces will contain grid lines.

AxesEdge is an option that specifies on which edges of the bounding box axes should be drawn.

- **AxesEdge** $\rightarrow$ **Automatic** (default) lets *Mathematica* decide on which edges axes should be drawn.
- **AxesEdge** $\rightarrow$ $\{\{y_1, z_1\}, \{x_2, z_2\}, \{x_3, y_3\}\}$ is where each of the x, y, and z values are either 1 or -1, to indicate on which edges of the bounding box the axes are to be drawn. 1 indicates that the axes will be drawn on the edge with the larger coordinate value, -1 indicates the smaller coordinate value. Any of the three lists $\{x, y\}$ can be replaced by **Automatic**, in which case *Mathematica* decides where to place the axis, or **None**, in which case the axis is not drawn.
- **BoxStyle** is an option that specifies how the bounding box is to be drawn. **BoxStyle** can be set to a list of style options such as **Dashing**, **Thickness**, **GrayLevel**, or **RGBColor**.
- **Mesh** is an option that determines whether a mesh should be drawn on the graphic surface. The default is **Mesh** $\rightarrow$ **True**; **Mesh** $\rightarrow$ **False** or **Mesh** $\rightarrow$ **None** eliminates the mesh.

The next example plots the parabolic cylinder $z = x^2$ using different options.

EXAMPLE 3 (Graphs are grouped together for easy comparison.)

```
Plot3D[x², {x, -2, 2}, {y, -2, 2}]
Plot3D[x², {x, -2, 2}, {y, -2, 2}, Mesh → False]
Plot3D[x², {x, -2, 2}, {y, -2, 2}, BoxRatios → 1]
Plot3D[x², {x, -2, 2}, {y, -2, 2}, FaceGrids → {{1, 0, 0}, {0, -1, 0}}]
Plot3D[x², {x, -2, 2}, {y, -2, 2}, AxesEdge → {{-1, 1}, {1, 1}, {1, -1}}]
```

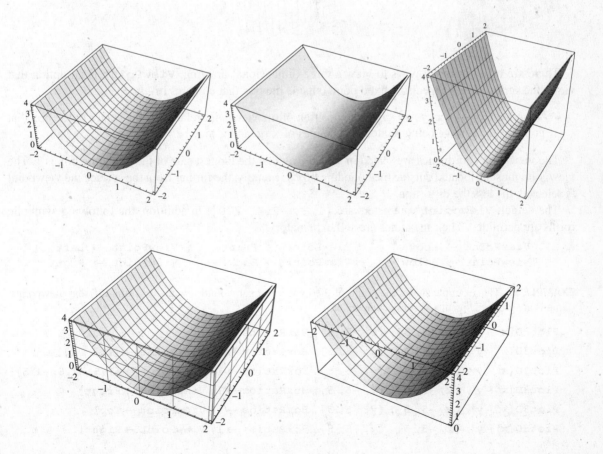

Three-dimensional graphics are generated as a sequence of polygons shaded to create a pleasing three-dimensional affect. The polygons are drawn opaque so that surfaces behind other surfaces are hidden. The following option can be used to draw the surface transparent.

- **PlotStyle → FaceForm[ ]** draws the polygons transparent (only the connecting lines are drawn) so that all surfaces are visible.

EXAMPLE 4   (Graphs are grouped together for easy comparison.)

```
Plot3D[1 - y², {x, -5, 5}, {y, -5, 5}, BoxRatios → {1, 1, 2}, Boxed → False,
    Axes → False]
Plot3D[1 - y², {x, -5, 5}, {y, -5, 5}, BoxRatios → {1, 1, 2}, Boxed → False,
    Axes → False, PlotStyle → FaceForm[]]
```

 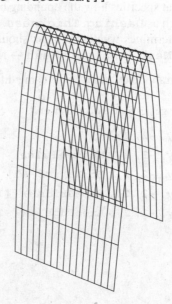

There are many different ways to view a three-dimensional drawing. **ViewPoint** is an option that views the surface from a specified fixed point outside the box that contains it.

- **ViewPoint → {x,y,z}** gives the position of the viewpoint relative to the center of the box that contains the surface being plotted. The values of x, y, and z may be ∞.

The viewpoint coordinates are scaled in such a way that the longest side of the bounding box is 1. The viewpoint must be located outside the bounding box. Generally, the further from the surface the viewpoint is selected, the less the distortion.

The default **Viewpoint** parameters are {1.3, -2.4, 2.0}. In addition, the following symbolic forms are permitted. Their meanings are self explanatory.

ViewPoint → Above	ViewPoint → Front	ViewPoint → Left
ViewPoint → Below	ViewPoint → Back	ViewPoint → Right

EXAMPLE 5   This example shows the graph of the hyperbolic paraboloid $z = x^2 - y^2$ from different viewpoints. (Graphs are grouped together for easy comparison.)

```
Plot3D[x² - y², {x, -5, 5}, {y, -5, 5}, BoxRatios → 1]
Plot3D[x² - y², {x, -5, 5}, {y, -5, 5}, BoxRatios → 1, ViewPoint → {2, 2, 2}]
Plot3D[x² - y², {x, -5, 5}, {y, -5, 5}, BoxRatios → 1, ViewPoint → {1.5, -2.6, -1.5}]
Plot3D[x² - y², {x, -5, 5}, {y, -5, 5}, BoxRatios → 1, ViewPoint → Front]
Plot3D[x² - y², {x, -5, 5}, {y, -5, 5}, BoxRatios → 1, ViewPoint → Top]
Plot3D[x² - y², {x, -5, 5}, {y, -5, 5}, BoxRatios → 1, ViewPoint → Right]
```

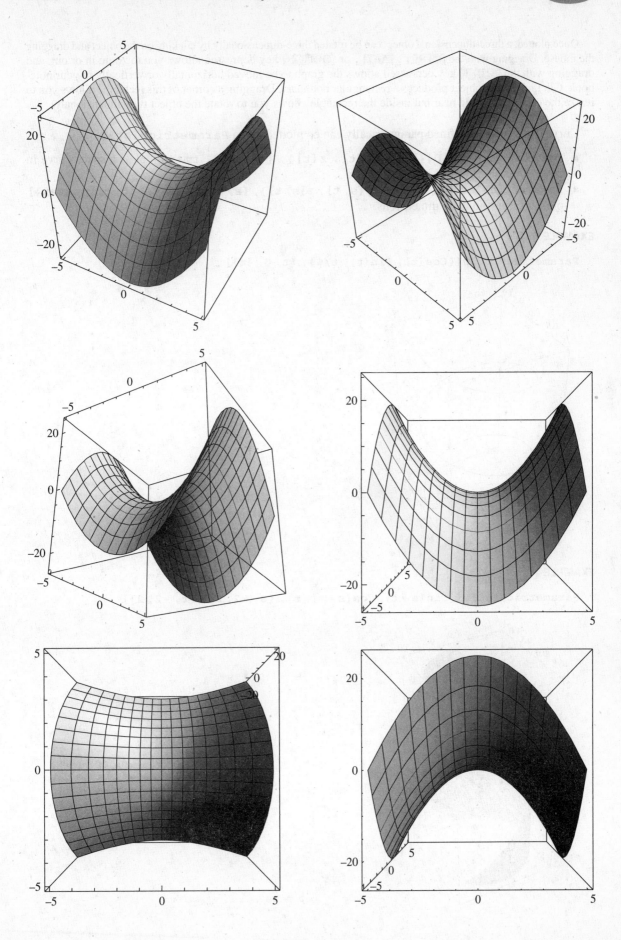

Once plotted, a three-dimensional object can be rotated three-dimensionally by clicking on the object and dragging the mouse. Dragging with the [CTRL] , [ALT] , or [OPTION] key depressed allows you to zoom in or out, and dragging with the [SHIFT] key depressed allows the graph to be moved horizontally or vertically in your notebook. Clicking on the object produces a rectangular boundary. Dragging a corner of this rectangle allows you to resize the object; dragging near but inside the rectangle allows you to rotate the object two-dimensionally.

Curves and surfaces defined parametrically can be plotted using **ParametricPlot3D**.

- **ParametricPlot3D[{x[t], y[t], z[t]}, {t, tmin, tmax}]** plots a space curve in three dimensions for $tmin \leq t \leq tmax$.
- **ParametricPlot3D[{x[s, t], y[s, t], z[s, t]}, {s, smin, smax}, {t, tmin, tmax}]** plots a surface in three dimensions.

EXAMPLE 6

```
ParametricPlot3D[{Cos[t], Sin[t], t/4}, {t, 0, 4π}]
```

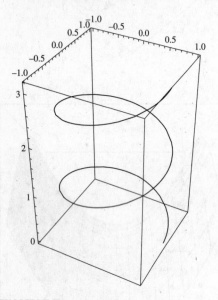

EXAMPLE 7

```
ParametricPlot3D[{Sin[s + t], Cos[s + t], s}, {s, -2, 2}, {t, -2, 2}]
```

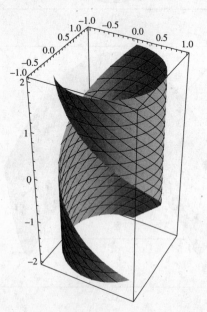

Plot3D allows you to plot surfaces expressed by equations in rectangular coordinates. Special surfaces, called surfaces of revolution, can be drawn using the command **RevolutionPlot3D**. (For additional options that provide more flexibility, please see **SurfaceOfRevolution**, which is discussed in Section 5.3.)

- **RevolutionPlot3D[f[x], {x, xmin, xmax}]** plots the surface generated by rotating the curve $z = f(x)$, xmin $\leq x \leq$ xmax, completely around the z-axis.
- **RevolutionPlot3D[f[x], {x, xmin, xmax}, {θ, θmin, θmax}]** plots the surface generated by rotating the curve $z = f(x)$, xmin $\leq x \leq$ xmax, around the z-axis for θmin $\leq θ \leq$ θmax where θ is the angle measured counterclockwise from the positive x-axis.
- **RevolutionPlot3D[{f[t],g[t]}, {t, tmin, tmax}]** generates a plot of the surface generated by rotating the curve $x = f(t)$, $z = g(t)$, tmin $\leq t \leq$ tmax, completely around the z-axis.
- **RevolutionPlot3D[{f[t],g[t]}, {t, tmin, tmax},{θ, θmin, θmax}]** generates a plot of the surface generated by the curve $x = f(t)$, $z = g(t)$, tmin $\leq t \leq$ tmax, around the z-axis for θmin $\leq θ \leq$ θmax where θ is the angle measured counterclockwise from the positive x-axis.

EXAMPLE 8 Sketch the surface of revolution generated when the curve $z = \sqrt{x}$, $0 \leq x \leq 4$, is rotated about the z-axis.

First we draw the two-dimensional generating curve and then the corresponding surface of revolution. (Graphs are placed side by side for easy comparison.)

```
Plot[√x , {x, 0, 4}, AspectRatio→1, AxesLabel→{"x", "z"}]
RevolutionPlot3D[√x , {x,0,4}, BoxRatios→1, ViewPoint→{1, -5, 2},
              AxesLabel→{"x", "y", "z"}]
```

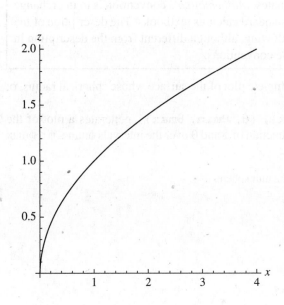
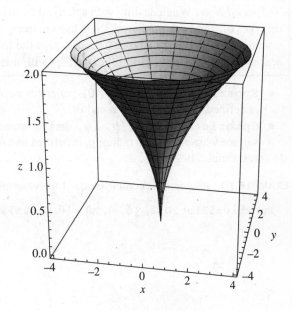

Cylindrical and spherical coordinate systems are useful for solving problems involving cylinders, spheres, and cones. Point P has cylindrical coordinates $(r, θ, z)$ where r and $θ$ are the polar coordinates of the projection of P in the x–y plane. Since the distance from P to the z-axis is r, the surface $z = z(r, θ)$ is a surface of revolution.

- **RevolutionPlot3D[z[r, θ], {r, rmin, rmax}]** generates a plot of the surface $z = z(r, θ)$ described in cylindrical coordinates for rmin $\leq r \leq$ rmax.
- **RevolutionPlot3D [z[r, θ], {r, rmin, rmax}, {θ, θmin, θmax}]** generates a plot of the surface $z = z(r, θ)$ for rmin $\leq r \leq$ rmax, θmin $\leq θ \leq$ θmax.

EXAMPLE 9   In cylindrical coordinates, the equation $z = r$ represents the cone $z = \sqrt{x^2 + y^2}$.

```
RevolutionPlot3D[r,{r,0,1},BoxRatios→1]
```

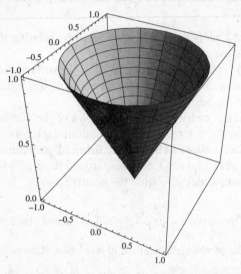

Point P has spherical coordinates (ρ, θ, ϕ) where ρ is the distance from P to the origin, θ is the angle formed by the positive *x*-axis and the line connecting the origin with the projection of P in the *x-y* plane, and ϕ is the angle formed by the positive *z*-axis and the line connecting P with the origin. The *Mathematica* command **SphericalPlot3D** allows the construction of surfaces given in spherical coordinates.

> *Special Note:* When dealing with spherical coordinates, *Mathematica*'s convention is to interchange the roles of θ and ϕ from that which is used in many standard calculus textbooks. The description of the command **SphericalPlot3D** described in the following, although different from the description in *Mathematica*'s documentation files, agrees with these conventions.

- **SphericalPlot3D[ρ, ϕ, θ]** generates a complete plot of the surface whose spherical radius, ρ, is defined as a function of ϕ and θ.
- **SphericalPlot3D[[ρ, {ϕ, ϕmin, ϕmax}, {θ, θmin, θmax}]** generates a plot of the surface whose spherical radius, ρ, is defined as a function of ϕ and θ over the intervals ϕmin $\leq \phi \leq \phi$max and θmin $\leq \theta \leq \theta$max.

EXAMPLE 10   In spherical coordinates, $\rho = 1$ represents the unit sphere.

```
SphericalPlot3D[1, {ϕ, 0, π}, {θ, 0, 2π}]
```

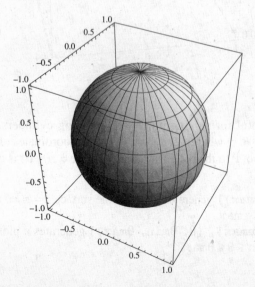

SOLVED PROBLEMS

5.1 Plot the graph of the function $e^{-x^2-y^2}$ above the rectangle $-2 \leq x \leq 2$, $-2 \leq y \leq 2$.

SOLUTION

```
Plot3D[Exp[-x² - y²], {x, -2, 2}, {y, -2, 2}]
```

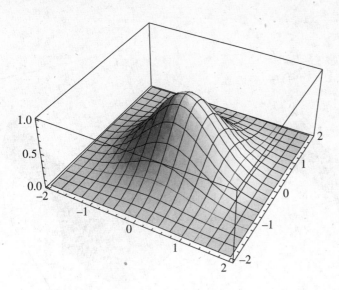

5.2 Show the intersection of the two paraboloids $f(x, y) = x^2 + y^2$ and $g(x, y) = 16 - x^2 - y^2$ above the square $-3 \leq x \leq 3$, $-3 \leq y \leq 3$.

SOLUTION

```
Plot3D[{x² + y², 16 - x² - y²}, {x, - 3, 3}, {y, - 3, 3}, BoxRatios → 1]
```

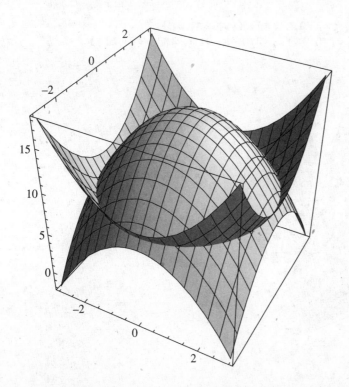

5.3 Obtain a graph of the "saddle-shaped" hyperboloid $z = x^2 - y^2$, $-5 \leq x \leq 5$, $-5 \leq y \leq 5$ in a cubic box. Draw the graph with and without a surface mesh.

SOLUTION   (Graphs are placed side by side for comparison purposes.)

```
Plot3D[x² - y², {x, -5, 5}, {y, -5, 5}, BoxRatios → 1]
Plot3D[x² - y², {x, -5, 5}, {y, -5, 5}, BoxRatios → 1, Mesh → False]
```

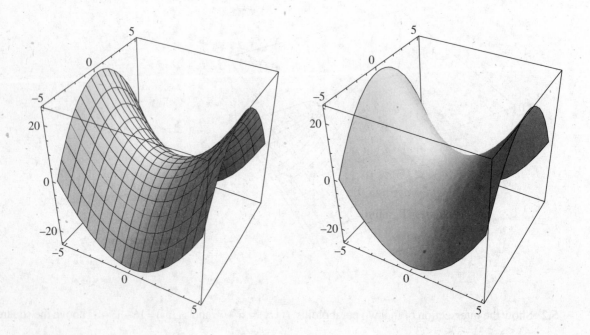

5.4 Draw the graph of the function $f(x, y) = \left| \sin x \sin y \right|$ for $-2\pi \leq x, y \leq 2\pi$. Label the x and y axes in terms of π.

SOLUTION

```
Plot3D[Abs[Sin[x]Sin[y]], {x, -2π, 2π}, {y, -2π, 2π},
       Ticks → {{-2π, -π, 0, π, 2π}, {-2π, -π, 0, π, 2π}, {0, 1}}]
```

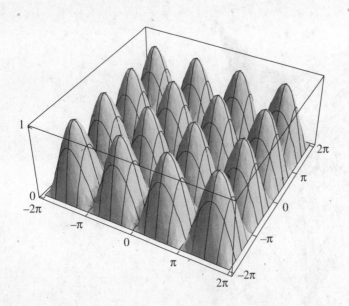

5.5 Draw the graph of the surface $z = |1 - x^2 - y^2|$ for $-1 \le x, y \le 1$. Do not draw axes or a surrounding box.

SOLUTION

```
Plot3D[Abs[1 - x² - y²], {x, -1, 1}, {y, -1, 1}, Axes → False, Boxed → False]
```

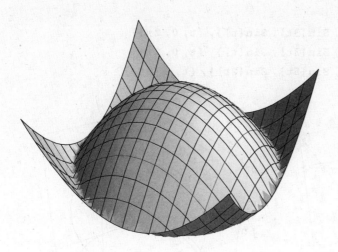

5.6 Graph the intersection of the paraboloid $z = x^2 + y^2$ with the plane $y + z = 12$. Obtain a front view and a side view.

SOLUTION (Graphs are placed side by side for easier comparison.)

```
paraboloid = Plot3D[x² + y², {x, -5, 5}, {y, -5, 5}];
plane = Plot3D[12 - y, {x, -5, 5}, {y, -5, 5}];
Show[paraboloid, plane, BoxRatios → 1, PlotRange → {0, 20},
    PlotLabel → "Default View"]
Show[paraboloid, plane, BoxRatios → 1, PlotRange → {0, 20}, ViewPoint → Front,
    PlotLabel → "Front View"]
Show[paraboloid, plane, BoxRatios → 1, PlotRange → {0, 20}, ViewPoint → Left,
    PlotLabel → "Left View"]
```

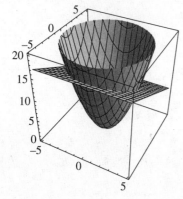
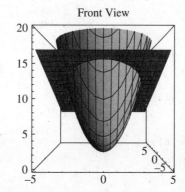
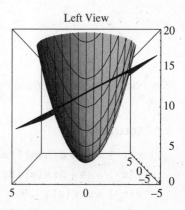

5.7 Sketch the space curves defined by $\begin{cases} x = \cos at \\ y = \sin bt \\ z = \sin ct \end{cases}$ $0 \le t \le 2\pi$ for

(i) $a = 5, b = 3, c = 1$;   (ii) $a = 3, b = 3, c = 1$;   (iii) $a = 2, b = 5, c = 2$.
These curves are known as Lissajous curves.

SOLUTION

```
ParametricPlot3D[{Cos[5t], Sin[3t], Sin[t]}, {t, 0, 2π}]
ParametricPlot3D[{Cos[3t], Sin[3t], Sin[t]}, {t, 0, 2π}]
ParametricPlot3D[{Cos[2t], Sin[5t], Sin[2t]}, {t, 0, 2π}]
```

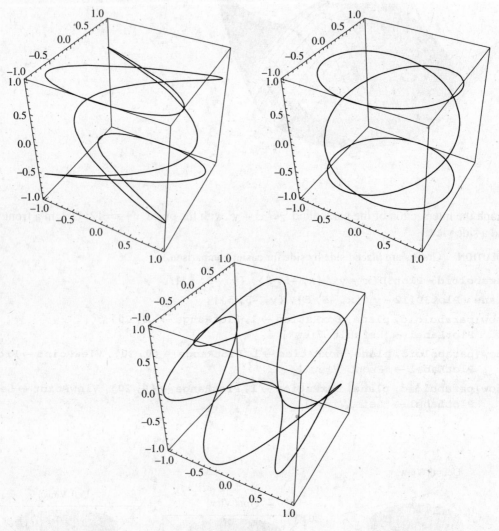

5.8 Sketch the torus defined by $\begin{cases} x = (4 + \sin s)\cos t \\ y = (4 + \sin s)\sin t \\ z = \cos s \end{cases}$ $0 \le s, t \le 2\pi$

SOLUTION

```
x[t_] = (4 + Sin[s])Cos[t];
y[t_] = (4 + Sin[s])Sin[t];
z[t_] = Cos[s];
g1 = ParametricPlot3D[{x[t], y[t], z[t]}, {s, 0, 2π}, {t, 0, 2π}, Mesh→False]
```

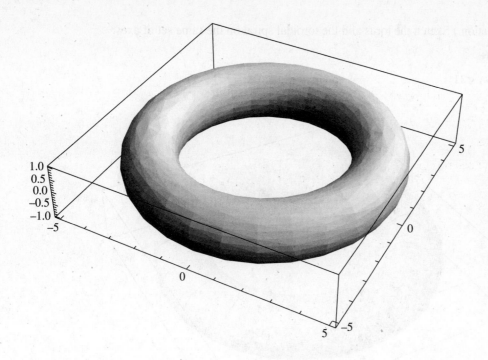

5.9 (Continuation.) Sketch the space curve $\begin{cases} x = (4 + \sin 20t)\cos t \\ y = (4 + \sin 20t)\sin t \\ z = \cos 20t \end{cases}$ $0 \le t \le 2\pi$

This curve is called a toroidal spiral since it lies on the surface of a torus (let $s = 20t$).

SOLUTION

```
x[t_] = (4 + Sin[20t]) Cos[t];
y[t_] = (4 + Sin[20t]) Sin[t];
z[t_] = Cos[20t];
g2 = ParametricPlot3D[{x[t], y[t], z[t]}, {t, 0, 2π}]
```

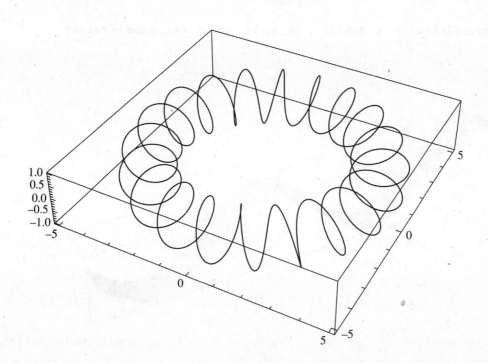

5.10 (Continuation.) Sketch the torus and the toroidal spiral on the same set of axes.

SOLUTION

`Show[g1, g2]`

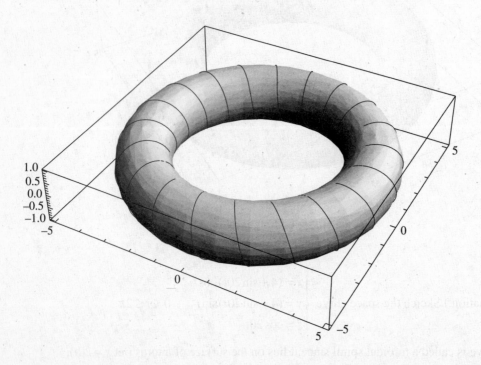

5.11 Sketch the graph of a "ribbon" one unit wide having the shape of a sine curve from 0 to 4π.

SOLUTION

We can represent this surface parametrically: $\begin{cases} x = t \\ y = s \\ z = \sin t \end{cases}$ $0 \le s \le 1, 0 \le t \le 4\pi.$

`ParametricPlot3D[{t, s, Sin[t]}, {s, 0, 1}, {t, 0, 4π}, Axes → False]`

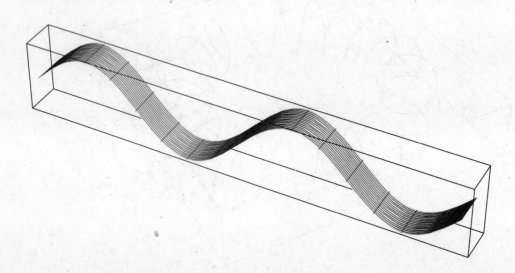

5.12 Draw the "ice cream cone" formed by the cone $z = 3\sqrt{x^2 + y^2}$ and the upper half of the sphere $x^2 + y^2 + (z-9)^2 = 9$. Use cylindrical coordinates.

SOLUTION

In cylindrical coordinates the cone has the equation $z = 3r$ and the hemisphere has the equation $z = 9 + \sqrt{9 - r^2}$.

```
cone = RevolutionPlot3D[3 r, {r, 0, 3}, BoxRatios → 1];
hemisphere = RevolutionPlot3D[9 + √(9 - r²), {r, 0, 3}];
Show[cone, hemisphere, PlotRange → All, BoxRatios → {1,1,2}]
```

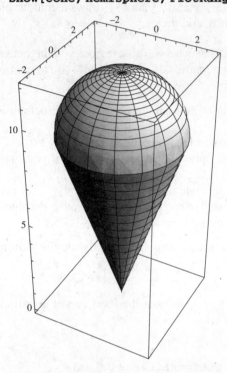

5.13 Sketch the graph of the following surface given in spherical coordinates:

$$\rho = 1 + \sin 4\theta \sin \phi, \qquad 0 \le \theta \le 2\pi, \qquad 0 \le \phi \le \pi$$

SOLUTION

```
SphericalPlot3D[1 + Sin[4 θ] Sin[φ], {φ, 0, π}, {θ, 0, 2π}]
```

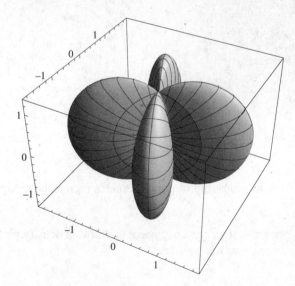

5.2 Other Graphics Commands

A level curve of a function of two variables, $f(x, y)$, is a two-dimensional graph of the equation $f(x, y) = k$ for some fixed value of k. A contour plot is a collection of level curves drawn on the same set of axes.

The *Mathematica* command **ContourPlot** draws contour plots of functions of two variables. The contours join points on the surface having the same height. The default is to have contours corresponding to a sequence of equally spaced values of the function.

- **ContourPlot[f[x, y], {x, xmin, xmax}, {y, ymin, ymax}]** draws a contour plot of $f(x, y)$ in a rectangle determined by xmin, xmax, ymin, and ymax.

Contour plots produced by *Mathematica* are drawn shaded, in such a way that regions with higher values of $f(x, y)$ are drawn lighter. As with all *Mathematica* graphics commands, options allow you to control the appearance of the graph.

- **Contours → n** allows you to determine the number of contours to be drawn. The default is ten equally spaced curves.
- **Contours → {k1, k2, ...}** draws contours corresponding to function values k1, k2, ...
- **ContourShading → False** turns off shading. This option is particularly useful if your monitor or printer does not handle grayscales well.
- **ContourLines → False** eliminates the lines that separate the shaded contours.
- **PlotPoints → n** controls how many points will be used in each direction in an adaptive algorithm to plot each curve. The default is 15. (The default for two-dimensional graphics is 25.)

A complete list of options and their default values can be obtained using the command **Options[ContourPlot]**.

EXAMPLE 11   Obtain contour plots of the paraboloid $z = x^2 + y^2$. Note that the level curves are all circles $x^2 + y^2 = k$. (Plots are placed side by side for easy comparison.)

ContourPlot[x² + y², {x, -10, 10}, {y, -10, 10}]

ContourPlot[x² + y², {x, -10, 10}, {y, -10, 10}, ContourLines → False]

ContourPlot[x² + y², {x, -10, 10}, {y, -10, 10}, ContourShading → False]

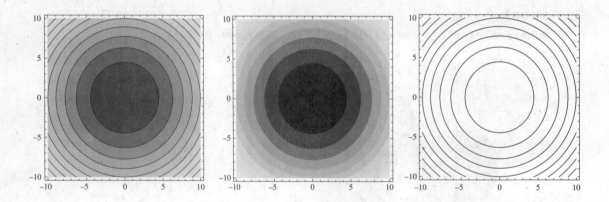

A density plot shows the values of a function at a regular array of points. Lighter regions have higher values.

- **DensityPlot[f[x, y], {x, xmin, xmax}, {y, ymin, ymax}]** draws a density plot of $f(x, y)$ in a rectangle determined by xmin, xmax, ymin, and ymax.

The option **Mesh** draws a rectangular mesh that subdivides the region.

- **Mesh → None** (default) draws no mesh.
- **Mesh → n** draws n equally spaced mesh divisions.
- **Mesh → Automatic** draws automatically chosen mesh divisions.
- **Mesh → All** draws mesh divisions between all elements.
- **Mesh → Full** draws mesh divisions through regular data points.

EXAMPLE 12

```
DensityPlot[x² + y², {x, -10, 10}, {y, -10, 10}]
```

```
DensityPlot[x² + y², {x, -10, 10}, {y, -10, 10}, Mesh → Automatic]
```

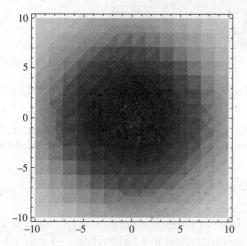 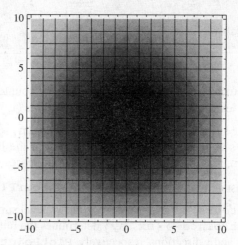

The commands **ListContourPlot** and **ListDensityPlot** are the analogs of **ContourPlot** and **DensityPlot** for lists of numbers. These commands are appropriate for use with functions defined on a lattice of integer coordinates.

- **ListContourPlot[*array*]** generates a contour plot from a two-dimensional array of numbers.
- **ListDensityPlot[*array*]** generates a density plot from a two-dimensional array of numbers.

array = { {z_{11}, z_{12}, ...}, {z_{21}, z_{22}, ...}, ...}, representing the heights of points in the *x*-*y* plane, must be a nested array of dimension 2×2 or larger. z_{ij} is the z-coordinate of the point (j, i). The options for **ListContourPlot** and **ListDensityPlot** are the same as for **ContourPlot** and **DensityPlot**, except that the axes are labeled, by default, with positive integers starting with 1. The option **DataRange** allows you to change the labeling of the axes to correspond to the actual values of the data.

- **DataRange → {{xmin, xmax}, {ymin, ymax}}** labels the x and y axes from xmin to xmax and from ymin to ymax, respectively.

EXAMPLE 13

```
list = Table[Random[], {x, 1, 10}, {y, 1, 10}];
ListContourPlot[list, DataRange → {{ -5, 5}, {3, 7}}]
ListDensityPlot[list, DataRange → {{ -5, 5}, {3, 7}}]
```

← Generates a 10×10 array of random numbers.

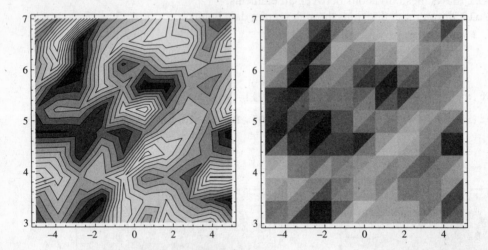

ContourPlot3D is the three-dimensional counterpart of **ContourPlot**. **ContourPlot3D** will sketch the level surfaces of f, i.e., the set of points (x, y, z) such that $f(x, y, z) = k$.

- **ContourPlot3D[f[x, y, z], {x, xmin, xmax}, {y, ymin, ymax}, {z, zmin, zmax}]** draws a three-dimensional contour plot of the level surface $f(x, y, z) = 0$ in a box determined by xmin, xmax, ymin, ymax, zmin, and zmax.

The most commonly used options for **ContourPlot3D** are

- **Contours → {k1, k2, ...}** draws level surfaces corresponding to k1, k2, ...
- **PlotPoints → {nx, ny}** determines the initial number of evaluation points that will be used in the x and y directions, respectively. **PlotPoints → n** is equivalent to **PlotPoints → {n, n}**.

EXAMPLE 14

```
ContourPlot3D[z - x² - y², {x, -5, 5}, {y, -5, 5}, {z, 0, 10}, Contours → {0, 5}, Mesh → None]
```

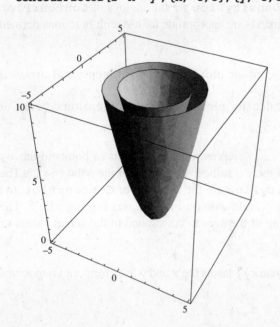

ListPlot3D is the three-dimensional analog of **ListPlot**.

- **ListPlot3D[{{z$_{11}$, z$_{12}$, ...}, {z$_{21}$, z$_{22}$, ...}, ...}]** generates a three-dimensional surface based upon a given array of heights, z$_{ij}$ (must be a nested array of dimension at least 2 × 2). The *x*- and *y*-coordinate values for each data point are taken to be consecutive integers beginning with 1.
- **ListPlot3D[{{x$_1$, y$_1$, z$_1$}, {x$_2$, y$_2$, z$_2$}, ...}]** generates a three-dimensional surface based upon a given array of heights, z$_j$ which are the z-coordinates corresponding to the points {x$_i$, y$_i$}.

Some options for **ListPlot3D** include:

- **MeshShading → *shades*** generates a surface shaded according to the descriptions in the array *shades* (**GrayLevel**, **Hue**, **RGBColor**, etc.). If *array* has dimensions m × n, then *shades* must have dimensions (m−1) × (n−1).
- **DataRange → {{xmin, xmax}, {ymin, ymax}}** labels the x- and y-axes from xmin to xmax and from ymin to ymax, respectively. The default is **DataRange → Automatic**, which assigns values starting with 1.

Mesh is an option that specifies how mesh divisions should be drawn. The default is **Mesh → Automatic**.

- **Mesh → n** specifies that n equally spaced mesh divisions (lines) should be drawn in each direction.
- **Mesh → All** specifies that mesh divisions should be drawn between all elements.
- **Mesh → None** eliminates all mesh divisions from being drawn.

EXAMPLE 15 (Graphs are grouped together for easy comparison.)

```
list = {{1, 5, 2, 2}, {3, 6, 1, 4}, {3, 1, 7, 2}};
ListPlot3D[list]
ListPlot3D[list, Mesh → None]
shades = {{Red, Orange, Green}, {Cyan, Yellow, Magenta}};
ListPlot3D[list, MeshShading → shades]
```

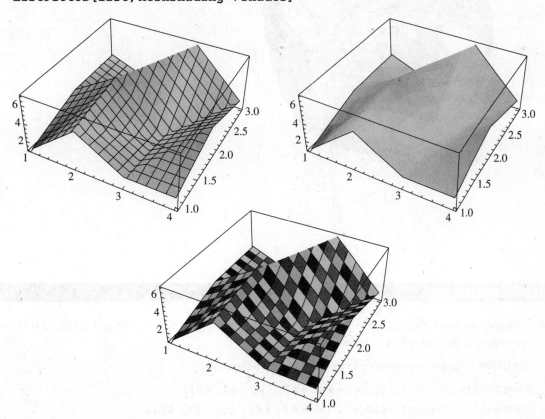

The discrete analog of **ContourPlot3D** is **ListContourPlot3D**.

- **ListContourPlot3D [*array*]** draws a contour plot of the values in *array*, a three-dimensional array of numbers representing the values of a function.
  - **Contours → n** is an option that draws contours at n equally spaced levels. The default is **Contours → 3. Contours → {k1, k2, . . .}** draws contours corresponding to function values k1, k2, ...
  - **DataRange → {{xmin, xmax}, {ymin, ymax}, {zmin, zmax}}** labels the x, y, and z axes from xmin to xmax, ymin to ymax, and zmin to zmax, respectively. The default is **DataRange → Automatic**, which assigns values starting with 1.

EXAMPLE 16   This example generates a *discrete* set of values of the function $f(x, y, z) = x^2 + y^2 + z^2$ and draws two contour plots of $f(x, y, z) = k$ for $k = .5$ and $k = 1.5$. The surfaces generated are spheres, but the larger sphere is drawn in a box that is too small to contain it completely. The result is that the inner sphere is partially visible in the picture.

```
list = Table[x² + y² + z², {x, -1, 1, .25}, {y, -1, 1, .25}, {z, -1, 1, .25}];

ListContourPlot3D[list, DataRange → {{-1, 1}, {-1, 1}, {-1, 1}},
                  Contours → {.5, 1.5}]
```

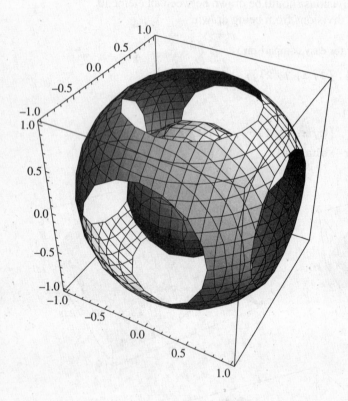

SOLVED PROBLEMS

5.14   Obtain a contour plot of $f(x, y) = \sin x + \sin y$ on the square $-4\pi \le x, y \le 4\pi$ and compare it to the three-dimensional graph of the function.

SOLUTION   (Graphs are placed side by side for easy comparison.)

```
Plot3D[Sin[x] + Sin[y], {x, -4π, 4π}, {y, -4π, 4π}]
ContourPlot[Sin[x] + Sin[y], {x, -4π, 4π}, {y, -4π, 4π}]
```

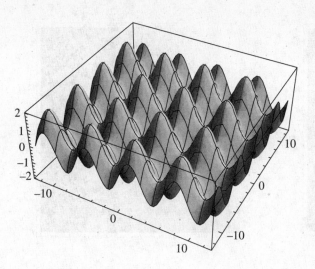
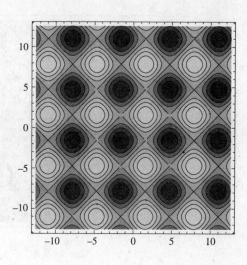

5.15 Compare a contour plot and a density plot for the function $f(x, y) = \sin xy$ over the rectangle $-\pi \le x$, $y \le \pi$.

SOLUTION   (Graphs are placed side by side for easy comparison.)

```
ContourPlot[Sin[x y], {x, -π, π}, {y, -π, π}]
DensityPlot[Sin[x y], {x, -π, π}, {y, -π, π}]
```

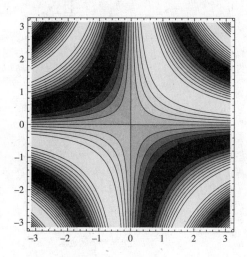
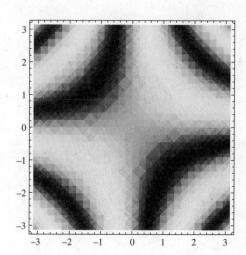

5.16 Obtain a contour plot and a density plot of the discrete function `Quotient[x, y]` as x and y range from 1 to 10.

SOLUTION   (Graphs are placed side by side for easy comparison.)

```
list = Table[Quotient[x, y], {x, 1, 10}, {y, 1, 10}];
ListContourPlot[list]
ListDensityPlot[list]
```

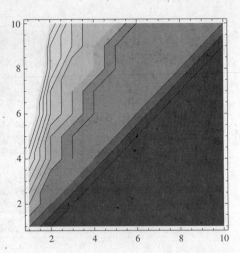 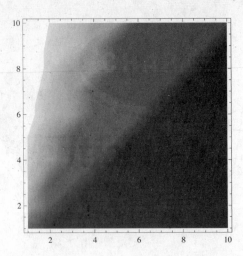

5.17 Let $f(x, y, z) = 5x^2 + 2y^2 + z^2$. Draw the level surfaces $f(x, y, z) = k$ for $k = 1, 4, 9, 16,$ and 25. Sketch the surfaces only for $y \geq 0$ so that all the surfaces will be visible.

SOLUTION

```
ContourPlot3D[5x² + 2y² + z², {x, -5, 5}, {y, 0, 5}, {z, -5, 5},
              Contours → {1, 4, 9, 16, 25}]
```

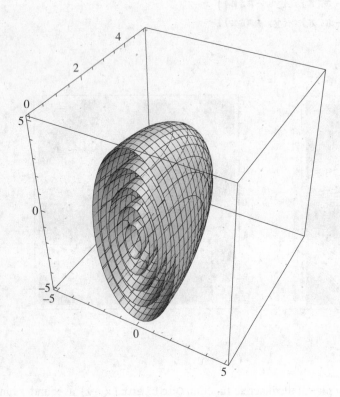

5.18 Generate a 5×5 array of random integers between 1 and 10 and construct a three-dimensional list plot of these values.

SOLUTION

```
list = Table[Random[Integer, {0, 10}], {x, 1, 5}, {y, 1, 5}];
ListPlot3D[list]
```

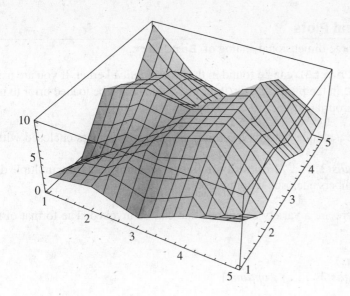

5.19 Draw hyperbolic cylinders $x^2 - y^2 = k$, $k = 0$, 2, and 5, by computing $f(x, y, z) = x^2 - y^2$ at integer values between -5 and 5 for each variable and using **ListContourPlot3D**.

SOLUTION

We use integer values of x, y, and z to construct our list.

```
list = Table[x² - y², {z, -5, 5}, {y, -5, 5}, {x, -5, 5}];

ListContourPlot3D[list, Contours → {0, 2, 5},
                  DataRange → {{-5, 5}, {-5, 5}, {-5, 5}}]
```

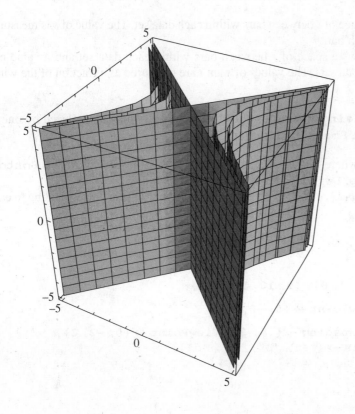

5.3 Special Three-Dimensional Plots

The command **BarChart3D** is the three-dimensional analog of **BarChart**.

Note: Starting with version 7, **BarChart3D** can be found in the *Mathematica* kernel. If you are using version 6, you will find **BarChart3D** in the package **BarCharts`** which must be loaded prior to use. See the Documentation Center for appropriate usage.

- **BarChart3D** [*datalist*] draws a simple bar graph. *datalist* is a set of numbers enclosed within braces.
- **BarChart3D** [{*datalist1*, *datalist2*, ...}] draws a bar graph containing data from multiple data sets. Each data list is a set of numbers enclosed within braces.

If a customized look is desired, there are a variety of options that can be invoked. The format of the command with options becomes

- **BarChart3D** [*datalist*, *options*]
- **BarChart3D** [{*datalist1*, *datalist2*, ...}, *options*]

Some of the more popular options are:

- **Chartstyle** → *g* specifies that style option *g* should be used to draw the bars. Examples of style options are **GrayLevel**, **Hue**, **Opacity**, **RGBColor**, and Colors (**Red**, **Blue**, etc.).
- **Chartstyle** → {*g1*, *g2*, ...} specifies that style options *g1*, *g2*, . . . should be used cyclically.
- **ChartLayout** → "*layout*" specifies that a layout of type *layout* should be used to draw the graph. Examples of layouts are "**Stacked**", which causes the bars to be stacked on top of each other rather than placed side by side, and "**Percentile**", which generates a stacked bar chart with the total height of each bar constant at 100%.

BarSpacing controls the spacing between bars and between groups of bars. The default is **BarSpacing** → **Automatic** which allows *Mathematica* to control the spacing.

- **BarSpacing** → *s* allows a space of *s* between bars within each data set. The value of *s* is measured as a fraction of the width of each bar.
- **BarSpacing** → {*s*, *t*} allows a space of *s* between bars within each data set and a value of *t* determines the space between data sets. The values of *s* ant *t* are measured as a fraction of the width of each bar.

In each of the preceding **BarSpacing** commands, the values of *s* and *t* may be replaced by one of the predefined symbols **None**, **Tiny**, **Small**, **Medium**, or **Large**.

- **BarOrigin** → *edge* controls where the bars originate from. The default value of *edge* is **Bottom**. Other acceptable values are **Top**, **Left**, and **Right**.
- **ChartLabels** → {*label1*, *label2*, ...} specifies the labeling for each bar corresponding to each value in the data list.

EXAMPLE 17

```
array = {{1, 2, 3, 4}, {5, 6, 7, 8}, {9, 10, 11, 12}};

g1 = BarChart3D[array, ViewPoint → {0, -2, 2}];

g2 = BarChart3D[array, BarSpacing → {.5, 2}, ViewPoint → {0, -2, 2},
                ChartLabels → {"a", "b", "c", "d"}];

g = GraphicsArray[{g1, g2}]
```

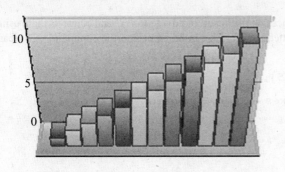
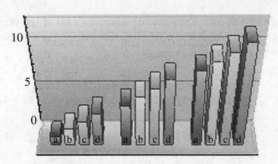

ListPointPlot3D is the three-dimensional analog of **ListPlot**, which plots discrete points in a two-dimensional plane.

- **ListPointPlot3D [*list*]** plots the points in *list* in a three-dimensional box. *list* must be a list of sublists, each of which contains three numbers, representing the coordinates of points to be plotted.

By default, **ListPointPlot3D** uses **BoxRatios → {1, 1, .4}** and accepts the **PlotStyle** option discussed in Chapter 4.

In the next example, we generate 50 random points and plot them in three-dimensional space.

EXAMPLE 18

```
list = Table[RandomInteger[{1,10}],{50},{3}]
```
← This generates a list of 50 three-element lists of random integers.

```
ListPointPlot3D[list, BoxRatios→1]
```

```
ListPointPlot3D[list, PlotStyle→PointSize[.02], BoxRatios → 1]
```

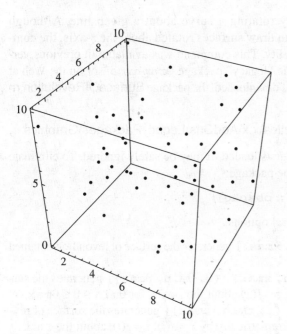
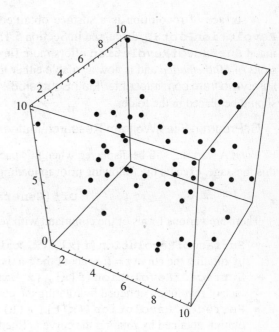

ListSurfacePlot3D creates a mesh of polygons constructed from the vertices specified in a list.

- **ListSurfacePlot3D [*list*]** creates a three-dimensional polygonal mesh from the vertices specified in *list*, which should be of the form

$$\{\{\{x_{11}, y_{11}, z_{11}\}, \{x_{12}, y_{12}, z_{12}\}, \ldots\}, \{\{x_{21}, y_{21}, z_{21}\}, \{x_{22}, y_{22}, z_{22}\}, \ldots\}, \ldots\}$$

EXAMPLE 19   The following generates a list of 169 vertices on the hyperboloid $z = x^2 - y^2$ and connects them using `ListSurfacePlot3D`. Note that the list must be flattened before it can be input into the command. (Compare with Problem 5.3.)

```
list = Table[{x, y, x² - y²}, {x, -3, 3, .5}, {y, -3, 3, .5}];

ListSurfacePlot3D[Flatten[list,1], Axes → True, BoxRatios → {1, 1, 1}]
```

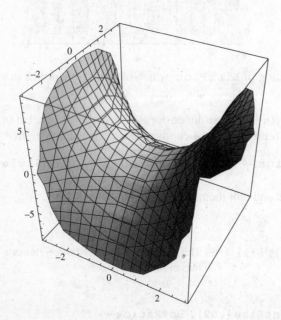

A surface of revolution is a surface obtained by rotating a curve about a given line. Although `RevolutionPlot3D`, discussed in Section 5.1, can draw surfaces rotated about the z-axis, the command `SurfaceOfRevolution` offers more flexibility. This command was available in previous versions of *Mathematica* and is now available either in the "legacy" package `Graphics`` or on the Web at library.wolfram.com/infocenter/MathSource/6824. If downloaded, the package SurfaceOfRevolution.m should be placed in the folder

     C:\Program Files\Wolfram Research\Mathematica\x.x\AddOns\LegacyPackages\Graphics

Note: A warning will be displayed when this package is loaded. It may be safely ignored. To eliminate this message, execute the following prior to loading the package:

`Off[General :: obspkg];`

There are various forms of the command with several options.

- `SurfaceOfRevolution[f[x], {x, xmin, xmax}]` generates the surface of revolution obtained by rotating the curve $z = f(x)$ about the z-axis.
- `SurfaceOfRevolution[f[x], {x, xmin, xmax}, {θ, θmin, θmax}]` generates the surface of revolution obtained by rotating the curve $z = f(x)$ about the z-axis, for $θmin \le θ \le θmax$.
- `SurfaceOfRevolution[{x[t], z[t]}, {t, tmin, tmax}]` generates the surface of revolution obtained by rotating the curve defined parametrically by $x = x(t)$, $z = z(t)$, about the z-axis.

The following example rotates the curve $z = x^2$ about the z-axis, completely and partially.

EXAMPLE 20

```
≪Graphics`
SurfaceOfRevolution[x², {x, 0, 3}];
SurfaceOfRevolution[x², {x, 0, 3}, {θ, 0, 3π/2}];
```

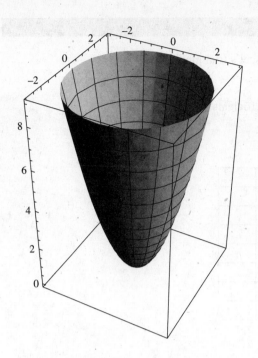
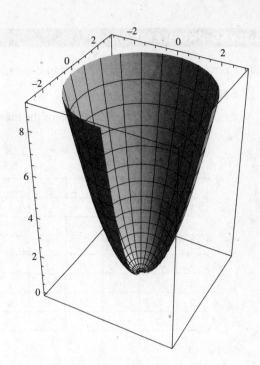

The option **RevolutionAxis** allows rotation about axes other than the *z*-axis.

- **RevolutionAxis → {x, z}** rotates the curve about an axis formed by connecting the origin to the point (x, z) in the *x-z* plane.
- **RevolutionAxis → {x, y, z}** rotates the curve about an axis formed by connecting the origin to the point (x, y, z) in space.

EXAMPLE 21

```
≪Graphics`
SurfaceOfRevolution[x², {x, 0, 3}, RevolutionAxis→{1, 0},
               BoxRatios→{1, 1, 1}, AxesLabel→{"x", "y", "z"}]
SurfaceOfRevolution[x², {x, 0, 3}, RevolutionAxis→{1, 1, 1},
               BoxRatios→{1, 1, 1}, AxesLabel→{"x", "y", "z"}]
```

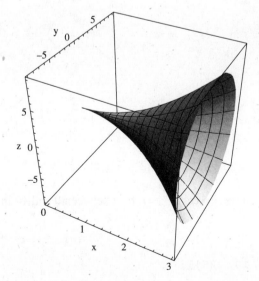

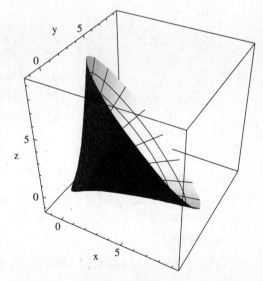

The curve $z = x^2$ is rotated about the line connecting the points $(0, 0, 0)$ and $(1, 0, 0)$.

The curve $z = x^2$ is rotated about the line connecting the points $(0, 0, 0)$ and $(1, 1, 1)$.

SOLVED PROBLEMS

5.20 Construct a 3 dimensional bar chart depiction of Pascal's triangle for $n = 7$.

SOLUTION

Pascal's triangle is a representation of the binomial coefficients $c(n,k) = \frac{n!}{k!(n-k)!}$.

n \ k	0	1	2	3	4	5	6	7
0	1							
1	1	1						
2	1	2	1					
3	1	3	3	1				
4	1	4	6	4	1			
5	1	5	10	10	5	1		
6	1	6	15	20	15	6	1	
7	1	7	21	35	35	21	7	1

```
c[n_, k_] = n!/(k! (n - k)!);

list = Table[c[n, k], {n, 0, 7}, {k, 0, n}];

g = BarChart3D[list, BarSpacing → {.5, 2}]
```

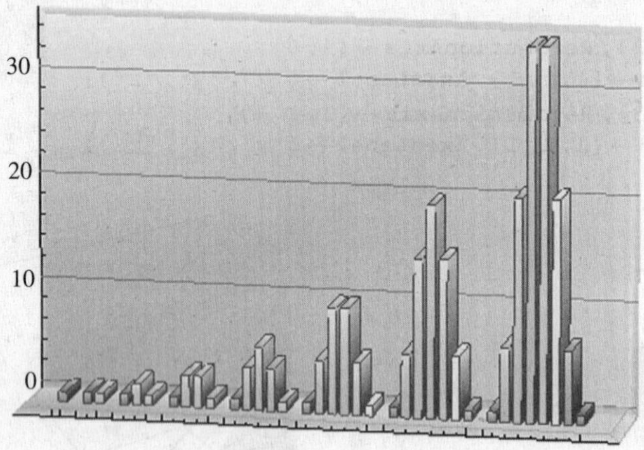

5.21 Construct a scatter plot of the points on the helix $x = \sin 2t$, $y = \cos 2t$, $z = t$ for t between 0 and 10 in increments of .25.

SOLUTION

```
list = Table[{Sin[2t], Cos[2t], t}, {t, 0, 10, .25}];

ListPointPlot3D[list, PlotStyle → PointSize[.03],
                BoxRatios → {.25, .25, 1}, Axes → False]
```

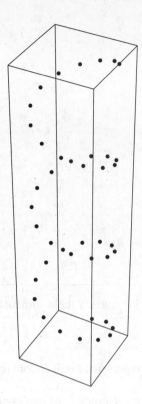

5.22 Construct the surface of revolution obtained by rotating the curve $z = \sin x$, $0 \le x \le 2\pi$, about (i) the z-axis and (ii) the x-axis.

SOLUTION

```
≪Graphics`
SurfaceOfRevolution[Sin[x], {x, 0, 2π}, Ticks → False,
                AxesLabel → {"x", "y", "z"}]

SurfaceOfRevolution[Sin[x], {x, 0, 2π}, RevolutionAxis → {1, 0}, Ticks → False,
                AxesLabel → {"x", "y", "z"}]
```

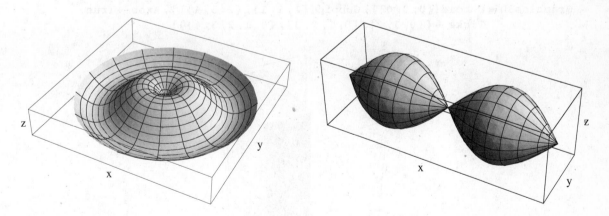

5.23 Sketch the surface obtained by rotating the curve $z = x^2$, $0 \le x \le 1$, about the line $z = x$.

SOLUTION

```
≪Graphics`
SurfaceOfRevolution[x², {x, 0, 1}, RevolutionAxis → {1, 1},
                AxesLabel → {"x", "y", "z"}, Ticks → False]
```

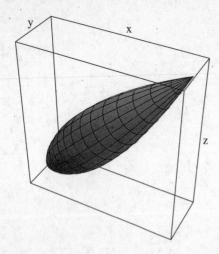

5.4   Standard Shapes—3D Graphics Primitives

- **Graphics3D [*primitives*]** or **Graphics3D [*primitives*, *options*]** creates a three-dimensional graphics object.

The standard primitives are

- **Cuboid[{x, y, z}]** is a three-dimensional graphics primitive that represents a unit cuboid (cube) with a corner at (x, y, z), with edges parallel to the axes.
- **Cuboid[{x_1, y_1, z_1}, {x_2, y_2, z_2}]** represents a cuboid (parallelepiped) whose opposite corners are (x_1, y_1, z_1) and (x_2, y_2, z_2).
- **Line[{x_1, y_1, z_1}, {x_2, y_2, z_2},...]** draws a sequence of line segments connecting the points (x_1, y_1, z_1), (x_2, y_2, z_2),...
- **Point[{x, y, z}]** plots a single point at coordinates (x, y, z).
- **Polygon[{x_1, y_1, z_1}, {x_2, y_2, z_2},...]** draws a filled polygon with coordinates (x_1, y_1, z_1), (x_2, y_2, z_2),...
- **Text[*expression*, {x, y, z}]** creates a graphics primitive representing the text *expression*, centered at position (x, y, z).

EXAMPLE 22

```
Graphics3D[{Cuboid[{0, 0, 0}], Cuboid[{1, 1, 1}, {2, 3, 4}]}, Axes → True,
          Ticks → {{0, 1, 2}, {0, 1, 2, 3}, {0, 1, 2, 3, 4}}]
```

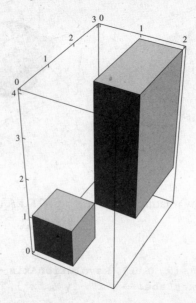

EXAMPLE 23

```
vertices = {{0, 0, 0}, {2, 2, 0}, {0, 2, 1}, {0, 0, 2}};
Graphics3D[Polygon[vertices], Axes → True, Ticks→{{0, 1, 2}, {0, 1, 2}, {0, 1, 2}}]
```

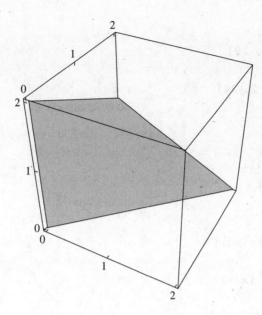

- **Sphere[{x, y, z}, r]** defines a sphere of radius r centered at {x, y, z}.
- **Cylinder[{{x1, y1, z1}, {x2, y2, z2}}, r]** defines a cylinder of radius r around the line from {x1, y1, z1} to {x2, y2, z2}.
- **Cone[{{x1, y1, z1}, {x2, y2, z2}}, r]** defines a cone with base radius r centered at {x1, y1, z1} and a tip at {x2, y2, z2}.

Additional three-dimensional graphics commands allow for convenient drawing of other standard shapes. Only **Cylinder**, **Cone** and **Sphere** are available in the *Mathematica* kernel. **DoubleHelix**, **Helix**, **OutlinePolygons**, **PerforatePolygons**, **RotateShape**, **ShrinkPolygons**, **Torus**, **TranslateShape**, and **WireFrame** were available in previous versions of *Mathematica* and are now available either in the "legacy" package **Graphics`** or on the Web at library.wolfram.com/infocenter/MathSource/6793. If downloaded, the package Shapes.m should be placed in the folder

C:\Program Files\Wolfram Research\Mathematica\x.x\AddOns\LegacyPackages\Graphics

Note: A warning will be displayed when this package is loaded. It may be safely ignored. To eliminate this message, execute the following prior to loading the package:

Off[General :: obspkg];

Torus and **MoebiusStrip** are also available in the kernel function **ExampleData**.

- **Cylinder[r, h, n]** draws a cylinder with radius r and half height h using n polygons.
- **Sphere[r, n, m]** draws a sphere of radius r using $n(m-2)+2$ polygons.
- **Cone[r, h, n]** draws a cone with radius r and half height h using n polygons.
- **Torus[r1, r2, n, m]** draws a torus with radii r1 and r2 using an $n \times m$ mesh.
- **MoebiusStrip[r1, r2, n]** draws a Moebius strip with radii r1 and r2 using 2n polygons.
- **Helix[r, h, m, n]** draws a helix with radius r, half height h, and m turns using an $n \times m$ mesh.
- **DoubleHelix[r, h, m, n]** draws a double helix with radius r, half height h, and m turns using an $n \times m$ mesh.

If the parameters are omitted, e.g., **Cone[]**, *Mathematica*'s defaults are used. The default values are

```
Cylinder[1, 1, 20]
Sphere[1, 20, 15]
Cone[1, 1, 20]
Torus[1, .5, 20, 10]
MoebiusStrip[1, .5, 20]
Helix[1, .5, 2, 20]
DoubleHelix[1, .5, 2, 20]
```

EXAMPLE 24

```
<<Graphics`
Graphics3D[Cylinder[]]
Graphics3D[Sphere[]]
Graphics3D[Cone[]]
Graphics3D[Torus[]]
Graphics3D[MoebiusStrip[]]
Graphics3D[Helix[]]
Graphics3D[DoubleHelix[]]
```

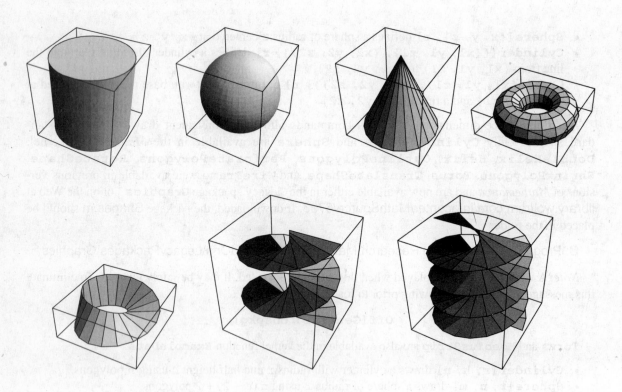

- **WireFrame[*object*]** shows all polygons used in the construction of *object* as transparent. It may be used on any **Graphics3D** object that contains the primitives **Polygon**, **Line**, and **Point**.
- **Opacity[a]** specifies the degree of transparency of a graphics object. The value of a must be between 0 and 1, with 0 representing perfect transparency and 1 representing complete opaqueness.

The **Opacity** directive should be placed within the **Graphics3D** directive as shown in the following example.

EXAMPLE 25

```
≪Graphics`
object = Torus[];
Graphics3D[object]
Graphics3D[{Opacity[.3], object}]
WireFrame[object]
```

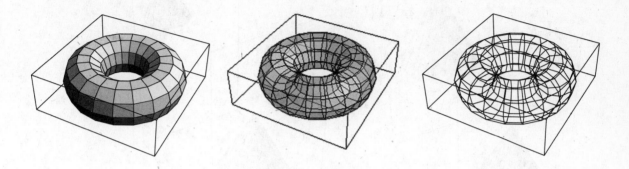

There are three commands in **Graphics`** that provide transformations in space:

- **RotateShape[*object*, ϕ, θ, ψ]** rotates *object* using the Euler angles[1] ϕ, θ, and ψ.
- **TranslateShape[*object*, {x, y, z}]** translates *object* by the vector {x, y, z}].
- **AffineShape[*object*, {*xscale*, *yscale*, *zscale*}]** scales the *x*-, *y*-, and *z*-coordinates by *xscale*, *yscale*, and *zscale*, respectively.

EXAMPLE 26

```
≪Graphics`
object = Graphics3D[Cone[]]
RotateShape[object, 0, π/2, 0]
RotateShape[object, 0, π/2, π/2]
```

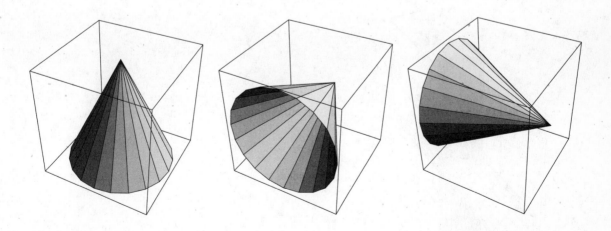

---

[1]Euler angles are a way of describing transformations in R³ by performing three rotations in a specified sequence. First we make a rotation ϕ about the z-axis. Then we perform a rotation θ about the *new* y-axis. Finally, we perform a rotation ψ about the (new) z-axis obtained from this rotation.

```
Show[object, TranslateShape[object, {1, 2, 3}]]
shrunkenobject = AffineShape[object, {.5, .5, .5}];
Show[object, TranslateShape[shrunkenobject, {1, 2, 3}]]
```

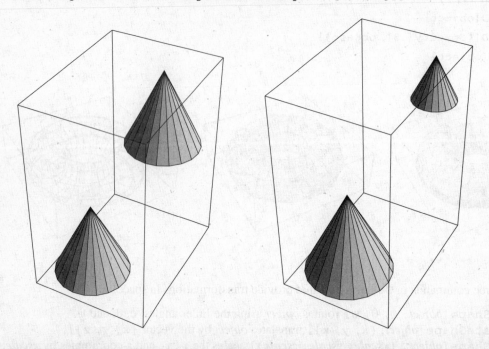

SOLVED PROBLEMS

5.24 Draw two cylinders intersecting at right angles.

SOLUTION

```
≪Graphics`
cyl1 = Graphics3D[Cylinder[1, 5, 20]];
cyl2 = Graphics3D[RotateShape[Cylinder[1, 5, 20], 0, π/2, 0]];
Show[cyl1, cyl2]
```

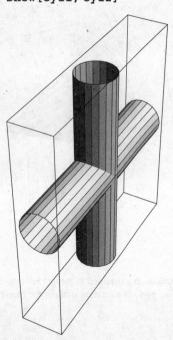

5.25 Construct a cylinder inscribed in a sphere of radius 1.

SOLUTION 1

Since the sphere has radius 1, we use the default parameters. In order for the cylinder to be inscribed in the sphere, r (radius) and h (half–height) must satisfy $r^2 + h^2 = 1$. We choose $r = 1/2$ and $h = \sqrt{3}/2$. In order for the cylinder to be visible, we draw the sphere as a wire frame.

```
≪Graphics`
sphere = WireFrame[Graphics3D[{Opacity[0.5], Sphere [{0, 0, 0}, 1]}]];
cylinder = Graphics3D[Cylinder [{{0, 0, -√3/2}, {0, 0, √3/2}}, 1/2]];
Show[sphere, cylinder, Boxed → False]
```

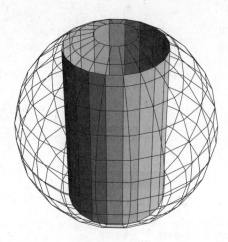

SOLUTION 2

Using the graphics Directive **Opacity**, we can make the sphere semitransparent so the cylinder is visible through the sphere.

```
sphere = Graphics3D[{Opacity[0.5], Sphere[]}];
cylinder = Graphics3D[Cylinder [{{0, 0, -√3/2}, {0, 0, √3/2}}, 1/2]];
Show[sphere, cylinder, Boxed → False]
```

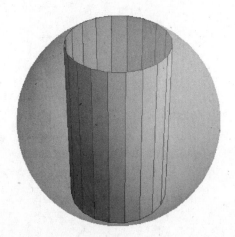

5.26　Draw two interlocking tori of default dimension ($r_1 = 1$, $r_2 = 0.5$).

SOLUTION

The second torus must be rotated 90° and translated one unit so that they interlock without intersecting.

```
<<Graphics`
torus1 = Graphics3D[Torus[]];
torus2 = Graphics3D[TranslateShape[RotateShape[Torus[], 0, π/2, π/2],
                {0, .5, 0}]];
Show[torus1, torus2, ViewPoint → {1.75, -2.8, 0.75}, Boxed → False]
```

5.27　Construct an animation showing a helix revolving about the z-axis.

SOLUTION

We use a default helix, **helix[]**. The helix makes one complete revolution as the Euler angle, ϕ, varies from 0 to 2π.

```
<<Graphics`
Animate[Graphics3D[RotateShape[Helix[], φ, 0, 0], Boxed → False], {φ, 0, 2π}]
```

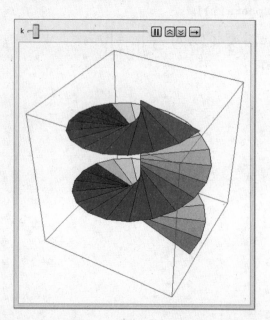

CHAPTER 6

Equations

6.1 Solving Algebraic Equations

Solutions of general algebraic equations may be found using the **Solve** command. The command is easy to use, but one must be careful to use a double equal sign, ==, between the left- and right-hand sides of the equation. (Recall that the double equal sign is a *logical* equality: lhs == rhs has a value of True if and only if lhs and rhs have the same value, False otherwise.)

- **Solve** [*equations, variables*] attempts to solve equations for *variables*.

The roots determined by **Solve** are expressed as of a list of the form

$$\{\{x \to x_1\}, \{x \to x_2\}, \ldots\}$$

The notation $x \to x_1$ indicates that the solution, x, is x_1, but x is not replaced by this value. If the equation has roots of multiplicity $m > 1$, each is repeated m times. If only one variable is present, *variables* may be omitted.

EXAMPLE 1   In this example, there is only one variable so the specification of *variables* is unnecessary.

```
Solve[7 x + 3 == 3 x + 8]
```

$$\left\{\left\{x \to \frac{5}{4}\right\}\right\}$$

If we solve the equation $ax = b$ for x, **Solve** tells us that $x = b/a$. However, if $a = b = 0$, then *every* number x is a solution. The command **Reduce** can be used to describe *all* possible solutions.

- **Reduce** [*equations, variables*] simplifies *equations*, attempting to solve for *variables*. If *equations* is an identity, **Reduce** returns the value True. If *equations* is a contradiction, the value False is returned.

In describing the solutions, **Reduce** uses the symbols && (logical and) and || (logical or). && takes precedence over ||.

EXAMPLE 2

```
Solve[a x == b, x]
```

$$\left\{\left\{x \to \frac{b}{a}\right\}\right\}$$

```
Reduce[a x == b, x]
```

$$(b == 0 \,\&\&\, a == 0) \,||\, \left(a \neq 0 \,\&\&\, x == \frac{b}{a}\right)$$ ← Either $a = b = 0$ *or* $a \neq 0$ and $x = b/a$.

```
Reduce[x² – 9 == (x + 3) (x – 3), x]
True
Reduce[x² – 10 == (x + 3) (x – 3), x]
False
```

If we try to solve an equation that contains two or more variables, we must specify which variable we are solving for.

EXAMPLE 3

> Note the space between a and y and between c and x. This is important. * may be used instead.

```
Solve[a y + b == c x + d]
```

Solve :: svars : Equations may not give solutions for all "solve" variables. >>

$\{\{b \rightarrow d + c\,x - a\,y\}\}$

We must specify which variable we wish to solve for:

```
Solve[a y + b == c x + d, x]
```

$$\left\{\left\{x \rightarrow \frac{b - d + a\,y}{c}\right\}\right\}$$

```
Solve[a y + b == c x + d, y]
```

$$\left\{\left\{y \rightarrow \frac{-b + d + c\,x}{a}\right\}\right\}$$

```
Solve[a y + b == c x + d, b]
```

$\{\{b \rightarrow d + c\,x - a\,y\}\}$

```
Solve[a y + b == c x + d, d]
```

$\{\{d \rightarrow b - c\,x + a\,y\}\}$

For systems of equations, *equations* is a list of the form {*equation1*, *equation2*, . . .} and *variables* represents either a single variable or a list of several. Alternatively, *equations* may be represented by the individual equations separated by && (logical and).

EXAMPLE 4　Here is an easy example that shows how to solve a simple system: $\begin{cases} 2x + 3y = 7 \\ 3x + 4y = 10 \end{cases}$

```
Solve[{2 x + 3 y == 7, 3 x + 4 y == 10}, {x, y}]   or   Solve[2 x + 3 y == 7 && 3 x + 4 y == 10, {x, y}]
```

$\{\{x \rightarrow 2, y \rightarrow 1\}\}$

In this example, the specification of {x, y} is not necessary because we do not have more variables than equations. If you have more unknown variables than equations, you must specify which variables you wish to solve for. Otherwise you get *Mathematica*'s default.

EXAMPLE 5

```
Solve[{x + 2 y + z == 5, 2 x + y + 3 z == 7}, {y, z}]
```

$$\left\{\left\{y \rightarrow \frac{8 - x}{5},\ z \rightarrow -\frac{3}{5}(-3 + x)\right\}\right\}$$

Of course, **Solve** is not limited to solving only linear equations.

EXAMPLE 6

```
Solve[a x² + b x + c == 0, x]
```

$$\left\{\left\{x \rightarrow \frac{-b - \sqrt{b^2 - 4ac}}{2a}\right\}, \left\{x \rightarrow \frac{-b + \sqrt{b^2 - 4ac}}{2a}\right\}\right\}$$

Observe that *Mathematica* gives the general solution in terms of arbitrary *a*, *b*, and *c* unless values are assigned to these variables.

EXAMPLE 7

$$\texttt{Solve[}x^3 + y^2 == 5 \&\& x + y == 3\texttt{]}$$

$$\left\{ \{y \to 2,\ x \to 1\},\ \left\{y \to 4 - \sqrt{5},\ x \to -1 + \sqrt{5}\right\},\ \left\{y \to 4 + \sqrt{5},\ x \to -1 - \sqrt{5}\right\} \right\}$$

Because *Mathematica* returns the solutions of equations as a nested list, they cannot be used directly as input to other mathematical structures. However, we can access their values without unnecessary typing or pasting by using **/.**

If we wish to compute the value of an expression using the solutions obtained from **Solve**, we can use the **/.** replacement operator and *Mathematica* will substitute the appropriate values.

EXAMPLE 8   Suppose we wish to solve the equations $\begin{cases} x^2 + y = 5 \\ x + y = 3 \end{cases}$ and compute the values of the expression $\sqrt{x^2 + y^2}$.
We use the **Solve** command and the object **solutions** for convenience.

$$\texttt{solutions = Solve[}\{x^2 + y == 5, x + y == 3\}, \{x, y\}\texttt{]}$$

$$\{\{y \to 1,\ x \to 2\},\ \{y \to 4,\ x \to -1\}\}$$

$$\sqrt{x^2 + y^2}\ \texttt{/. solutions} \qquad \leftarrow \textit{Mathematica} \text{ produces a list containing}$$
$$\left\{\sqrt{5},\ \sqrt{17}\right\} \qquad\qquad\qquad \text{both values of the expression.}$$

EXAMPLE 9   Suppose we wish to find the *sum of the squares* of the roots of

$$x^6 - 21x^5 + 175x^4 - 735x^3 + 1{,}624x^2 - 1{,}764x + 720 = 0$$

We use the **Solve** command:

$$\texttt{solutions = Solve[}\{x^6 - 21x^5 + 175x^4 - 735x^3 + 1624x^2 - 1764x + 720 == 0\texttt{]}$$
$$\{\{x \to 1\},\ \{x \to 2\},\ \{x \to 3\},\ \{x \to 4\},\ \{x \to 5\},\ \{x \to 6\}\}$$

Now we can define a list containing the solutions listed above.

$$\texttt{list = x /. solutions}$$
$$\{1, 2, 3, 4, 5, 6\}$$

Now we can easily compute the sum of the squares of the elements of the list.

$$\texttt{Total[list}^2\texttt{]}$$
$$91$$

Solve is designed to solve algebraic equations, but can sometimes be used to find limited solutions of transcendental equations. A warning message is given to indicate that not all solutions can be found.

EXAMPLE 10

$$\texttt{Solve[Sin[x] == 1/2, x]}$$

Solve :: ifun : Inverse functions are being used by Solve, so some solutions may not be found;
 use Reduce for complete solution information. ≫

$$\left\{\left\{x \to \frac{\pi}{6}\right\}\right\}$$

To get a more general solution to this equation, use **Reduce**.

$$\texttt{Reduce[Sin[x] == 1/2, x]}$$

C[1]∈ Integers &&

$$\left(x = \frac{\pi}{6} + 2\pi\,C[1] \ || \ x = \frac{5\pi}{6} + 2\pi\,C[1]\right)$$

> $x = \dfrac{\pi}{6}$ or $\dfrac{5\pi}{6}$ plus any integer multiple of 2π

If the equations to be solved are inconsistent, *Mathematica* returns an empty list.

EXAMPLE 11

```
Solve[{2x + 3y == 5, 4x +6y == 11}]
{}
```

If the roots of an equation involve complex numbers, they are represented as rational powers of -1. However, if a more traditional expression is desired, the function **ComplexExpand** can be used.

EXAMPLE 12

```
Solve[x³ == 1]
```

$$\{\{x \to 1\}, \{x \to -(-1)^{1/3}\}, \{x \to (-1)^{2/3}\}\}$$

```
Solve[x³ == 1] //ComplexExpand
```

$$\left\{\{x \to 1\}, \left\{x \to -\frac{1}{2} - \frac{i\sqrt{3}}{2}\right\}, \left\{x \to -\frac{1}{2} + \frac{i\sqrt{3}}{2}\right\}\right\}$$

A system of equations need not have a unique solution. For example, a system of two equations in three unknowns will either be inconsistent or have an infinite number of solutions. In the latter case it is possible to eliminate one or more variables from the system.

- **Eliminate[*equations*, *variables*]** eliminates *variables* from a set of simultaneous equations.

equations is a list of simultaneous equations, and *variables* may be a single variable or a list of two or more.

EXAMPLE 13   (a) Eliminate the variable z; (b) eliminate the variables y and z from the following equations:

$$w + x + y + z = 3$$
$$2w + 2x + 5y + z = 6$$
$$3w + 6x + 2y + 2z = 1$$

(a) `Eliminate[{w+x+y+z == 3, 2w+2x+5y+z == 6, 3w+6x+2y+3z == 1}, z]`
      $w == 3 - x - 4y \&\& 3x == -8 + y$

(b) `Eliminate[{w+x+y+z == 3, 2w+2x+5y+z == 6, 3w+6x+2y+3z == 1}, {y, z}]`
      $-29 - 13x == w$

Not all algebraic equations are solvable by *Mathematica*, even if theoretical solutions exist. If *Mathematica* is unable to solve an equation, it will represent the solution in a symbolic form. For the most part, such solutions are useless and a numerical approximation is more appropriate. Numerical approximations are obtained with the command **NSolve**.

- **NSolve[*equations*, *variables*]** solves *equations* numerically for *variables*.
- **NSolve[*equations*, *variables*, n]** solves *equations* numerically for *variables* to n digits of precision.

As with **Solve**, the list of variables may be omitted if there is no ambiguity.

EXAMPLE 14   Solve the equation $x^5 + x^4 + x^3 + x^2 + x + 2 = 0$.

   SOLUTION

```
Solve[x⁵ + x⁴ + x³ + x² + x + 2 == 0]
```

```
{ {x → Root [2 + #1 + #1² + #1³ + #1⁴ + #1⁵&, 1] },
  {x → Root [2 + #1 + #1² + #1³ + #1⁴ + #1⁵&, 2] },
  {x → Root [2 + #1 + #1² + #1³ + #1⁴ + #1⁵&, 3] },
  {x → Root [2 + #1 + #1² + #1³ + #1⁴ + #1⁵&, 4] },
  {x → Root [2 + #1 + #1² + #1³ + #1⁴ + #1⁵&, 5] } }
```

Mathematica cannot solve this equation exactly, so it returns a symbolic solution. However, we can obtain a numerical approximation.

```
NSolve [x⁵ + x⁴ + x³ + x² + x + 2 == 0]
```

```
{ {x → -1.21486}, {x → -0.522092-1.06118 i}, {x → -0.522092 + 1.06118 i},
    {x → 0.629523 - 0.883585 i}, {x → 0.629523 + 0.883585 i} }
```

An *extraneous solution* is a number that is technically not a solution of the equation, but evolves from the solution process. When solving radical equations, one typically encounters extraneous solutions. For example, when solving $\sqrt{x} = -3$, which has no real solution, the squaring process yields $x = 9$.

- **VerifySolutions** is an option that determines whether *Mathematica* should verify if solutions obtained are extraneous. The default, **VerifySolutions → True**, eliminates extraneous solutions from the solution list. If such solutions are desired, the option **VerifySolutions → False** should be used.

EXAMPLE 15

```
Solve [x + √x == 5]
```

$$\left\{\left\{x \rightarrow \frac{1}{2}\left(11 - \sqrt{21}\right)\right\}\right\}$$

```
Solve [x + √x == 5, VerifySolutions → False]
```

$$\left\{\left\{x \rightarrow \frac{1}{2}\left(11 - \sqrt{21}\right)\right\}, \left\{x \rightarrow \frac{1}{2}\left(11 + \sqrt{21}\right)\right\}\right\}$$

$\boxed{\frac{1}{2}\left(11 + \sqrt{21}\right) \text{ is extraneous.}}$

SOLVED PROBLEMS

6.1 Find an equation of the line passing through (2, 5) and (7, 9).

SOLUTION

The general equation of a line is $y = ax + b$. Substituting the coordinates of the given points leads to the equations $2a + b = 5$ and $7a + b = 9$.

```
Solve [2 a + b == 5 && 7 a + b == 9]
```

$$\left\{\left\{a \rightarrow \frac{4}{5}, b \rightarrow \frac{17}{5}\right\}\right\}$$

The line has equation $y = \frac{4}{5}x + \frac{17}{5}$.

6.2 Find an equation of the circle passing through (1, 4), (2, 7), and (4, 11).

SOLUTION

The general equation of a circle is $x^2 + y^2 + ax + by + c = 0$. We substitute the coordinates of the given points into the equation to obtain $17 + a + 4b + c = 0$, $53 + 2a + 7b + c = 0$, and $137 + 4a + 11b + c = 0$.

```
Solve[{17 + a + 4b + c == 0, 53 + 2a + 7b + c == 0,137 + 4a + 11b + c == 0}]
```
$$\{\{a \to -54,\ b \to 6,\ c \to 13\}\}$$

The equation of the circle is $x^2 + y^2 - 54x + 6y + 13 = 0$.

6.3 Solve the equation $x^4 - 16x^3 + 61x^2 - 22x - 12 = 0$, exactly and numerically.

SOLUTION

```
equation = x⁴ - 16x³ + 61x² - 22x - 12 == 0;
Solve[equation]
```
$$\left\{\left\{x \to 3 - \sqrt{5}\right\},\left\{x \to 3 + \sqrt{5}\right\},\left\{x \to 5 - 2\sqrt{7}\right\},\left\{x \to 5 + 2\sqrt{7}\right\}\right\}$$

```
NSolve[equation]
```
$$\{\{x \to -0.291503\},\ \{x \to 0.763932\},\ \{x \to 5.23607\},\ \{x \to 10.2915\}\}$$

6.4 Solve the following system for w, x, and y and then determine the solution when $z = 1$, $z = 2$, and $z = 3$.
$$\begin{aligned} w + x + y + z &= 3 \\ 2w + 3x + 4y + 5z &= 10 \\ w - x + y - z &= 4 \end{aligned}$$

SOLUTION

```
equations = {w + x + y + z == 3, 2w + 3x + 4y + 5z == 10,w - x + y - z == 4};
solution = Solve[equations, {w, x, y}]
```
$$\left\{\left\{w \to \tfrac{1}{4}(5 + 4z),\ x \to -\tfrac{1}{2} - z),\ y \to \tfrac{9}{4} - z\right\}\right\}$$

```
solution /. z → 1
```
$$\left\{\left\{w \to \tfrac{9}{4},\ x \to -\tfrac{3}{2},\ y \to \tfrac{5}{4}\right\}\right\}$$

```
solution /. z → 2
```
$$\left\{\left\{w \to \tfrac{13}{4},\ x \to -\tfrac{5}{2},\ y \to \tfrac{1}{4}\right\}\right\}$$

```
solution /. z → 3
```
$$\left\{\left\{w \to \tfrac{17}{4},\ x \to -\tfrac{7}{2},\ y \to -\tfrac{3}{4}\right\}\right\}$$

6.5 Find, to 20 significant digits, a real number such that the sum of itself, its square, and its cube is 30.

SOLUTION

```
NSolve[x + x² + x³ == 30, x, 20]
```
$$\{\{x \to -1.8557621138713175532 - 2.7604410593413850003\,i\},$$
$$\{x \to -1.8557621138713175532 + 2.7604410593413850003\,i\},$$
$$\{x \to 2.7115242277426351064\}\}$$

The only real solution is $x = 2.7115242277426351064$.

6.6 Solve the trigonometric equation $2\sin^2 x + 1 = 3\sin x$ for $\sin x$ and then for x.

SOLUTION

To solve for $\sin x$, we can write

```
Solve[2 Sin[x]² + 1 == 3 Sin[x]]
```
$$\left\{\left\{\mathrm{Sin}[x] \to \tfrac{1}{2}\right\},\{\mathrm{Sin}[x] \to 1\}\right\}$$

If we solve for *x*, only the principal solutions (using inverse functions) are obtained.

`Solve[2 Sin[x]² + 1 == 3 Sin[x], x]`

Solve:: ifun : Inverse functions are being used by Solve, so some solutions may not be found; use Reduce for complete solution information. »

$$\left\{\left\{x \rightarrow \frac{\pi}{6}\right\}, \left\{x \rightarrow \frac{\pi}{2}\right\}\right\}$$

Using **Reduce** we can get all the solutions.

`Reduce[2 Sin[x]² + 1 == 3 Sin[x], x]`

C[1]∈ Integers &&

$$\left(x = \frac{\pi}{2} + 2\pi C[1] \;||\; x = \frac{\pi}{6} + 2\pi C[1] \;||\; x = \frac{5\pi}{6} + 2\pi C[1]\right)$$

$x = \frac{\pi}{2}, \frac{\pi}{6},$ or $\frac{5\pi}{6}$ plus any integer multiple of 2π

6.7 Solve for *x*: $e^{2x} + e^x = 3$.

SOLUTION

`Solve[Exp[2 x] + Exp[x] == 3, x]`

Solve :: ifun : Inverse functions are being used by Solve, so some solutions may not be found; use Reduce for complete solution information. »

$$\left\{\left\{x \rightarrow \text{Log}\left[\frac{1}{2}\left(-1 + \sqrt{13}\right)\right]\right\}, \left\{x \rightarrow i\pi + \text{Log}\left[\frac{1}{2}\left(-1 - \sqrt{13}\right)\right]\right\}\right\}$$

`Reduce[Exp[2 x] + Exp[x] == 3, x]`

C[1]∈ Integers &&

$$\left(x = i\pi + 2 i\pi C[1] + \text{Log}\left[\frac{1}{2}\left(1 + \sqrt{13}\right)\right] \;||\; x = 2 i\pi C[1] + \text{Log}\left[\frac{1}{2}\left(-1 + \sqrt{13}\right)\right]\right)$$

6.8 Sketch the graphs of $f(x) = x^3 - 7x^2 + 2x + 20$ and $g(x) = x^2$ on the same set of axes and find their points of intersection exactly and approximately.

SOLUTION

`f[x_] = x³ - 7x² + 2x + 20;`

`g[x_] = x²;`

`Plot[{f[x], g[x]}, {x, -10, 10}, PlotRange → {-100, 100}]`

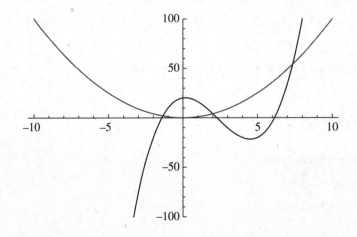

```
xvalues = Solve[f[x] == g[x], x];
{x, f[x]} /. xvalues //Expand
```

$$\left\{\{2,4\}, \left\{3-\sqrt{19}, 28-6\sqrt{19}\right\}, \left\{3+\sqrt{19}, 28+6\sqrt{19}\right\}\right\}$$

```
% // N
```

```
{{2., 4.}, {-1.3589, 1.84661}, {7.3589, 54.1534}}
```

6.9 A theorem from algebra says that if $p(x) = a_n x^n + a_{n-1}x^{n-1} + a_{n-2}x^{n-2} + \cdots + a_1 x + a_0$, the sum of the roots of the equation $p(x) = 0$ is $-\dfrac{a_{n-1}}{a_n}$ and their product is $(-1)^n \dfrac{a_0}{a_n}$. Verify this for the equation

$$20x^7 + 32x^6 - 221x^5 - 118x^4 + 725x^3 - 18x^2 - 726x + 252 = 0$$

SOLUTION

```
solution = Solve[20x^7 + 32x^6 - 221x^5 - 118x^4 + 725x^3 - 18x^2 - 726x + 252 ==0]
```

$$\left\{\left\{x \to -\tfrac{7}{2}\right\}, \left\{x \to \tfrac{2}{5}\right\}, \left\{x \to \tfrac{3}{2}\right\}, \left\{x \to -\sqrt{2}\right\}, \left\{x \to \sqrt{2}\right\}, \left\{x \to -\sqrt{3}\right\}, \left\{x \to \sqrt{3}\right\}\right\}$$

```
list = x /. solution
```

$$\left\{-\tfrac{7}{2}, \tfrac{2}{5}, \tfrac{3}{2}, -\sqrt{2}, \sqrt{2}, -\sqrt{3}, \sqrt{3}\right\}$$

$$\sum_{k=1}^{7} \text{list[[k]]} \quad \text{or} \quad \textbf{Sum[list[[k]], \{k, 1, 7\}]} \quad \text{or} \quad \textbf{Total[list]}$$

$$-\frac{8}{5}$$

$$n = 7; \quad -\frac{a_{n-1}}{a_n} = -\frac{32}{20} = -\frac{8}{5}$$

$$\prod_{k=1}^{7} \text{list[[k]]} \quad \text{or} \quad \textbf{Product[list[[k]], \{k, 1, 7\}]}$$

$$-\frac{63}{5}$$

$$n = 7; \quad (-1)^n \frac{a_0}{a_n} = (-1)^7 \frac{252}{20} = -\frac{63}{5}$$

6.10 Find all possible solutions, x, for the equation $ax + b = cx + d$.

SOLUTION

```
Solve[a x + b == c x + d, x]
```

$$\left\{\left\{x \to \frac{-b+d}{a-c}\right\}\right\}$$

This solution presumes $a \neq c$. A more general solution is obtained using **Reduce**.

```
Reduce[a x + b == c x + d, x]
```

$$(b == d \,\&\&\, a == c) \;||\; \left(a - c \neq 0 \,\&\&\, x == \frac{-b+d}{a-c}\right)$$

6.11 Eliminate the variable x from the nonlinear system

$$x^3 + y^2 + z = 1$$
$$x + y + z = 3$$

SOLUTION

```
Eliminate[{x³ + y² + z == 1, x + y + z == 3}, x]
```

$(26 - 18y + 3y^2) z + (-9 + 3y) z^2 + z^3 == 26 - 27y + 10y^2 - y^3$

6.2 Solving Transcendental Equations

A transcendental equation is one that is non-algebraic. Although **Solve** and **NSolve** can be used in a limited way to handle simple trigonometric or exponential equations, it was not designed to handle equations involving more complicated transcendental functions. The *Mathematica* command **FindRoot** is better equipped to handle these.

FindRoot uses iterative methods to find solutions. A starting value, sometimes called the *initial guess*, must be specified. For best results, the initial guess should be as close to the desired root as possible.

- **FindRoot[lhs == rhs, {x, x0}]** solves the equation lhs = rhs using Newton's method with starting value x0.
- **FindRoot[lhs == rhs, {x, {x0, x1}]** solves the equation lhs = rhs using (a variation of) the secant method[1] with starting values x0 and x1.
- **FindRoot[lhs == rhs, {x, x0, xmin, xmax}]** attempts to solve the equation, but stops if the iteration goes outside the interval [xmin, xmax].

If a *function* is specified in place of the equation lhs == rhs, **FindRoot** will compute a zero of the function. A zero of f is a number x such that $f(x) = 0$.

EXAMPLE 16 The equation $\sin x = x^2 - 1$ has two solutions.

```
Plot[{Sin[x], x² - 1}, {x, -π, π}]
```

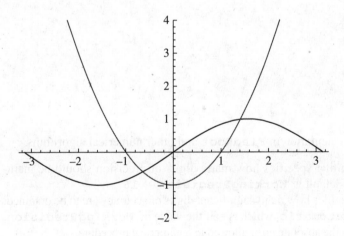

The graph of the two functions shows that they intersect near $x = -1$ and $x = 1$.

```
FindRoot[Sin[x] == x² - 1, {x, -1}]
```
$\{x \rightarrow -0.636733\}$

```
FindRoot[Sin[x] == x² - 1, {x, 1}]
```
$\{x \rightarrow 1.40962\}$

---

[1]Newton's method uses the *x*-intercept of the tangent line to improve the accuracy of the initial guess. Thus, Newton's method fails if the derivative of the function cannot be computed. The secant method, although a bit slower, uses the values of the function at two distinct points, computing the *x*-intercept of the *secant* line.

By default, 100 iterations are performed before **FindRoot** is aborted. The number of iterations performed before quitting is controlled by the option **MaxIterations**.

- **MaxIterations → n** instructs *Mathematica* to use a maximum of n iterations in the iterative process before aborting.

EXAMPLE 17   The equation $e^{2x} - 2e^x + 1 = 0$ has $x = 0$ as its only root. However, because its multiplicity is 2, Newton's method converges very slowly.

```
FindRoot[Exp[2x] - 2 Exp[x] + 1 == 0, {x, 100}]
```

FindRoot::cvmit : Failed to converge to the requested accuracy or precision within 100 iterations. ≫

$\{x \to 50.\}$

```
FindRoot[Exp[2x] - 2 Exp[x] + 1 == 0, {x, 100}, MaxIterations → 300]
```

$\{x \to 4.54676 \times 10^{-9}\}$

FindRoot attempts to find real solutions. However, if a complex initial value is specified, or if the equation contains complex numbers, complex solutions will be sought. The equation in the next example has no real solutions.

EXAMPLE 18

```
FindRoot[x² + x + 1 == 0, {x, 2}]
```

FindRoot :: lstol:

The line search decreased the step size to within tolerance specified by AccuracyGoal and PrecisionGoal but was unable to find a sufficient decrease in the merit function. You may need more than MachinePrecision digits of working precision to meet these tolerances. ≫

$\{x \to -0.500002\}$

```
FindRoot[x² + x + 1 == 0, {x, I}]
```

$\{x \to -0.5 + 0.866025\,i\}$

```
FindRoot[x² + x + 1 == 0, {x, -I}]
```

$\{x \to -0.5 - 0.866025\,i\}$

There are three options that control the calculation in **FindRoot** and other numerical algorithms.

- **WorkingPrecision** is an option that specifies how many digits of precision should be maintained internally in computation. The default is **WorkingPrecision → 16**.
- **AccuracyGoal** is an option that specifies how many significant digits of accuracy are to be obtained. The default is **AccuracyGoal → Automatic**, which is half the value of **WorkingPrecision**. **AccuracyGoal** effectively specifies the absolute error allowed in a numerical procedure.
- **PrecisionGoal** is an option that specifies how many effective digits of precision should be sought in the final result. The default is **PrecisionGoal → Automatic**, which is half the value of **WorkingPrecision**. **PrecisionGoal** effectively specifies the relative error allowed in a numerical procedure.

EXAMPLE 19   We wish to obtain a 10-decimal place approximation to the solution of the equation $\cos\left(\dfrac{100}{x}\right) = \dfrac{x}{x+1}$, nearest to 5,000.

```
FindRoot[Cos[100/x] == x/(x+1), {x, 5000}]
```

$\{x \to 5000.83\}$

Mathematica's defaults are insufficient to give the required accuracy. By increasing **WorkingPrecision**, we can obtain the desired result.

$$\texttt{FindRoot}\left[\texttt{Cos}\left[\frac{100}{x}\right] == \frac{x}{x+1}, \{x, 5000\}, \texttt{WorkingPrecision} \rightarrow 28\right]$$

```
{x → 5000.8331911595560 9589817}
```

Since **AccuracyGoal** is, by default, half the value of **WorkingPrecision**, only the first 14 significant digits can be trusted. Thus, $x \approx 5000.8331911595$ (accurate to ten decimal places).

- **EvaluationMonitor** can be used to show intermediate calculations to be performed and displayed. The format is **EvaluationMonitor :→ *expression***.

The symbol :→ can be found on the **Basic Math Input** palette or can be created by typing :>. This symbol is used instead of → to avoid *expression* being immediately evaluated. This technique is illustrated in the next two examples.

EXAMPLE 20   To see how quickly the sequence of approximations converges when we solve the equation $e^{-x} = x$, we can use **EvaluationMonitor** to print the results of intermediate calculations.

```
n = -1;

FindRoot[Exp[-x] == x, {x, 2}, EvaluationMonitor :→ {n++, Print[n, "     ", x]}]

0    2.
1    0.357609
2    0.558708
3    0.56713
4    0.567143
5    0.567143
{x → 0.567143}
```

EXAMPLE 21   To obtain a comparison between Newton's method and the secant method, we can ask **EvaluationMonitor** to print the number of iterations needed to converge to 100 significant digits.

<u>Newton's Method</u>

```
n = 0;

FindRoot[Exp[-x] == x, {x, 1}, WorkingPrecision → 100,
        AccuracyGoal → 100, EvaluationMonitor :→ n++]

Print[n, " iterations"]

{x → 0.56714329040978387299996866221035554975381578718651250813513107922304
        5793086684566693219446961752946}
8 iterations
```

<u>Secant Method</u>

```
n = 0;

FindRoot[Exp[-x] == x, {x, 1, 2}, WorkingPrecision → 100,
        AccuracyGoal → 100, EvaluationMonitor :→ n++]

Print[n, " iterations"]

{x → 0.56714329040978387299996866221035554975381578718651250813513107922304
        5793086684566693219446961752946}
24 iterations
```

If the equation to be solved has a root of multiplicity 2 or greater, Newton's method may converge slowly or not at all. In this situation, convergence can sometimes be improved by a judicious choice of `DampingFactor`.

- `DampingFactor` → *factor* is an option that controls the behavior of convergence in Newton's method. The size of each step taken in Newton's method is multiplied by the value of *factor*. The default is `DampingFactor → 1`.

EXAMPLE 22

```
n = 0;
FindRoot[(Exp[x] - 1)², {x, 2}, EvaluationMonitor :→ n++]
Print[n," iterations"]
```

$\{x \to 6.95942 \times 10^{-9}\}$

```
32 iterations
```

```
n = 0;
FindRoot[(Exp[x] - 1)², {x, 2}, DampingFactor → 2, EvaluationMonitor :→ n++]
Print[n," iterations"]
```

$\{x \to 6.6703 \times 10^{-17}\}$

```
8 iterations
```

FindRoot can also be used to determine the solution of simultaneous equations.

- **FindRoot** [*equations*, {*var1*, *a1*}, {*var2*, *a2*}, . . .] attempts to solve *equations* using initial values *a1*, *a2*, . . . for *var1*, *var2*, . . . , respectively. The equations are enclosed in a list: {*equation1*, *equation2*, . . . }. Alternatively, the equations may be separated by && (logical and).

Convergence of Newton's method for functions of several variables is much more sensitive to choice of starting values than its counterpart for single variables. Therefore, a good graph of the functions involved is quite helpful.

EXAMPLE 23   Solve the system of equations $\begin{cases} e^x + \ln y = 2 \\ \sin x + \cos y = 1 \end{cases}$

First we graph the equations.

```
ContourPlot[{Exp[x] + Log[y] == 2, Sin[x] + Cos[y] == 1}, {x, 0, 2}, {y, 0, 3},
            Frame → False, Axes → True]
```

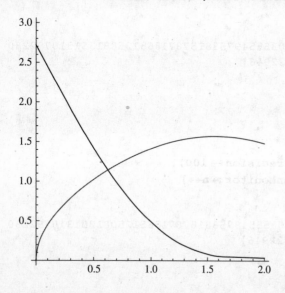

It appears that there is only one solution. We use $x = 1$, $y = 1$ for our initial guess.

```
FindRoot[{Exp[x] + Log[y] == 2, Sin[x] + Cos[y] == 1}, {x, 1}, {y, 1}]
```

$\{x \rightarrow 0.624295, y \rightarrow 1.14233\}$

If the function in an equation is such that its evaluation is costly, particularly if high precision is desired, there is another procedure that may be beneficial.

- **InterpolateRoot[lhs == rhs, {x, a, b}]** solves the equation lhs = rhs using initial values a and b.

Whereas **FindRoot** uses linear functions (straight lines) to approximate the root of the equation, **InterpolateRoot** uses polynomials of degree 3 or less. The result is that higher precision can be achieved with fewer function evaluations. **InterpolateRoot** is contained within the package **FunctionApproximations`** and must be loaded prior to use.
As with **FindRoot**, the equation may be replaced by a function, in which case its zero is computed.

EXAMPLE 24 This example computes the zero (between 2 and 3) of the Bessel function[2] $J_0(x)$, using a working precision of 1000 significant digits. For comparison purposes, the *Mathematica* function **Timing** is used. The actual numerical approximation is suppressed to save space. As a result, the value Null is returned. Delete the semicolon and run the command to see the actual result of the calculation.

```
FindRoot[BesselJ[0, x], {x, 2}, WorkingPrecision → 1000]; //Timing
```

$\{0.219, \text{Null}\}$

```
≪FunctionApproximations`
```

```
InterpolateRoot[BesselJ[0, x], {x, 2, 3}, WorkingPrecision → 1000]; //Timing
```

$\{0.046, \text{Null}\}$

SOLVED PROBLEMS

6.12 Solve the equation $5\cos x = 4 - x^3$. Make sure you find all solutions.

SOLUTION

Since $5\cos x = 4 - x^3$ if and only if $5\cos x - 4 + x^3 = 0$, we introduce the function $f(x) = 5\cos x - 4 + x^3$ and look for x-intercepts. (Although we could look for the intersection of two curves, it is easier to approximate where points intercept an axis.)

```
f[x_] = 5 Cos[x] - 4 + x³;
Plot[f[x], {x, -1, 2}]
```

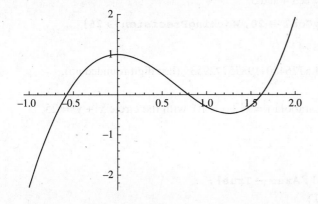

---

[2] $J_0(x)$ is a solution of the differential equation $x^2 y'' + xy' + x^2 y = 0$.

It appears that there are three solutions, near −0.5, 0.8, and 1.6.

```
FindRoot[f[x], {x, -0.5}]
```
$\{x \rightarrow -0.576574\}$
```
FindRoot[f[x], {x, 0.8}]
```
$\{x \rightarrow 0.797323\}$
```
FindRoot[f[x], {x, 1.6}]
```
$\{x \rightarrow 1.61805\}$

6.13 Find a solution of the equation $\sin x = 2$. (This problem may be omitted by those unfamiliar with functions of a complex variable.)

SOLUTION

Since $-1 \leq \sin x \leq 1$ for all real x, this problem has no real solutions. We can force **FindRoot** to search for a complex solution by using a complex initial guess.

```
FindRoot[Sin[x] == 2, {x, I}]
```
$\{x \rightarrow 1.5708 + 1.31696\, i\}$

6.14 Find a 20 significant digit approximation to the equation $x + \mid \sin (x-1) \mid = 5$.

SOLUTION

First we plot the function $f(x) = x + \mid \sin (x-1) \mid - 5$.

```
f[x_] = x + Abs[Sin[x - 1]] - 5;
Plot[f[x], {x, -10, 10}]
```

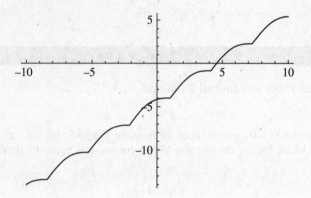

It appears that the only solution lies between 4 and 5.

```
FindRoot[f[x], {x, 5}, AccuracyGoal → 20, WorkingPrecision → 25]
```
$\{x \rightarrow 4.577640011987577295259374\}$

To 20 significant digits, the solution is 4.5776400119875772953 (last digit rounded up).

6.15 Find the points of intersection of the parabola $y = x^2 + x - 10$ with the circle $x^2 + y^2 = 25$.

SOLUTION

First, plot the two graphs.

```
g1 = Graphics[Circle[{0, 0}, 5], Axes → True];
g2 = Plot[x² + x - 10, {x, -5, 5}];
Show[g1, g2, AspectRatio → Automatic, PlotRange → {-10, 10}]
```

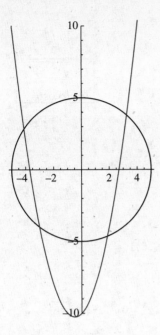

The parabola $y = x^2 + x + 1$ intersects the circle $x^2 + y^2 = 25$ at four points. Now solve for the intersection points. Because of the complicated structure of the exact solution, we obtain a numerical approximation.

```
NSolve[y == x² + x - 10 && x² + y² == 25]
```

```
{{y → -4.63752, x → 1.86907}, {y → 2.83654, x → -4.11753},
     {y → -4., x → -3.}, {y → 3.80098, x → 3.24846}}
```

6.16 Find the points of intersection of the limacon $r = 5 - 4\cos\theta$ and the parabola $y = x^2$.

SOLUTION

First we plot both curves on the same set of axes.

```
limacon = PolarPlot[5 - 4 Cos[t], {t, 0, 2π}];
parabola = Plot[x², {x, -3, 3}];
Show[limacon, parabola, PlotRange → All]
```

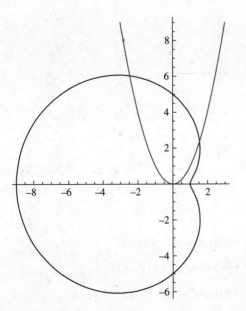

We convert the equation of the limacon to rectangular coordinates:

$$r = 5 - 4\cos\theta$$

$$r^2 = 5r - 4r\cos\theta$$

$$x^2 + y^2 = 5\sqrt{x^2 + y^2} - 4x$$

$$\boxed{\begin{aligned} r &= \sqrt{x^2 + y^2} \\ x &= r\cos\theta \end{aligned}}$$

The first intersection point appears to be near (2, 2):

`FindRoot [{y == x², x² + y² == 5 √(x² + y²) - 4x}, {x, 2}, {y, 2}]`

$\{x \to 1.53711, y \to 2.3627\}$

The second point lies near (−3, 6):

`FindRoot [{y == x², x² + y² == 5 √(x² + y²) - 4x}, {x, -3}, {y, 6}]`

$\{x \to -2.4552, y \to 6.02802\}$

6.17 Where does the Spiral of Archimedes, $r = \theta$, intersect the ellipse $4x^2 + 9y^2 = 400$?

SOLUTION

```
spiral = PolarPlot[θ, {θ, 0, 6π}];
ellipse = ContourPlot[4x² + 9y² == 400, {x, -10, 10}, {y, -20, 20},
                  ContourStyle → Dashing[.02]];
Show[spiral, ellipse]
```

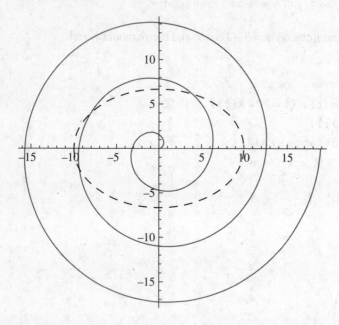

The graph shows three points of intersection that appear to be near (4, 6), (−8, 4), and (−9, −2). To convert the polar equation to rectangular, we use the transformations $r = \sqrt{x^2 + y^2}$ and $\theta = \tan^{-1}(y/x)$. However, Newton's method is more stable if we write this as $\tan(\sqrt{x^2 + y^2}) = y/x$.

`FindRoot [{Tan[√(x² + y²)] == y/x, 4x² + 9y² == 400}, {x, 4}, {y, 6}]`

`FindRoot [{Tan[√(x² + y²)] == y/x, 4x² + 9y² == 400}, {x, -8}, {y, 4}]`

`FindRoot [{Tan[√(x² + y²)] == y/x, 4x² + 9y² == 400}, {x, -9}, {y, -2}]`

{x → 3.93476, y → 6.1289}

{x → -8.04703, y → 3.95785}

{x → -9.38786, y → -2.29668}

6.18 Find a solution of the system of equations

$$x + y + z = 6$$

$$\sin x + \cos y + \tan z = 1$$

$$e^x + \sqrt{y} + \frac{1}{z} = 5$$

near the point (1, 2, 3).

SOLUTION

```
FindRoot [{x + y + z == 6, Sin[x] + Cos[y] + Tan[z] == 1,
          Exp[x] + Sqrt[y] + 1/z == 5}, {x, 1}, {y, 2}, {z, 3}]
```

{x → 1.23382, y → 1.5696, z → 3.19658}

CHAPTER 7

Algebra and Trigonometry

7.1 Polynomials

Because they are so prevalent in algebra, *Mathematica* offers commands that are devoted exclusively to polynomials.

- **PolynomialQ [*expression*, *variable*]** yields True if *expression* is a polynomial in *variable*, and False otherwise.
- **Variables [*polynomial*]** gives a list of all independent variables in *polynomial*.
- **Coefficient [*polynomial*, *form*]** gives the coefficient of *form* in *polynomial*.
- **Coefficient [*polynomial*, *form*, n]** gives the coefficient of *form* to the nth power in *polynomial*.
- **CoefficientList [*polynomial*, *variable*]** gives a list of the coefficients of powers of *variable* in *polynomial*, starting with the 0th power.

EXAMPLE 1

```
PolynomialQ [x² + 3x + 2, x]
```
True
```
PolynomialQ [x² + 3x + 2/x, x]
```
False
```
PolynomialQ [x² + 3x + 2/y, x]
```    ←2/y is treated as a constant with respect to x.
True
```
PolynomialQ [x² + 3x + 2/y, y]
```
False

EXAMPLE 2

```
poly1 = (x + 1)¹⁰;
poly2 = x³ – 5x² y + 3x y² – 7y³;
Variables [poly2]
```
{x, y}
```
Coefficient [poly1, x, 5]
```
252
```
Coefficient [poly2, x]
```
3 y²
```
Coefficient [poly2, y, 2]
```
3 x
```
Coefficient [poly2, x y²]
```
3
```
CoefficientList [poly1, x]
```
{1, 10, 45, 120, 210, 252, 210, 120, 45, 10, 1}

```
CoefficientList[poly2, x]
```

$\{-7y^3, 3y^2, -5y, 1\}$

```
CoefficientList[poly2, y]
```

$\{x^3, -5x^2, 3x, -7\}$

Often it is convenient to write the solution of a polynomial equation as a _logical_ expression. For example, if $x^2 - 4 = 0$, then $x = -2$ _or_ $x = 2$. Roots of polynomial equations can be expressed in this form using two specialized commands, **Roots** and **NRoots**. The solutions are given in disjunctive form separated by the symbol || (logical _or_).

- **Roots[lhs == rhs,** _variable_**]** produces the solutions of a polynomial equation.
- **NRoots[lhs == rhs,** _variable_**]** produces numerical approximations of the solutions of a polynomial equation.

EXAMPLE 3 Find all the solutions of $x^4 + x^3 - 8x^2 - 5x + 15 = 0$ that are greater than 2.

```
solutions = Roots[x⁴ + x³ - 8x² - 5x + 15 == 0, x]
```

$x == \frac{1}{2}(-1 - \sqrt{13})\,||\,x == \frac{1}{2}(-1 + \sqrt{13})\,||\,x == \sqrt{5}\,||\,x == -\sqrt{5}$

```
solutions && x > 2 //Simplify
```

$x == \sqrt{5}$

> **&&** is _Mathematica_'s logical _and_.
> See Section 7.4 for a discussion of **Simplify**.

```
numericalsolutions = NRoots[x⁴ + x³ - 8x² - 5x + 15 == 0, x]
```

$x == -2.30278\,||\,x == -2.23607\,||\,x == 1.30278\,||\,x == 2.23607$

```
numericalsolutions && x > 2 //Simplify
```

$x == 2.23607$

The division algorithm for polynomials guarantees that given two polynomials, p and s, for which degree(p) $\geq$ degree(s), there exist uniquely determined polynomials, q and r, such that

$$p(x) = q(x)\,s(x) + r(x), \qquad \text{where} \qquad \deg(r) < \deg(s)$$

The _Mathematica_ commands that produce the quotient and remainder are

- **PolynomialQuotient[p, s, x]** gives the quotient upon division of p by s expressed as a function of x. Any remainder is ignored.
- **PolynomialRemainder[p, s, x]** returns the remainder when p is divided by s. The degree of the remainder is less than the degree of s.

EXAMPLE 4

```
p = x⁵ - 7x⁴ + 3x² - 5x + 9;
s = x² + 1;
q = PolynomialQuotient[p, s, x]
```

$10 - x - 7x^2 + x^3$

```
r = PolynomialRemainder[p, s, x]
```

$-1 - 4x$

- **Expand[**_poly_**]** expands products and powers, writing _poly_ as a sum of individual terms.
- **Factor[**_poly_**]** attempts to factor _poly_ over the integers. If factoring is unsuccessful, _poly_ is unchanged.
- **FactorTerms[**_poly_**]** factors out common constants that appear in the terms of _poly_.
- **FactorTerms[**_poly_**,** _var_**]** factors out any common monomials containing variables other than _var_.
- **Collect[**_poly_**,** _var_**]** takes a polynomial having two or more variables and expresses it as a polynomial in _var_.

EXAMPLE 5

```
poly = 6 x² y³ z⁴ + 8 x³ y² z⁵ + 10 x² y⁴ z³;
Factor[poly]
```

$2 x^2 y^2 z^3 (5 y^2 + 3 y z + 4 x z^2)$ ← poly is factored completely.

```
FactorTerms[poly]
```

$2 (5 x^2 y^4 z^3 + 3 x^2 y^3 z^4 + 4 x^3 y^2 z^5)$ ← Only the constants are factored.

```
FactorTerms[poly, x]
```

$2 y^2 z^3 (5 x^2 y^2 + 3 x^2 y z + 4 x^3 z^2)$ ← Only the common factors *not* involving x are factored.

```
FactorTerms[poly, y]
```

$2 x^2 z^3 (5 y^4 + 3 y^3 z + 4 x y^2 z^2)$ ← Only the common factors *not* involving y are factored.

```
FactorTerms[poly, z]
```

$2 x^2 y^2 (5 y^2 z^3 + 3 y z^4 + 4 x z^5)$ ← Only the common factors *not* involving z are factored.

EXAMPLE 6

```
poly = 1 + 2 x + 3 y + 4 x y + 5 x² y + 6 x y² + 7 x² y²;
Collect[poly, x]
```

$1 + 3 y + x (2 + 4 y + 6 y^2) + x^2 (5 y + 7 y^2)$ ← Powers of x are factored out.

```
Collect[poly, y]
```

$1 + 2 x + (3 + 4 x + 5 x^2) y + (6 x + 7 x^2) y^2$ ← Powers of y are factored out.

EXAMPLE 7   The following **Manipulate** command expands $(x + 1)^n$ to any power between 1 and 10, controlled by radio buttons.

```
Manipulate[Expand[(x + 1)ⁿ] //TraditionalForm, {n, Range[10]},
          ControlType → RadioButton]
```

n ○1 ○2 ○3 ○4 ○5 ○6 ○7 ●8 ○9 ○10

$x^8 + 8 x^7 + 28 x^6 + 56 x^5 + 70 x^4 + 56 x^3 + 28 x^2 + 8 x + 1$

By default, **Factor** allows factorization only over the integers. There are options that allow this default to be overridden.

- **Extension** → {*extension1, extension2, . . .*} can be used to specify a list of algebraic numbers that may be included as well. (The brackets, { }, are not needed if only one extension is used.)
- **Extension** → **Automatic** extends the field to include any algebraic numbers that appear in the polynomial.
- **GaussianIntegers** → **True** allows the factorization to take place over the set of integers with i adjoined. Alternatively, i or I may be included in the list of extensions.

EXAMPLE 8

```
Factor[x⁸ - 41x⁴ + 400]
```

$(-2 + x) (2 + x) (-5 + x^2) (4 + x^2) (5 + x^2)$

```
Factor[x⁸ - 41x⁴ + 400, GaussianIntegers → True]
```

$(-2 + x) (-2 i + x) (2 i + x) (2 + x) (-5 + x^2) (5 + x^2)$

```
Factor[x⁸ - 41x⁴ + 400, Extension → √5 ]
```

$-(\sqrt{5} - x) (-2 + x) (2 + x) (\sqrt{5} + x) (4 + x^2) (5 + x^2)$

```
Factor[x⁸ - 41x⁴ + 400, Extension → {I, √5}]
```

$-(\sqrt{5} - x) (\sqrt{5} - i x) (\sqrt{5} + i x) (-2 + x) (-2 i + x) (2 i + x) (2 + x) (\sqrt{5} + x)$

The greatest common divisor (GCD) of polynomials, $p_1, p_2, \ldots$ is the polynomial of largest degree that can be divided evenly (remainder = 0) into $p_1, p_2, \ldots$. The least common multiple (LCM) of polynomials $p_1, p_2, \ldots$ is the polynomial of smallest degree that can be divided evenly by $p_1, p_2, \ldots$.

- **PolynomialGCD[p1, p2, ...]** computes the greatest common divisor of the polynomials p1, p2, ...
- **PolynomialLCM[p1, p2, ...]** computes the least common multiple of the polynomials p1, p2, ...

EXAMPLE 9

```
p = (x - 1) (x - 2)² (x - 3)³;
q = (x - 1)² (x - 2) (x - 3)⁴;
PolynomialGCD[p, q]
```
$(-3 + x)^3 (-2 + x) (-1 + x)$
```
PolynomialLCM[p, q]
```
$(-3 + x)^4 (-2 + x)^2 (-1 + x)^2$

By default, both **PolynomialGCD** and **PolynomialLCM** assume the coefficients of the polynomials to be rational numbers. As with **Factor**, the option **Extension** can be used to specify a list of algebraic numbers (and/or I) that may be allowed.

EXAMPLE 10

```
p = x² - 5;
q = x + √5
PolynomialGCD[p, q]
```
1
```
PolynomialGCD[p, q, Extension → Automatic]
```
$\sqrt{5} + x$
```
PolynomialLCM[p, q]
```
$(\sqrt{5} + x)(-5 + x^2)$
```
PolynomialLCM[p, q, Extension → Automatic]
```
$-5 + x^2$

Although *Mathematica* will automatically expand integer exponents of products and quotients, if the exponent is non-integer, the expression will be left unexpanded. To force the "distribution" of the exponent, the command **PowerExpand** is available.

- **PowerExpand[*expression*]** expands nested powers, powers of products and quotients, roots of products and quotients, and their logarithms.

EXAMPLE 11

```
(a b)⁵
```
$a^5 b^5$ ← *Mathematica* distributes the exponent because it is an integer.
```
(a b)ˣ
```
$(a\,b)^x$ ← *Mathematica* does nothing because the exponent is undefined.
```
PowerExpand[(a b)ˣ]
```
$a^x b^x$ ← We force the expansion with **PowerExpand**.

One must be very careful with **PowerExpand** when multi-valued functions are involved.

EXAMPLE 12

$\sqrt{ab}$ **/. {a → -1, b → -1}**

1

$\boxed{\sqrt{(-1)(-1)} = \sqrt{1} = 1}$

PowerExpand$\left[\sqrt{ab}\right]$ **/. {a → -1, b → -1}**

-1

$\boxed{\textbf{PowerExpand} \text{ expands and then replaces the values of } a \text{ and } b \text{ by } -1.}$

Here are a few additional examples illustrating **PowerExpand**:

EXAMPLE 13

(a^x)^y // PowerExpand

a^{xy}

(a/b)^x // PowerExpand

a^{x}b^{-x}

Log[x y] // PowerExpand

Log[x] + Log[y]

Log[x/y] // PowerExpand

Log[x] - Log[y]

Log[x^y] // PowerExpand

y Log[x]

SOLVED PROBLEMS

7.1 Test to see if $1 + x \sin y + x^2 \cos y + x^5 e^y$ is a polynomial in x. Is it a polynomial in y?

SOLUTION

PolynomialQ[1 + x Sin[y] + x^2 Cos[y] + x^5 Exp[y], x]
True
PolynomialQ[1 + x Sin[y] + x^2 Cos[y] + x^5 Exp[y], y]
False

$\boxed{y \text{ is treated as a constant in this expression.}}$

7.2 What are the coefficients of the polynomial expansion of $(2x+3)^5$?

SOLUTION

poly = (2x + 3)^5;
CoefficientList[poly, x]
{243, 810, 1080, 720, 240, 32}

7.3 What is the coefficient of xy^2z^3 in the expansion of $(x+y+z)^6$?

SOLUTION

poly = (x + y + z)^6;
Coefficient[poly, xy^{2}z^3]
60

7.4 Expand $(x + a + 1)^4$ completely.

SOLUTION

```
Expand[(x + a + 1)⁴]
```

$1 + 4a + 6a^2 + 4a^3 + a^4 + 4x + 12ax + 12a^2x + 4a^3x + 6x^2 + 12ax^2 + 6a^2x^2 + 4x^3 + 4ax^3 + x^4$

7.5 Express $(x + a + 1)^4$ as a polynomial in x.

SOLUTION

```
Collect[(x + a + 1)⁴, x]
```

$1 + 4a + 6a^2 + 4a^3 + a^4 + (4 + 12a + 12a^2 + 4a^3)x + (6 + 12a + 6a^2)x^2 + (4 + 4a)x^3 + x^4$

7.6 Factor the polynomial

$$poly = 6x^3 + x^2y - 11xy^2 - 6y^3 - 5x^2z + 11xyz + 11y^2z - 2xz^2 - 6yz^2 + z^3$$

and solve for z so that $poly = 0$.

SOLUTION

```
poly = 6x³ + x²y - 11xy² - 6y³ - 5x²z + 11xyz + 11y²z - 2xz² - 6yz² + z³;
Factor[poly]
```

$(x + y - z)(3x + 2y - z)(2x - 3y + z)$

SOLUTION using `Solve`

```
Solve[poly == 0, z]
```

$\{\{z \to x + y\}, \{z \to 3x + 2y\}, \{z \to -2x + 3y\}\}$

SOLUTION using `Roots`

```
Roots[poly == 0, z]
```

$z == x + y \,||\, z == 3x + 2y \,||\, z == -2x + 3y$

7.7 Find the quotient and remainder when $x^5 + 2x^4 - 3x^3 + 7x^2 - 10x + 5$ is divided by $x^2 - 4$ and verify that the answer is correct.

SOLUTION

```
p = x⁵ + 2x⁴ - 3x³ + 7x² - 10x + 5;
s = x² - 4;
q = PolynomialQuotient[p, s, x]
```

$15 + x + 2x^2 + x^3$

```
r = PolynomialRemainder[p, s, x]
```

$65 - 6x$

```
checkpoly = q * s + r//Expand
```

$5 - 10x + 7x^2 - 3x^3 + 2x^4 + x^5$

```
checkpoly == p
```

True

7.8 Express $(x + y + z)^3$ as a polynomial in z.

SOLUTION

```
Collect[(x + y + z)³, z]
```

$x^3 + 3x^2y + 3xy^2 + y^3 + (3x^2 + 6xy + 3y^2)z + (3x + 3y)z^2 + z^3$

7.9 Let $p = 2x^4 - 15x^3 + 39x^2 - 40x + 12$ and $q = 4x^4 - 24x^3 + 45x^2 - 29x + 6$. Compute their GCD and LCM and show that their product is equal to pq.

SOLUTION

```
p = 2x⁴ - 15x³ + 39x² - 40x + 12;
q = 4x⁴ - 24x³ + 45x² - 29x + 6;
a = PolynomialGCD[p, q]
```
$-6 + 17x - 11x^2 + 2x^3$
```
b = PolynomialLCM[p, q]
```
$(-2 + x)(6 - 29x + 45x^2 - 24x^3 + 4x^4)$
```
Expand[a * b] == Expand[p * q]
```
True

7.10 Factor $x^4 - 25$ over the integers and then over the field containing $\sqrt{5}$ and i.

SOLUTION

```
Factor[x⁴ - 25]
```
$(-5 + x^2)(5 + x^2)$
```
Factor[x⁴ - 25, Extension → {√5, I}]
```
$-(\sqrt{5} - x)(\sqrt{5} - ix)(\sqrt{5} + ix)(\sqrt{5} + x)$

7.11 Expand $\ln\left[\sqrt{\dfrac{x^a y^b}{z^c}}\right]$.

SOLUTION

```
Log[√(xᵃ yᵇ / zᶜ)] //PowerExpand
```
$\frac{1}{2}(a \operatorname{Log}[x] + b \operatorname{Log}[y] - c \operatorname{Log}[z])$

7.2   Rational and Algebraic Functions

There are a few commands appropriate for use with rational functions (fractions).

- **Numerator[***fraction***]** returns the numerator of *fraction*.
- **Denominator[***fraction***]** returns the denominator of *fraction*.
- **Cancel[***fraction***]** cancels out common factors in the numerator and denominator of *fraction*. The option **Extension → Automatic** allows operations to be performed on algebraic numbers that appear in *fraction*.
- **Together[***expression***]** combines the terms of *expression* using a common denominator. Any common factors in numerator and denominator are cancelled.
- **Apart[***fraction***]** writes *fraction* as a sum of partial fractions.

EXAMPLE 14

```
Cancel[ (x² + 5x + 6) / (x² + 3x + 2) ]
```
$\frac{3 + x}{1 + x}$

EXAMPLE 15

```
Together[ 1/(x+1) + 2/(x²-1) ]
```
$\frac{1}{-1 + x}$

EXAMPLE 16

$$\text{Apart}\left[\frac{x^2 + 5\,x}{x^4 + x^3 - x - 1}\right]$$

$$\frac{1}{-1+x} + \frac{2}{1+x} + \frac{-1-3\,x}{1+x+x^2}$$

Since *Mathematica*, by default, converts factors with negative exponents to their positive exponent equivalents, the result of **Numerator** or **Denominator** may be different than expected.

EXAMPLE 17

$$\text{fraction} = \frac{x^{-1}\,y^{-2}}{z^{-3}};$$

Numerator[fraction]

$$z^3$$

Denominator[fraction]

$$x\,y^2$$

- **ExpandNumerator[*expression*]** expands the numerator of *expression* but leaves the denominator alone.
- **ExpandDenominator[*expression*]** expands the denominator of *expression* but leaves the numerator alone.
- **ExpandAll[*expression*]** expands both numerator and denominator of *expression*, writing the result as a sum of fractions with a common denominator.

EXAMPLE 18

$$\text{expression} = \frac{(x+1)\,(x+2)}{(x+3)\,(x+4)};$$

ExpandNumerator[expression]

$$\frac{2+3\,x+x^2}{(3+x)\,(4+x)}$$

ExpandDenominator[expression]

$$\frac{(1+x)\,(2+x)}{12+7\,x+x^2}$$

ExpandAll[expression]

$$\frac{2}{12+7\,x+x^2} + \frac{3\,x}{12+7\,x+x^2} + \frac{x^2}{12+7\,x+x^2}$$

ExpandNumerator[ExpandDenominator[expression]]

$$\frac{2+3\,x+x^2}{12+7\,x+x^2}$$

The commands described in this section are not limited to rational functions (quotients of polynomials) but will work for both algebraic expressions involving radicals and non-algebraic expressions involving functions or undefined objects. In addition, if the option **Trig → True** is set within the command, *Mathematica* will use standard trigonometric identities to simplify the expression. This will be discussed further in Section 7.3.

EXAMPLE 19

$$\text{Expand}\left[\left(1 + \sqrt{x}\right)^6\right]$$

$$1 + 6\,\sqrt{x} + 15\,x + 20\,x^{3/2} + 15\,x^2 + 6\,x^{5/2} + x^3$$

EXAMPLE 20

$$\text{Apart}\left[\frac{1}{(\sqrt{x}+1)\,(\sqrt{x}+2)}\right]$$

$$\frac{1}{1+\sqrt{x}}-\frac{1}{2+\sqrt{x}}$$

SOLVED PROBLEMS

7.12 The expression $\dfrac{f(x)-f(a)}{x-a}$ appears in calculus in connection with the derivative. Simplify this expression for $f(x)=x^9$, $a=-3$.

SOLUTION

$$f[x\_]=x^9;$$
$$a=-3;$$

$$\text{Cancel}\left[\frac{f[x]-f[a]}{x-a}\right]$$

$$6561-2187\,x+729\,x^2-243\,x^3+81\,x^4-27\,x^5+9\,x^6-3\,x^7+x^8$$

7.13 Express the sum of $\dfrac{a}{b}$, $\dfrac{c}{d}$, and $\dfrac{e}{f}$ as a single fraction.

SOLUTION

$$\text{Together}[a/b+c/d+e/f]$$

$$\frac{b\,d\,e+b\,c\,f+a\,d\,f}{b\,d\,f}$$

7.14 Write $\dfrac{(x+2)(x^2+3)(2x-7)}{(x^2+5x+2)(x-5)(x+6)}$ with expanded numerator and denominator.

SOLUTION 1

$$\text{ExpandNumerator}\left[\text{ExpandDenominator}\left[\frac{(x+2)\,(x^2+3)\,(2x-7)}{(x^2+5x+2)\,(x-5)\,(x+6)}\right]\right]$$

$$\frac{-42-9\,x-8\,x^2-3\,x^3+2\,x^4}{-60-148\,x-23\,x^2+6\,x^3+x^4}$$

SOLUTION 2

$$\text{ExpandAll}\left[\frac{(x+2)\,(x^2+3)\,(2x-7)}{(x^2+5x+2)\,(x-5)\,(x+6)}\right]\,//\text{Together}$$

$$\frac{-42-9\,x-8\,x^2-3\,x^3+2\,x^4}{-60-148\,x-23\,x^2+6\,x^3+x^4}$$

7.15 Add $\dfrac{2x+3}{5x-7}$, $\dfrac{7x-2}{3x+1}$, and $\dfrac{x^2}{x^2+1}$ and express as a single fraction with expanded numerator and denominator.

SOLUTION

$$p=\frac{2x+3}{5x-7};$$

$$q=\frac{7x-2}{3x+1};$$

$$r=\frac{x^2}{x^2+1};$$

```
Together[p + q + r] //ExpandDenominator
```

$$\frac{17 - 48x + 51x^2 - 64x^3 + 56x^4}{-7 - 16x + 8x^2 - 16x^3 + 15x^4}$$

> Without //**ExpandDenominator**, the denominator would be expressed in factored form.

7.16 What is the partial fraction expansion of $\dfrac{(x-1)^6}{(x^2+1)(x+1)^2(x-4)}$?

SOLUTION

```
Apart[    (x - 1)⁶
      ─────────────────────── ]
      (x² + 1) (x + 1)²(x - 4)
```

$$-4 + \frac{729}{425(-4+x)} + x - \frac{32}{5(1+x)^2} + \frac{288}{25(1+x)} - \frac{4(4+x)}{17(1+x^2)}$$

7.17 Find the partial fraction expansion of the function in the previous problem with linear complex denominators.

SOLUTION

```
Apart[         (x - 1)⁶
      ─────────────────────────── ]
      (x + I) (x - I) (x + 1)²(x - 4)
```

> To force *Mathematica* to express the result using linear complex denominators, we factor $x^2 + 1$ as $(x + I)(x - I)$.

$$-4 + \frac{729}{425(-4+x)} + x - \frac{\frac{2}{17} - \frac{8i}{17}}{-i+x} - \frac{\frac{2}{17} + \frac{8i}{17}}{i+x} - \frac{32}{5(1+x)^2} + \frac{288}{25(1+x)}$$

7.18 Express $(e^x + e^{2x})^4$ as a sum of exponentials.

SOLUTION

```
Expand[ (Eˣ + E²ˣ)⁴]
```

$$e^{4x} + 4\,e^{5x} + 6\,e^{6x} + 4\,e^{7x} + e^{8x}$$

7.3 Trigonometric Functions

Although the commands discussed in the previous section may be applied to trigonometric functions, doing so does not take advantage of the simplification offered by trigonometric identities. To incorporate these into the calculation, the option **Trig → True** must be set. (The default is **Trig → False** for all but the **Simplify** command.) The following examples show the difference.

EXAMPLE 21

```
Cancel[   Sin[x]
        ──────────── ]
        1 - Cos [x]²
```

$$\frac{Sin[x]}{1 - Cos[x]^2}$$

```
Cancel[   Sin[x]
        ────────────, Trig → True ]
        1 - Cos [x]²
```

```
Csc[x]
```

EXAMPLE 22

```
Together[  Cos [x]²       Sin [x]²
          ──────────── + ──────────── ]
          1 - Sin [x]²    1 - Cos [x]²
```

$$\frac{Cos[x]^2 - Cos[x]^4 + Sin[x]^2 - Sin[x]^4}{(-1 + Cos[x]^2)(-1 + Sin[x]^2)}$$

Together$\left[\dfrac{\text{Cos}[x]^2}{1 - \text{Sin}[x]^2} + \dfrac{\text{Sin}[x]^2}{1 - \text{Cos}[x]^2}, \text{Trig} \rightarrow \text{True} \right]$

2

Trig → True applies to hyperbolic as well as circular functions.

EXAMPLE 23

Expand[(Cosh[x]^2 + Sinh[x]^2)(Cosh[x]^2 – Sinh[x]^2)]

Cosh[x]^4 – Sinh[x]^4

Expand[(Cosh[x]^2 + Sinh[x]^2)(Cosh[x]^2 – Sinh[x]^2), Trig → True]

Cosh[x]^2 + Sinh[x]^2

To allow additional manipulation of trigonometric expressions, *Mathematica* offers the following specialized commands, which apply to both circular and hyperbolic functions:

- **TrigExpand[*expression*]** expands *expression*, splitting up sums and multiples that appear in arguments of trigonometric functions and expanding out products of trigonometric functions into sums and powers, taking advantage of trigonometric identities whenever possible.
- **TrigReduce[*expression*]** rewrites products and powers of trig functions in *expression* as trigonometric expressions with combined arguments, reducing *expression* to a linear trig function (i.e., without powers or products).
- **TrigFactor[*expression*]** converts *expression* into a factored expression of trigonometric functions of a single argument.

The next example shows the difference between **Expand** and **TrigExpand**.

EXAMPLE 24

Expand[(Sin[x] + Cos[x])^2]

Cos[x]^2 + 2 Cos[x] Sin[x] + Sin[x]^2

TrigExpand[(Sin[x] + Cos[x])^2]

1 + 2 Cos[x] Sin[x]

EXAMPLE 25

TrigExpand[Sin[x + y]]

Cos[y] Sin[x] + Cos[x] Sin[y]

TrigExpand[Sin[2 x]]

2 Cos[x] Sin[x]

TrigExpand[Sin[2 x + y]]

2 Cos[x] Cos[y] Sin[x] + Cos[x]^2 Sin[y] – Sin[x]^2 Sin[y]

TrigExpand can also be applied to hyperbolic functions.

EXAMPLE 26

TrigExpand[Cosh[x + y]]

Cosh[x] Cosh[y] + Sinh[x] Sinh[y]

EXAMPLE 27

TrigReduce[Sin[2 x]^2 + Sin[x] Cos[3 x]^3]

$\dfrac{1}{8}$ (4 – 4 Cos[4x] – 3 Sin[2x] + 3 Sin[4x] – Sin[8x] + Sin[10x])

TrigReduce rewrites the original expression as a linear trig expression.

```
TrigReduce[Sinh[2x]² + Sinh[x] Cosh[3x]³]
```

$$\frac{1}{8}(-4 + 4\,\text{Cosh}[4x] - 3\,\text{Sinh}[2x] + 3\,\text{Sinh}[4x] - \text{Sinh}[8x] + \text{Sinh}[10x])$$

The next example shows the difference between **TrigFactor** and **TrigReduce**. Notice that **TrigFactor** writes the expression as a product, while **TrigReduce** writes the expression as a sum of linear trig functions.

EXAMPLE 28

```
expression = 24 Sin[x]² Cos[x]² + 16 Cos[x]⁴;
TrigFactor[expression]
```

$-4\,\text{Cos}[x]^2\,(-5 + \text{Cos}[2x])$

```
TrigReduce[expression]
```

$9 + 8\,\text{Cos}[2x] - \text{Cos}[4x]$

The **Solve** command can be used to solve trigonometric equations. However, because only principal values of inverse trigonometric functions are returned, not all solutions will be obtained.

EXAMPLE 29 Consider the equation $1 - 2\cos x - \sin x + \sin 2x = 0$.

```
equation = 1 - 2 Cos[x] - Sin[x] + Sin[2x] == 0
Solve[equation , x]
```

Solve::ifun: Inverse functions are being used by Solve, so some solutions may not be found; use Reduce for complete solution information. ≫

$$\left\{\left\{x \to -\frac{\pi}{3}\right\}, \left\{x \to \frac{\pi}{3}\right\}, \left\{x \to \frac{\pi}{2}\right\}\right\}$$

Since trigonometric and hyperbolic functions can be represented in terms of exponential functions (complex exponentials in the case of circular trig functions), *Mathematica* offers two conversion functions:

- **TrigToExp[*expression*]** converts trigonometric and hyperbolic functions to exponential form.
- **ExpToTrig[*expression*]** converts exponential functions to trigonometric and/or hyperbolic functions.

TrigToExp and **ExpToTrig** may also be used to convert *inverse* trigonometric and hyperbolic functions.

EXAMPLE 30

```
TrigToExp[Cos[x]]
```

$\dfrac{e^{-ix}}{2} + \dfrac{e^{ix}}{2}$

```
TrigToExp[Sinh[x]]
```

$-\dfrac{e^{-x}}{2} + \dfrac{e^{x}}{2}$

```
ExpToTrig[Exp[x]]
```

$\text{Cosh}[x] + \text{Sinh}[x]$

```
ExpToTrig[Exp[I x]]
```

$\text{Cos}[x] + i\,\text{Sin}[x]$

SOLVED PROBLEMS

7.19 Simplify the trigonometric function $\dfrac{1}{\cos^2 x - \sin^2 x}$.

SOLUTION

$$\texttt{TrigReduce}\left[\frac{1}{\texttt{Cos[x]}^2 - \texttt{Sin[x]}^2}\right]$$

$$\texttt{Sec[2 x]}$$

7.20 Factor and simplify: $\sin^2 x \cos^2 x + \cos^4 x$.

SOLUTION

$$\texttt{TrigFactor[Sin[x]}^2\,\texttt{Cos[x]}^2 + \texttt{Cos[x]}^4]$$

$$\texttt{Cos[x]}^2$$

7.21 Solve the trigonometric equation $1 - 2 \cos x - 2 \sin x + 4 \sin 2x = 0$.

SOLUTION

$$\texttt{equation = 1 - 2 Cos[x] - 2 Sin[x] + 4 Sin[2 x] == 0}$$
$$\texttt{Solve[equation, x]}$$

Solve::ifun : Inverse functions are being used by Solve, so some solutions may not be found; use Reduce for complete solution information. >>

$$\left\{\left\{x \to \text{ArcCos}\left[\frac{1}{8} + \frac{\sqrt{13}}{8} - \frac{1}{4}\sqrt{\frac{1}{2}(9 - \sqrt{13})}\right]\right\}, \left\{x \to \text{ArcCos}\left[\frac{1}{8} + \frac{\sqrt{13}}{8} + \frac{1}{4}\sqrt{\frac{1}{2}(9 - \sqrt{13})}\right]\right\},\right.$$
$$\left.\left\{x \to \text{ArcCos}\left[\frac{1}{8}\left(1 - \sqrt{13} - 4\sqrt{\frac{9}{8} + \frac{\sqrt{13}}{8}}\right)\right]\right\}, \left\{x \to -\text{ArcCos}\left[\frac{1}{8}\left(1 - \sqrt{13} + 4\sqrt{\frac{9}{8} + \frac{\sqrt{13}}{8}}\right)\right]\right\}\right\}$$

A numerical solution would probably be more useful.

$$\texttt{\% //N}$$

$$\{\{x \to 1.40492\}, \{x \to 0.165873\}, \{x \to 2.83487\}, \{x \to -1.26407\}\}$$

7.22 Add and simplify: $\dfrac{\cos x}{1 + \sin x} + \tan x$.

SOLUTION

$$\texttt{Together}\left[\frac{\texttt{Cos[x]}}{1 + \texttt{Sin[x]}} + \texttt{Tan[x]}, \texttt{Trig} \to \texttt{True}\right]$$

$$\frac{1}{\left(\texttt{Cos}\left[\frac{x}{2}\right] - \texttt{Sin}\left[\frac{x}{2}\right]\right)\left(\texttt{Cos}\left[\frac{x}{2}\right] + \texttt{Sin}\left[\frac{x}{2}\right]\right)}$$

> Sometimes you have to apply two or more trig commands to simplify completely.

$$\texttt{TrigReduce[\%]}$$
$$\texttt{Sec[x]}$$

7.23 Combine and simplify: $\dfrac{\sinh x}{\cosh x - \sinh x} + \dfrac{\cosh x}{\cosh x + \sinh x}$

SOLUTION

$$\texttt{Together}\left[\frac{\texttt{Sinh[x]}}{\texttt{Cosh[x]} - \texttt{Sinh[x]}} + \frac{\texttt{Cosh[x]}}{\texttt{Cosh[x]} + \texttt{Sinh[x]}}, \texttt{Trig} \to \texttt{True}\right]$$

$$\texttt{Cosh[2 x]}$$

7.24 Construct a table of multiple angle formulas for sin nx and cos nx, $n = 2, 3, 4$, and 5.

SOLUTION

```
trigtable = Table[{n, TrigExpand[Sin[n x]], TrigExpand[Cos[n x]]},
            {n, 2, 5}];
TableForm[trigtable, TableHeadings →
        {None, {"n", "    sin nx", "    cos nx"}}]
```

n	sin nx	cos nx
2	$2 \cos[x] \sin[x]$	$\cos[x]^2 - \sin[x]^2$
3	$3 \cos[x]^2 \sin[x] - \sin[x]^3$	$\cos[x]^3 - 3 \cos[x] \sin[x]^2$
4	$4 \cos[x]^3 \sin[x] - 4 \cos[x] \sin[x]^3$	$\cos[x]^4 - 6 \cos[x]^2 \sin[x]^2 + \sin[x]^4$
5	$5 \cos[x]^4 \sin[x] - 10 \cos[x]^2 \sin[x]^3 + \sin[x]^5$	$\cos[x]^5 - 10 \cos[x]^3 \sin[x]^2 + 5 \cos[x] \sin[x]^4$

7.25 Construct a table of linear trig formulas for $\sin^n x$ and $\cos^n x$, $n = 2, 3, 4$, and 5.

SOLUTION

```
trigtable = Table[{n, TrigReduce[Sin[x]^n], TrigReduce[Cos[x]^n]},
            {n, 2, 5}];
TableForm[trigtable, TableHeadings →
        {None, {"n", "    sinⁿ x", "    cosⁿ x"}}]
```

n	$\sin^n x$	$\cos^n x$
2	$\frac{1}{2}(1 - \cos[2x])$	$\frac{1}{2}(1 + \cos[2x])$
3	$\frac{1}{4}(3 \sin[x] - \sin[3x])$	$\frac{1}{4}(3 \cos[x] + \cos[3x])$
4	$\frac{1}{8}(3 - 4 \cos[2x] + \cos[4x])$	$\frac{1}{8}(3 + 4 \cos[2x] + \cos[4x])$
5	$\frac{1}{16}(10 \sin[x] - 5 \sin[3x] + \sin[5x])$	$\frac{1}{16}(10 \cos[x] + 5 \cos[3x] + \cos[5x])$

7.26 Express e^{x+y} in terms of hyperbolic functions and expand.

SOLUTION

```
ExpToTrig[E^(x+y)]
```

$\cosh[x + y] + \sinh[x + y]$

```
TrigExpand[%]
```

$\cosh[x] \cosh[y] + \cosh[y] \sinh[x] + \cosh[x] \sinh[y] + \sinh[x] \sinh[y]$

7.27 Express $\sinh^{-1} x$ and $\tanh^{-1} x$ in logarithmic form.

SOLUTION

```
TrigToExp[ArcSinh[x]]
```

$\log[x + \sqrt{1 + x^2}]$

```
TrigToExp[ArcTanh[x]]
```

$-\frac{1}{2} \log[1 - x] + \frac{1}{2} \log[1 + x]$

7.28 Use **Manipulate** to control the graph of $f(x) = a \sin(bx + c)$, $0 \le x < 2\pi$, with controls for a, b, and c varying between 1 and 10. Move the sliders and observe the effect upon the graph.

SOLUTION

```
Manipulate[Plot[a Sin[bx + c], {x, 0, 2π},
        PlotRange → {-10, 10}], {a, 1, 10}, {b, 1, 10}, {c, 1, 10}]
```

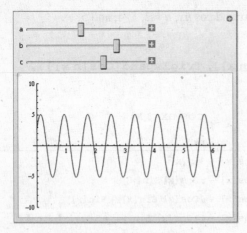

7.4 The Art of Simplification

There are many different ways to write any particular algebraic or trigonometric expression. Obviously one person's interpretation of "simple" may not agree with another's. For example, in dealing with rational functions, $(x + 3)^2$ may be preferable to $x^2 + 6x + 9$, but when manipulating polynomials, the latter is clearly more desirable.

As you have seen from reading this chapter, *Mathematica* offers a variety of commands that allow full control of how an expression will appear. With practice, you will learn to use these commands to reshape appearances to suit your needs.

As a step in the direction toward simplification, *Mathematica* offers two commands that can be used to simplify complex structures.

- **Simplify[*expression*]** performs a sequence of transformations on *expression* and returns the simplest form it finds.
- **FullSimplify[*expression*]** tries a wider range of transformations on *expression* including elementary and special functions and returns the simplest form it finds.

Simplify tries expanding, factoring, and other standard mathematical transformations to reduce the complexity of *expression*. Because of its general nature, **Simplify** tends to be quite slow in comparison to more direct instructions. **FullSimplify** always produces an expression at least as simple as **Simplify**, but may take somewhat longer.

You can specify a time limitation (in seconds) with the option **TimeConstraint**. The default for **Simplify** is **TimeConstraint → 300** and for **FullSimplify**, **TimeConstraint → Infinity**. For both commands, **Trig → True** is the default for trigonometric evaluation.

EXAMPLE 31   First let us generate a messy algebraic expression.

$$\texttt{messyexpression = Expand}\left[\left(\frac{1}{x+1} + \frac{1}{x+2} + \frac{1}{x+3}\right)^5\right]$$

$$\frac{1}{(1+x)^5} + \frac{1}{(2+x)^5} + \frac{5}{(1+x)(2+x)^4} + \frac{10}{(1+x)^2(2+x)^3} + \frac{10}{(1+x)^3(2+x)^2} +$$

$$\frac{5}{(1+x)^4(2+x)} + \frac{1}{(3+x)^5} + \frac{5}{(1+x)(3+x)^4} + \frac{5}{(2+x)(3+x)^4} + \frac{10}{(1+x)^2(3+x)^3} +$$

$$\frac{10}{(2+x)^2(3+x)^3} + \frac{20}{(1+x)(2+x)(3+x)^3} + \frac{10}{(1+x)^3(3+x)^2} + \frac{10}{(2+x)^3(3+x)^2} +$$

$$\frac{30}{(1+x)(2+x)^2(3+x)^2} + \frac{30}{(1+x)^2(2+x)(3+x)^2} + \frac{5}{(1+x)^4(3+x)} + \frac{5}{(2+x)^4(3+x)} +$$

$$\frac{20}{(1+x)(2+x)^3(3+x)} + \frac{30}{(1+x)^2(2+x)^2(3+x)} + \frac{20}{(1+x)^3(2+x)(3+x)}$$

Now we will simplify. Of course, *Mathematica* does not "remember" how **messyexpression** was generated.

```
Simplify[messyexpression]
```

$$\frac{(11 + 12\,x + 3\,x^2)^5}{(6 + 11\,x + 6\,x^2 + x^3)^5}$$

```
FullSimplify[messyexpression]
```

$$\frac{(11 + 3\,x\,(4 + x)\,)^5}{(1 + x)^5\,(2 + x)^5\,(3 + x)^5}$$

EXAMPLE 32

```
messytrigexpression = Expand[(Tan[x]² + Sin[x]² + Cos[x]²)⁵]
```

$Cos[x]^{10} + 5\,Cos[x]^6\,Sin[x]^2 + 5\,Cos[x]^8\,Sin[x]^2 + 10\,Cos[x]^2\,Sin[x]^4 + 20\,Cos[x]^4\,Sin[x]^4 +$
$10\,Cos[x]^6\,Sin[x]^4 + 30\,Sin[x]^6 + 30\,Cos[x]^2\,Sin[x]^6 + 10\,Cos[x]^4\,Sin[x]^6 +$
$20\,Sin[x]^8 + 5\,Cos[x]^2\,Sin[x]^8 + Sin[x]^{10} + 10\,Sin[x]^4\,Tan[x]^2 + 30\,Sin[x]^6\,Tan[x]^2 +$
$5\,Sin[x]^8\,Tan[x]^2 + 20\,Sin[x]^4\,Tan[x]^4 + 10\,Sin[x]^6\,Tan[x]^4 + 5\,Sin[x]^2\,Tan[x]^6 +$
$10\,Sin[x]^4\,Tan[x]^6 + 5\,Sin[x]^2\,Tan[x]^8 + Tan[x]^{10}$

```
Simplify[messytrigexpression]
```

$Sec[x]^{10}$

CHAPTER 8

Differential Calculus

8.1 Limits

The limit of a function is the foundation stone of differential calculus. For a complicated function, the calculation of a limit can be quite difficult and can require specialized techniques for its evaluation. *Mathematica* has built-in procedures for accomplishing this task and always attempts to determine the *exact* value of the limit.

- **Limit[f[x], x → a]** computes the value of $\lim_{x \to a} f(x)$.

EXAMPLE 1   We wish to compute $\lim_{x \to 2} \dfrac{x^5 - 32}{x^3 - 8}$. Because both numerator and denominator approach zero as x → 2, the limit is not immediately obvious.

$$\text{Limit}\left[\frac{x^5 - 32}{x^3 - 8},\ x \to 2\right]$$

$\dfrac{20}{3}$

Left- and right-hand limits can be computed with the **Direction** option.

- **Direction → 1** causes the limit to be computed as a *left-hand limit* with values of x approaching *a* from below.
- **Direction → −1** causes the limit to be computed as a *right-hand limit* with values of x approaching *a* from above.

The default for the **Limit** command is **Direction → Automatic**, which provides **Direction → −1** except for limits at ∞. Thus, *Mathematica* may give a misleading representation of the limit of a discontinuous function if the **Direction** option is omitted.

EXAMPLE 2   Evaluate $\lim_{x \to 0} \dfrac{|x|}{x}$.

$$\text{Limit}\left[\frac{\text{Abs}[x]}{x},\ x \to 0\right]$$

1

By default, only the right-hand limit has been computed, since no direction was specified. To fully analyze the limit we must compute the left-hand limit as well.

$$\text{Limit}\left[\frac{\text{Abs}[x]}{x},\ x \to 0,\ \text{Direction} \to 1\right]$$

−1

The limit does not exist since the left- and right-hand limits are different numbers.

Mathematica can compute infinite limits and limits at ∞.

EXAMPLE 3

```
Limit[1/x, x → 0, Direction → -1]
∞

Limit[1/x, x → 0, Direction → 1]
-∞

Limit[ 2 x² + 3 x + 4 / x² + 1 , x → ∞]
2
```

The functions in the next example exhibit a different behavior. As $x \to 0$, the function oscillates an infinite number of times. *Mathematica* returns the limit as an **Interval** object. **Interval[{*min*, *max*}]** represents the range of values between *min* and *max*.

EXAMPLE 4

```
Limit[Sin[1/x], x → 0]
Interval[{-1, 1}]

Limit[Tan[1/x], x → 0]
Interval[{ -∞, ∞}]
```

SOLVED PROBLEMS

8.1 Compute $\lim\limits_{x \to 0} \dfrac{2^x + x - 1}{3x}$.

SOLUTION

```
Limit[ 2^x + x - 1 / 3 x , x → 0]
1/3 (1 + Log[2])
```

8.2 Compute $\lim\limits_{x \to 0} \dfrac{\tan x - x}{x^3}$

SOLUTION

```
Limit[ Tan[x] - x / x³ , x → 0]
1/3
```

8.3 Compute $\lim\limits_{x \to 0} (1 + \sin x)^{\cot 2x}$

SOLUTION

```
Limit[(1 + Sin[x])^Cot[2x], x → 0]
√e
```

8.4 Compute $\lim_{x \to \infty}(e^x + x)^{1/x}$ and $\lim_{x \to -\infty}(e^x + x)^{1/x}$

SOLUTION

```
Limit[(Exp[x] + x)^(1/x), x → ∞]
e
Limit[(Exp[x] + x)^(1/x), x → -∞]
1
```

8.5 Compute $\lim_{x \to 1}(2 - x)^{\tan(\frac{\pi}{2}x)}$

SOLUTION

```
Limit[(2 - x)^Tan[π/2 x], x → 1]
e^(2/π)
```

8.6 If p dollars is compounded n times per year at an annual interest rate of r, the money will be worth $p\left(1 + \dfrac{r}{n}\right)^{nt}$ dollars after t years. How much will the money be worth after t years if it is compounded continuously $(n \to \infty)$?

SOLUTION

```
Limit[p(1 + r/n)^(nt), n → ∞]
e^(rt)p
```

8.7 The derivative of a function is defined to be $\lim_{h \to 0}\dfrac{f(x+h) - f(x)}{h}$. Use this definition to compute the derivative of $f(x) = \ln x + x^5 + \sin x$.

SOLUTION

```
f[x_] = Log[x] + x^5 + Sin[x];
Limit[ (f[x + h] - f[x])/h, h → 0]
1/x + 5 x^4 + Cos[x]
```

8.8 The second derivative of a function can be computed as the limit

$$\lim_{h \to 0}\frac{f(x+h) - 2f(x) + f(x-h)}{h^2}$$

Use this limit to compute the second derivative of $f(x) = \ln x + x^5 + \sin x$.

SOLUTION

```
f[x_] = Log[x] + x^5 + Sin[x];
Limit[ (f[x + h] - 2 f[x] + f[x - h])/h^2, h → 0]
-1/x^2 + 20 x^3 - Sin[x]
```

8.2 Derivatives

There are several ways derivatives can be computed in *Mathematica*. Each has its advantages and disadvantages, so the proper choice for a particular situation must be determined.

- If **f[x]** represents a function, its derivative is represented by **f'[x]**. Higher order derivatives are represented by **f''[x]**, **f'''[x]**, and so on.

EXAMPLE 5

```
f[x_] = x^5 + x^4 + x^3 + x^2 + x + 1;
f'[x]
1 + 2x + 3x^2 + 4x^3 + 5x^4
f''[x]
2 + 6x + 12x^2 + 20x^3
f'''[x]
6 + 24x + 60x^2
```

If a more traditional formatting of the derivatives is desired, the command **TraditionalForm** can be used.

EXAMPLE 6

```
f[x_] = x^5 + x^4 + x^3 + x^2 + x + 1;
f'[x] // TraditionalForm
```
$5x^4 + 4x^3 + 3x^2 + 2x + 1$
```
f''[x] // TraditionalForm
```
$20x^3 + 12x^2 + 6x + 2$
```
f'''[x] // TraditionalForm
```
$60x^2 + 24x + 6$

The prime notation can also be used for "built-in" functions, as illustrated in the next example. If the argument is omitted, *Mathematica* returns a *pure* function representing the required derivative. (Pure functions are discussed in the appendix.)

EXAMPLE 7

```
Sqrt'
```
$\dfrac{1}{2\sqrt{\#1}}$ &

```
Sqrt'[x]
```
$\dfrac{1}{2\sqrt{x}}$ ← The variable x replaces the symbol #1.

```
Sqrt''
```
$-\dfrac{1}{4\ \#1^{3/2}}$ &

```
Sqrt''[x]
```
$-\dfrac{1}{4\ x^{3/2}}$

- **D[f[x], x]** returns the derivative of f with respect to x.
- **D[f[x], {x, n}]** returns the nth derivative of f with respect to x.

EXAMPLE 8

```
D[x⁵ + x⁴ + x³ + x² + x + 1, x]
```

$$1 + 2x + 3x^2 + 4x^3 + 5x^4$$

```
D[x⁵ + x⁴ + x³ + x² + x + 1, {x, 2}]
```

$$2 + 6x + 12x^2 + 20x^3$$

```
D[x⁵ + x⁴ + x³ + x² + x + 1, {x, 3}]
```

$$6 + 24x + 60x^2$$

- $\partial_{\square}$, which can be found on the Basic Math Input palette, is equivalent to **D**. ∂_x will return the derivative with respect to x. The *n*th derivative is represented by $\partial_{\{x, n\}}$.

EXAMPLE 9

$$\partial_x (x^5 + x^4 + x^3 + x^2 + x + 1)$$

$$1 + 2x + 3x^2 + 4x^3 + 5x^4$$

$$\partial_{\{x, 2\}} (x^5 + x^4 + x^3 + x^2 + x + 1)$$

$$2 + 6x + 12x^2 + 20x^3$$

$$\partial_{\{x, 3\}} (x^5 + x^4 + x^3 + x^2 + x + 1)$$

$$6 + 24x + 60x^2$$

- **Derivative[n]** is a *functional operator* that acts on a function to produce a new function, namely, its nth derivative. **Derivative[n][f]** gives the nth derivative of f as a *pure* function and **Derivative[n][f][x]** evaluates the nth derivative of f at x.

It is useful to remember that **f'** is converted to **Derivative[1]**. Thus, **f'[x]** becomes **Derivative[1][x]**. Higher order derivatives **f''**, **f'''**, etc. are handled in a similar manner.

EXAMPLE 10

```
f[x_] = x⁵ + x⁴ + x³ + x² + x + 1;
```

```
Derivative[1][f]
```

$$1 + 2\#1 + 3\#1^2 + 4\#1^3 + 5\#1^4 \&$$ ← *Mathematica* returns a pure function representing the derivative of f. Pure functions are discussed in the appendix.

```
Derivative[1][f][x]
```

$$1 + 2x + 3x^2 + 4x^3 + 5x^4$$ ← #1 is replaced by x.

The numerical *value* of a derivative at a specific point can be computed several different ways, depending upon how the derivative is computed. The next example illustrates the most common techniques.

EXAMPLE 11

```
f[x_] = (x² - x + 1)⁵;
f''[1]
```
30   ← In each of the first three parts of this example, the second derivative is computed and then x is replaced by 1.

```
D[f[x], {x, 2}] /. x → 1
```
30

$$\partial_{\{x, 2\}} f[x] /. x \to 1$$
30

```
g := Derivative[2][f]
```
     ← Here we have defined a new function, g, as the second derivative of f. If f is changed, g will be the second derivative of the new function. Note the use of := here. This is crucial if g is to reflect the change in f.

```
g[1]
```
30

```
f[x_] = x³
```

```
g[1]
```
6

Mathematica computes derivatives of combinations of functions, sums, differences, products, quotients, and composites by "memorizing" the various rules. If we do not define the functions, we can see what the rules are.

EXAMPLE 12

```
Clear[f, g]
D[f[x] + g[x], x]
```
$f'[x] + g'[x]$ ← The derivative of a sum is the sum of
```
D[f[x] g[x], x]
```
the derivatives of its terms.

$g[x] f'[x] + f[x] g'[x]$ ← This is the familiar product rule.
```
D[f[x]/g[x], x]//Together
```
$$\frac{g[x] f'[x] - f[x] g'[x]}{g[x]^2}$$ ← Quotient rule.
```
D[f[g[x]], x]
```
$f'[g[x]] g'[x]$ ← Chain rule.

We can use *Mathematica* to investigate some basic theory from a graphical perspective. Rolle's Theorem guarantees, under certain conditions, the existence of a point where the derivative of a function is 0:

> *Let f be continuous on the closed interval* $[a, b]$ *and differentiable on the open interval* (a, b) *and suppose* $f(a) = f(b) = 0$. *Then there exists a number, c, between a and b, such that* $f'(c) = 0$.

In other words, if a smooth (differentiable) function vanishes (has a value of 0) at two distinct locations, its derivative must vanish somewhere in between.

EXAMPLE 13   Show that the function $f(x) = (x^3 + 2x^2 + 15x + 2)\sin \pi x$ satisfies Rolle's Theorem on the interval $[0, 1]$ and find the value of c referred to in the theorem.

Since f is the product of a polynomial and a trigonometric sine function, f is continuous and differentiable everywhere.

```
f[x_] = (x³ + 2x² + 15x + 2) Sin[π x];
f[0]
```
0
```
f[1]
```
0
```
FindRoot[f'[c] == 0, {c, 0.5}]
```
← We used 0.5 as our initial guess since it is
$\{c \rightarrow 0.640241\}$ halfway between 0 and 1.
```
Plot[{f[x], f[.640241]}, {x, 0, 1}]
```

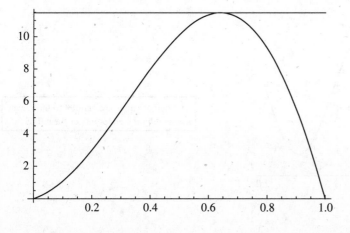

The Mean Value Theorem is similar to Rolle's Theorem and does not require f to be 0 at each endpoint of the interval:

> *Let f be continuous on the closed interval $[a, b]$ and differentiable on the open interval (a, b). Then there exists a number, c, between a and b such that $f(b) - f(a) = f'(c)(b - a)$.*

If we write the conclusion of the theorem in the form $\dfrac{f(b) - f(a)}{b - a} = f'(c)$, we see that the Mean Value Theorem guarantees the existence of a number, c, between a and b, such that the tangent line at $(c, f(c))$ is parallel to the line segment connecting the endpoints of the curve.

Note: Rolle's Theorem and the Mean Value Theorem guarantee the existence of *at least* one number c. In actuality, there may be several.

EXAMPLE 14 Find the value(s), c, guaranteed by the Mean Value Theorem for the function $f(x) = \sqrt{x} + \sin 2\pi x$ on the interval $[0, 2]$.

```
f[x_] = √x + Sin[2π x]; a = 0; b = 2; m = f[b] - f[a] / b - a ;
Plot[f'[x] - m, {x, 0, 2}, PlotRange → {-8, 8}]]
```

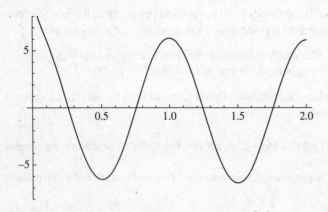

> We estimate the zeros of the function $f'(x) - m$ to determine the approximate locations of c. There appear to be four values: 0.3, 0.7, 1.3, and 1.7 (approximately).

```
FindRoot[f'[c] == m, {c, {.3, .7, 1.3, 1.7}}]
{c → {0.257071, 0.753319, 1.24344, 1.75836}}
c1 = .257071; c2 = .753319; c3 = 1.24344; c4 = 1.75836;
l1[x_] := f[c1] + f'[c1] (x - c1) /; c1 - .25 ≤ x ≤ c1 + .25
l2[x_] := f[c2] + f'[c2] (x - c2) /; c2 - .25 ≤ x ≤ c2 + .25
l3[x_] := f[c3] + f'[c3] (x - c3) /; c3 - .25 ≤ x ≤ c3 + .25
l4[x_] := f[c4] + f'[c4] (x - c4) /; c4 - .25 ≤ x ≤ c4 + .25
l[x_] := f[a] + m (x - a)
Plot[{f[x], l[x], l1[x], l2[x], l3[x], l4[x]}, {x, a, b}]
```

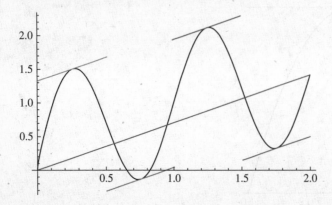

> The tangent lines are parallel to the secant connecting the endpoints of the curve.

SOLVED PROBLEMS

8.9   Compute the 3rd derivative of tan x.

SOLUTION

```
Tan'''[x]
```
$$2 \operatorname{Sec}[x]^4 + 4 \operatorname{Sec}[x]^2 \operatorname{Tan}[x]^2$$

8.10   Compute the values of the first ten derivatives of $f(x) = e^{x^2}$ at $x = 0$. Put the results in tabular form.

SOLUTION

```
f[x_] = Exp[x²]
derivtable = Table[{n, D[f[x], {x, n}] /. x → 0}, {n, 1, 10}];
TableForm[derivtable, TableAlignments → Right, TableSpacing → {1,5},
        TableHeadings → {None, {"n", "f⁽ⁿ⁾(0)"}}]
```

n	$f^{(n)}(0)$
1	0
2	2
3	0
4	12
5	0
6	120
7	0
8	1680
9	0
10	30 240

8.11   Sketch the graph of $f(x) = x^4 - 50x^2 + 300$ and its derivative, on one set of axes, for $-10 \le x \le 10$.

SOLUTION

```
≪PlotLegends`
f[x_] = x⁴ - 50x² + 300;
Plot[{f[x], f'[x]}, {x, -10, 10}, PlotRange → {-1000, 1000},
     PlotStyle → {GrayLevel[0], Dashing[{.015}]},
     PlotLegend → {"f(x)", "f'(x)"}]
```

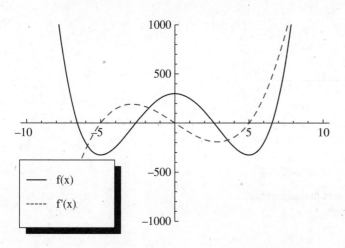

Observe that $f'(x) = 0$ precisely where $f(x)$ has a relative (local) maximum or minimum.

8.12 Given $f(x)$ whose graph is C, the slope of the line tangent to C at a is $f'(a)$. Let $f(x) = \sin x$. Sketch the graph and its tangent line at $a = \pi/3$.

SOLUTION

```
f[x_] = Sin[x];
a = π/3;
l[x_] = f[a] + f'[a] (x - a);
Plot[{f[x], l[x]}, {x, 0, 2π}]
```

> Recall that the equation of a line having slope m, passing through (x_1, y_1) is
> $$y - y_1 = m(x - x_1)$$
> or $$y = y_1 + m(x - x_1)$$
> Here, $x_1 = a$, $y_1 = f(a)$, and $m = f'(a)$ so
> $$y = f(a) + f'(a)(x - a)$$

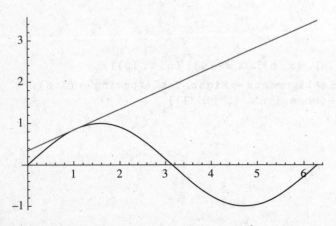

8.13 Use **Manipulate** to show the tangent line at various positions along the curve $y = \sin x$, $0 \le x \le 2\pi$.

SOLUTION

The tangent line has equation $y = f(a) + f'(a)(x - a)$.

```
f[x_] = Sin[x];
```

This guarantees that the tangent line will have a constant length of 2.

$$l[x\_, a\_] := f[a] + f'[a](x-a) \; /; \; a - \frac{1}{\sqrt{1 + f'[a]^2}} \le x \le a + \frac{1}{\sqrt{1 + f'[a]^2}}$$

```
Manipulate[Plot[{f[x], l[x, a]}, {x, 0, 2π},
        PlotRange → {-1.5, 1.5}], {a, 0, 2π}]
```

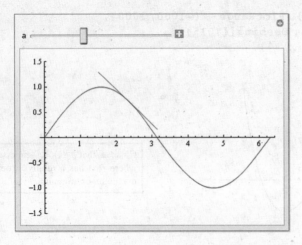

Move the slider to change the location of the tangent line.

8.14 Find the value(s) of c guaranteed by Rolle's Theorem for the function
$f(x) = 4x + 39x^2 - 46x^3 + 17x^4 - 2x^5$ on the interval $[0, 4]$.

SOLUTION

Since $f(x)$ is a polynomial, it is continuous and differentiable everywhere. First we verify that $f(0) = f(4) = 0$.

```
f[x_] = 4 x + 39 x² - 46 x³ + 17 x⁴ - 2 x⁵;

f[0]

0

f[4]

0
```

Now we look to see where $f'(c) = 0$. Since f' is a polynomial, we can use **NSolve**.

```
NSolve[f'[c] == 0]

{{c → -0.0472411}, {c → 1.05962}, {c → 2.27466}, {c → 3.51296}}
```

There are three values of c between 0 and 4 (Rolle's Theorem guarantees *at least* one). A plot of the graph confirms our result.

```
Plot[f[x], {x, -1, 4}]
```

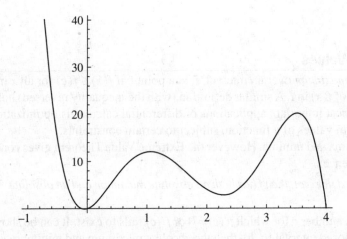

8.15 Verify the Mean Value Theorem for the function $f(x) = x + \sin 2x$ on the interval $[0, \pi]$.

SOLUTION

$f(x)$ is continuous and differentiable everywhere. Define $a = 0$, $b = \pi$ and solve the equation $f(b) - f(a) = f'(c)(b - a)$ for c. To approximate their values, we look at the graph with the endpoints connected by a line segment.

```
f[x_] = x + Sin[2 x];

a = 0; b = π;

m = f[b] - f[a] / b - a ;          ← Slope of the secant connecting the endpoints.

l[x_] = f[a] + m (x - a);          ← Function representing the secant line.
```

```
Plot[{f[x], l[x]}, {x, a, b}]
```

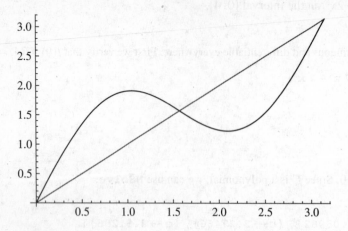

It looks like the tangent line will be parallel to the secant when $x \approx 1$ or $x \approx 2.5$. Clearly both values lie between 0 and π.

```
FindRoot[f[b] - f[a] == f'[c] (b - a), {c, 1}]

FindRoot[f[b] - f[a] == f'[c] (b - a), {c, 2.5}]
```

$\{c \to 0.785398\}$

$\{c \to 2.35619\}$

8.3 Maximum and Minimum Values

A function f has an *absolute (global) maximum* over an interval, I, at a point c if $f(x) \le f(c)$ for all x in I. In other words, $f(c)$ is the largest value of $f(x)$ in I. A similar definition (with the inequality reversed) holds for an *absolute minimum*. One of the most important applications of differential calculus is optimization, i.e., finding the maximum and minimum values of a function, subject to certain constraints.

Not all functions have absolute maxima and minima. However the Extreme Value Theorem gives conditions sufficient to guarantee their existence:

If f is continuous on a closed bounded interval, then f has both an absolute maximum and an absolute minimum in that interval.

A critical number of a function f is a number c for which $f'(c) = 0$ or $f'(c)$ fails to exist. It can be shown that if a function is continuous on the closed interval $[a, b]$, then the absolute maximum and minimum will be found either at a critical number or at an endpoint of the interval. We can use *Mathematica* to help us find the maximum and/or minimum values.

EXAMPLE 15   We wish to find the absolute maximum and minimum values of the function $f(x) = x^4 - 4x^3 + 2x^2 + 4x + 2$ on the interval $[0, 4]$. First we find the critical numbers.

```
f[x_] = x^4 - 4x^3 + 2x^2 + 4x + 2;

Solve[f'[x] == 0]
```

$\left\{ \{x \to 1\}, \left\{x \to 1 - \sqrt{2}\right\}, \left\{x \to 1 + \sqrt{2}\right\} \right\}$

Of these three numbers, only two lie in the interval $[0, 4]$. We compute the value of the function at these numbers as well as the endpoints of the interval.

```
c1 = 0; c2 = 1; c3 = 1 + √2 ; c4 = 4;

points = {{c1, f[c1]}, {c2, f[c2]}, {c3, f[c3]}, {c4, f[c4]}} //Expand;
```

```
TableForm[points, TableHeadings → {None, {"x", "f[x]"}}]
```

x	f[x]
0	2
1	5
$1 + \sqrt{2}$	1
4	50

```
Max[{f[c1], f[c2], f[c3], f[c4]}] //Expand
```
50

```
Min[{f[c1], f[c2], f[c3], f[c4]}] //Expand
```
1

The absolute maximum of *f* is 50 and the absolute minimum is 1.

EXAMPLE 16   A wire, 100 in. long, is to be used to form a square and a circle. Determine how the wire should be distributed in order for the combined area of the two figures to be (a) as large as possible and (b) as small as possible.

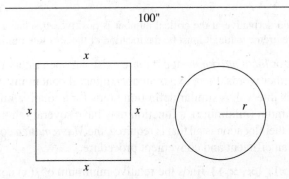

The combined area of the two figures is $A(x) = x^2 + \pi r^2$. The circle has a circumference of $2\pi r$, so it follows that $4x + 2\pi r = 100$. Since the wire is 100 in. long, $0 \leq x \leq 25$.

```
Solve[4 x + 2 π r == 100, r]
```

$$\left\{\left\{r \to -\frac{2(-25 + x)}{\pi}\right\}\right\}$$

```
a[x_] = x² + π r² /. r → - 2 (-25 + x)
                          _____
                               π
```
← Replace r in terms of x.

$$\frac{4(-25 + x)^2}{\pi} + x^2$$

```
Solve[a'[x] == 0]
```
← Find critical value(s).

$$\left\{\left\{x \to \frac{100}{4 + \pi}\right\}\right\}$$

```
x1 = 0;
```

```
x2 = 100 ;
     _____
     4 + π
```

Compute the values of $a(x)$ at these three points. The values are placed in a table with numerical approximations for comparison.

```
x3 = 25;
```

```
points = {{x1, a[x1], N[a[x1]]}, {x2, a[x2], N[a[x2]]},
                {x3, a[x3], N[a[x3]]}} //Together;
TableForm[points, TableAlignments → Center, TableSpacing → {2, 5},
        TableHeadings → {None, {"x", "a[x]", "N[a[x]]"}}]
```

x	a[x]	N[a[x]]
0	$\dfrac{2500}{\pi}$	795.775
$\dfrac{100}{4+\pi}$	$\dfrac{2500}{4+\pi}$	350.062
25	625	625.

The largest combined area occurs when $x = 0$ (all the wire is used to form the circle). The smallest area occurs when one side of the square is $\dfrac{100}{4+\pi}$ (cut the wire $\dfrac{400}{4+\pi}$ from one end). To further confirm that $x = \dfrac{100}{4+\pi}$ gives a minimum area, we can apply the second derivative test.

```
Sign[a''[100/(4+π)]]
```

```
1
```

Since the sign of the second derivative at the critical number is positive, $A(x)$ has a relative minimum at $\dfrac{100}{4+\pi}$. Since this is the only relative extreme value, it must be the location of the absolute minimum.

A function has a *relative* or *local maximum* at c if there exists an open interval, I, containing c such that $f(x) \leq f(c)$ for all x in I. In other words, there exists an open interval containing c such that $f(c)$ is the largest value of f for all x in this interval. A similar definition holds for a relative minimum.

Unlike an absolute maximum (minimum), a function may have several relative maxima (minima). If a numerical approximation of their location is all that is required, the *Mathematica* commands **FindMinimum** and **FindMaximum** offer an efficient and convenient procedure.

- **FindMinimum[f[x], {x, x₀}]** finds the relative minimum of $f(x)$ near x_0.
- **FindMaximum[f[x], {x, x₀}]** finds the relative maximum of $f(x)$ near x_0.

As with **FindRoot**, the options **AccuracyGoal** and **WorkingPrecision** can be set if greater accuracy is desired. In addition, **PrecisionGoal** can be set to determine the precision in the value of the function at the maximum or minimum point. (Precision is the number of significant digits in the answer; accuracy is the number of significant digits to the right of the decimal point.)

EXAMPLE 17   The function $f(x) = x + \sin(5x)$ has three relative maxima and two relative minima in the interval $[0, \pi]$. A quick look at its graph gives good approximations to their locations.

```
f[x_] = x + Sin[5x];
Plot[f[x], {x, 0, π}]
```

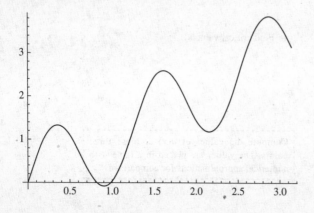

```
FindMinimum[f[x], {x, 1}]
```
$\{-0.0775897, \{x \to 0.902206\}\}$ ← The value of the function comes first,
                                            followed by the value of *x*.
```
FindMinimum[f[x], {x, 2}]
```
$\{1.17905, \{x \to 2.15884\}\}$
```
FindMaximum[f[x], {x, 0.4}]
```
$\{1.33423, \{x \to 0.354431\}\}$
```
FindMaximum[f[x], {x, 1.5}]
```
$\{2.59086, \{x \to 1.61107\}\}$
```
FindMaximum[f[x], {x, 3}]
```
$\{3.8475, \{x \to 2.8677\}\}$

The relative maximum points are (0.354431, 1.33423), (1.61107, 2.59086), and (2.8677, 3.8475). The relative minimum points are (0.902206, – 0.0775897) and (2.15884, 1.17905).

Note: Caution must be taken to examine the results of the calculation. The value obtained is not necessarily the one closest to the initial guess. For example,

```
FindMaximum[f[x], {x, 2.8}]
```
$\{5.10414, \{x \to 4.12434\}\}$

but the value of *x* is not between 0 and π.

SOLVED PROBLEMS

8.16 Find two positive numbers whose sum is 50, such that the square root of the first added to the cube root of the second is as large as possible.

SOLUTION

```
y = 50 - x;
```
$f[x\_] = \sqrt{x} + \sqrt[3]{y};$
```
Plot[f[x],{x, 50}];
```

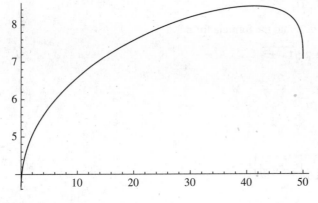

```
NSolve[f'[x] == 0]
```
$\{\{x \to 41.1553\}\}$
```
y /. x → 41.1553
```
8.8447
```
f[41.1553]
```
8.48329

The two numbers are *x* = 41.1553 and *y* = 8.8447. The maximum sum is 8.48329.

8.17 A right circular cylinder is inscribed in a unit sphere.
   (a) Find the largest possible volume.
   (b) Find the largest possible surface area.

SOLUTION

(a) We consider a two-dimensional perspective of the problem. Label the radius and height of the inscribed cylinder r and h, respectively. The volume of the inscribed cylinder is $V = \pi r^2 h$ and, by the Theorem of Pythagoras, $r^2 + \left(\dfrac{h}{2}\right)^2 = 1$. It is easily seen (even without *Mathematica*) that $r^2 = 1 - \left(\dfrac{h}{2}\right)^2$. Thus, the volume, as a function of h, becomes $V(h) = \pi\left[1 - \left(\dfrac{h}{2}\right)^2\right]h$.

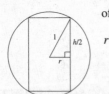

```
v[h_] = π (1 - (h / 2)²) h;
```

```
Solve[v'[h] == 0, h]
```

$$\left\{\left\{h \to -\frac{2}{\sqrt{3}}\right\}, \left\{h \to \frac{2}{\sqrt{3}}\right\}\right\}$$

```
vmax = v[2 / √3]
```

$$\frac{4\pi}{3\sqrt{3}}$$

> Obviously only the positive value of h is appropriate.
>
> Since the sign of the second derivative at the critical point is negative, we have a relative maximum at $2/\sqrt{3}$. Since this is the only relative extremum, it must be the absolute maximum.

```
Sign[v''[2 / √3]]
```

-1

(b) The surface area of the cylinder (including top and bottom) is $S = 2\pi r h + 2\pi r^2$. As in part (a), $r^2 + \left(\dfrac{h}{2}\right)^2 = 1$, but because r and r^2 both appear in the equation for S, it is easier to solve for h in terms of r.

```
Solve[r² + (h / 2)² == 1, h]
```

$$\left\{\left\{h \to -2\sqrt{1-r^2}\right\}, \left\{h \to 2\sqrt{1-r^2}\right\}\right\}$$

Now substitute the (positive) value of h into the formula for s:

```
s[r_] = 2 π r h + 2 π r² /. h → 2 √(1 - r²)
```

$$2\pi r^2 + 4\pi r\sqrt{1-r^2}$$

Solve for the critical value of r:

```
Solve[s'[r] == 0, r]
```

$$\left\{\left\{r \to \sqrt{\tfrac{1}{10}\left(5 + \sqrt{5}\right)}\right\}, \left\{r \to -\sqrt{\tfrac{1}{10}\left(5 - \sqrt{5}\right)}\right\}\right\}$$

Only the positive value of r is acceptable. We use it to compute the maximum surface area.

```
s[√(1/10 (5 + √5))] //Simplify
```

$$\left(1 + \sqrt{5}\right)\pi$$

```
Sign[s''[√(1/10 (5 + √5))]]
```
                                     ← Confirmation of a maximum.

-1

8.18 Find the points on the circle $x^2 + y^2 - 2x - 4y = 0$ closest to and furthest from P(4, 4).

SOLUTION

First we draw a diagram.

```
circle = ContourPlot[x² + y² - 2 x - 4 y == 0, {x, -5, 5}, {y, -2, 5}];
point = Graphics[{PointSize[Medium], Point[{4,4}]}];
Show[circle, point, Frame → False, AspectRatio → Automatic, Axes → True]
```

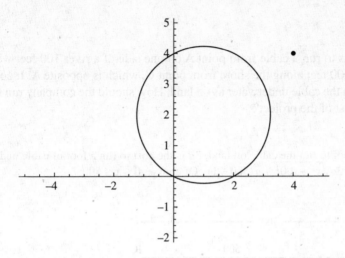

Let (*x*, *y*) represent a point on the circle. First, we need to solve for *y* in terms of *x*.

```
Solve[x² + y² - 2 x - 4 y == 0, y] //Simplify
```

$$\left\{\left\{y \to 2 - \sqrt{4 + 2x - x^2}\right\}, \left\{y \to 2 + \sqrt{4 + 2x - x^2}\right\}\right\}$$

We shall minimize the *square* of the distance from (*x*, *y*) to (4, 4). We call this d2. It is clear from the picture that the point closest to P lies on the upper semicircle.

```
y = 2 + √(4 + 2 x - x²);
d2[x_] = (x - 4)² + (y - 4)²;
Solve[d2'[x] == 0]
```

$$\left\{\left\{x \to \frac{1}{13}\left(13 + 3\sqrt{65}\right)\right\}\right\}$$

```
{x, y} /. x → 1/13(13 + 3 √65) //Simplify
```

$$\left\{1 + 3\sqrt{\frac{5}{13}}, \; 2 + 2\sqrt{\frac{5}{13}}\right\}$$

```
%//N
```

```
{2.86052, 3.24035}
```

The point furthest from P lies on the lower semicircle.

```
y = 2 - √(4 + 2 x - x²);
d2[x_] = (x - 4)² + (y - 4)²;
Solve[d2'[x] == 0]
```

$$\left\{\left\{x \to \frac{1}{13}\left(13 - 3\sqrt{65}\right)\right\}\right\}$$

`{x,y} /. x →` $\frac{1}{13}\left(13 - 3\sqrt{65}\right)$ `//Simplify`

$$\left\{1 - 3\sqrt{\frac{5}{13}},\ 2 - 2\sqrt{\frac{5}{13}}\right\}$$

`%//N`

`{-0.860521, 0.759653}`

8.19 A local telephone company wants to run a cable from point A on one side of a river 100 feet wide to point B on the opposite side, 500 feet along the shore from point C, which is opposite A. It costs three times as much money to run the cable underwater as on land. How should the company run the cable in order to minimize the cost of the project?

SOLUTION

If we let a represent the cost per foot to run the cable on land, $3a$ is the cost to run a foot of cable underwater. The total cost is then $c(x) = 3a\sqrt{x^2 + 100^2} + a(500 - x)$. Of course, $0 \le x \le 500$.

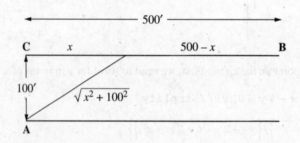

`c[x_]=3 a` $\sqrt{x^2 + 100^2}$ `+ a (500 - x);`

`Solve[c'[x]==0,x]`

$$\left\{\left\{x \to 25\sqrt{2}\right\}\right\}$$

Now we must compare the cost corresponding to this solution with the cost at the endpoints of the interval.

`c[0] //N`

`800. a`

`c[25` $\sqrt{2}$ `] //N`

`782.843 a`

`c[500] //N`

`1529.71 a`

> For minimum cost, run the cable underwater to the point $25\sqrt{2}$ feet from C, then on land to B.

8.4 Power Series

The nicest functions to work with are polynomials. They are continuous and can easily be differentiated and integrated. If a difficult function is encountered in a problem, one approach is to approximate it by a polynomial.

If the value of the function and its derivatives are known at a single point, a, the function can often be represented by a power series. This, however, is usually an infinite series that must be truncated for practical application. The trick is to truncate it in such a way that it accurately approximates the given function, at least in some neighborhood of a.

The following series, known as a Taylor series, gives a representation of an analytic[1] function, $f(x)$. If $a = 0$, the series is known as a Maclaurin series.

$$f(x) = \sum_{k=0}^{\infty} \frac{f^{(k)}(a)}{k!}(x-a)^k$$

$f^{(k)}(a)$ represents the kth derivative of f evaluated at a. If $k = 0$, it represents $f(a)$.

If we truncate this infinite series by omitting all terms of degree greater than n, we obtain the nth degree Taylor polynomial of f about a. We shall represent this polynomial as $p_n(x)$. If $a = 0$, the polynomial is called a Maclaurin polynomial.

EXAMPLE 18   To obtain the Maclaurin polynomial of degree 5 for the function $f(x) = e^x$, we can use the **Sum** command or the Σ symbol from the **Basic Math Input** palette. Here are three different ways the polynomial can be generated:

```
f[x_] = Exp[x];
```
(a) `Sum[(D[f[x], {x, k}]/.x→0)/k! * x^k, {k, 0, 5}]`

$$1 + x + \frac{x^2}{2} + \frac{x^3}{6} + \frac{x^4}{24} + \frac{x^5}{120}$$

(b) $\displaystyle\sum_{k=0}^{5} \frac{\partial_{\{x,k\}} f[x] \ /. \ x \to 0}{k!} x^k$

$$1 + x + \frac{x^2}{2} + \frac{x^3}{6} + \frac{x^4}{24} + \frac{x^5}{120}$$

(c) $\displaystyle\sum_{k=0}^{5} \frac{\text{Derivative}[k][f][0]}{k!} x^k$

$$1 + x + \frac{x^2}{2} + \frac{x^3}{6} + \frac{x^4}{24} + \frac{x^5}{120}$$

Mathematica includes a convenient command for constructing the Taylor polynomial.

- **Series[f[x], {x, a, n}]** generates a **SeriesData** object[2] representing the nth degree Taylor polynomial of $f(x)$ about a.

EXAMPLE 19

```
f[x_] = Exp[x];
Series[f[x], {x, 0, 5}]
```

$$1 + x + \frac{x^2}{2} + \frac{x^3}{6} + \frac{x^4}{24} + \frac{x^5}{120} + O[x]^6$$

The symbol $O[x]^6$ in the above expansion represents the "order" of the omitted terms in the (infinite) expansion. $O[x]^6$ means that the omitted terms have powers of x of degree ≥ 6.

We can see what a **SeriesData** object looks like by using the command **InputForm**.

- **InputForm[*expression*]** prints *expression* in a form suitable for input to *Mathematica*.

---

[1]An analytic function of a real variable is one that has a Taylor series expansion. Most functions encountered in applications are of this type; however, even if a function has derivatives of all orders, it may not be analytic.

[2]A **SeriesData** object is a *representation* of a power series but does not have a numerical value.

EXAMPLE 20

```
f[x_] = Exp[x];
s = Series[f[x], {x, 0, 5}];
InputForm[s]
SeriesData[x, 0, {1, 1, 1/2, 1/6, 1/24, 1/120}, 0, 6, 1]
```

A **SeriesData** object is non-numerical and therefore cannot be evaluated numerically.

EXAMPLE 21

```
f[x_] = Exp[x];
p[x_] = Series[f[x], {x, 0, 5}]
```

$$1 + x + \frac{x^2}{2} + \frac{x^3}{6} + \frac{x^4}{24} + \frac{x^5}{120} + O[x]^6$$

```
p[1]
```

SeriesData::ssdn :

> Attempt to evaluate a series at the number 1. Returning Indeterminate. »

```
Indeterminate
```

In order to convert the series into one that can be evaluated, the function **Normal** can be used to transform it into an ordinary polynomial.

- **Normal [*series*]** returns a polynomial representation of the **SeriesData** object *series*, which can then be evaluated numerically. The $O[x]^n$ term is omitted.

EXAMPLE 22

```
f[x_] = Exp[x];
s = Series[f[x], {x, 0, 5}]
```

$$1 + x + \frac{x^2}{2} + \frac{x^3}{6} + \frac{x^4}{24} + \frac{x^5}{120} + O[x]^6$$

```
p[x_] = Normal[s]
```

$$1 + x + \frac{x^2}{2} + \frac{x^3}{6} + \frac{x^4}{24} + \frac{x^5}{120}$$

> **Normal** has converted the **SeriesData** object into an ordinary polynomial, whose value can now be computed.

```
p[1]
```

$$\frac{163}{60}$$

The number obtained in the previous example, 163/60, is approximately 2.71667. If we compare this to the (known) value of $e \approx 2.71828$, we see a small error in our approximation. We would expect the error to diminish as the degree of the polynomial increases. This is shown to be the case in the next example.

EXAMPLE 23

```
f[x_] = Exp[x];
exactvalue = f[1];
p[n_] := Normal[Series[f[x], {x, 0, n}]] /. x → 1
data = Table[{n, N[p[n]], N[Abs[p[n] - exactvalue]]}, {n, 1, 10}];
```

```
TableForm[data, TableSpacing → {1,10},
        TableHeadings → {None, {"n"," p(1)"," Error"}}]
```

n	p(1)	Error
1	2.	0.718282
2	2.5	0.218282
3	2.66667	0.0516152
4	2.70833	0.0099485
5	2.71667	0.00161516
6	2.71806	0.000226273
7	2.71825	0.0000278602
8	2.71828	3.05862×10^{-6}
9	2.71828	3.02886×10^{-7}
10	2.71828	2.73127×10^{-8}

EXAMPLE 24   To see the convergence of a power series even more dramatically, we can construct an animation showing the sequence of Maclaurin polynomials converging to e^x. We consider the interval [0, 5].

```
f[x_] = Exp[x];
p[n_, x_] := Normal[Series[f[t], {t, 0, n}]] /.t → x
Animate[Plot[{p[n, x], f[x]}, {x, 0, 5},
        PlotRange → {0,Exp[5]}], {n, 1, 10, 1}]
```

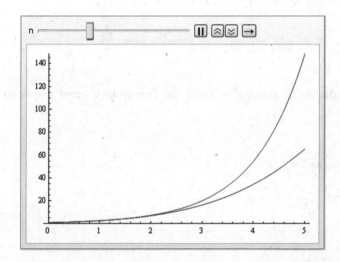

If only the *coefficient* of a particular term of a series is needed, the command **SeriesCoefficient** may be used. The actual series, which may be quite long, need not be printed in its entirety. **SeriesCoefficient** is the **SeriesData** equivalent of **Coefficient** for polynomials.

- **SeriesCoefficient** [*series*, **n**] returns the coefficient of the nth degree term of a **SeriesData** object.

EXAMPLE 25

```
f[x_] = Exp[x];
s = Series[f[x], {x, 0, 10}];
SeriesCoefficient[s, 10]
```

$$\frac{1}{3\,628\,800}$$

SOLVED PROBLEMS

8.20 Obtain the Maclaurin polynomial of degree 10 for the function $f(x) = \tan^{-1}x$ by using a direct summation and then by using the **Series** command.

SOLUTION

```
f[x_] = ArcTan[x];
```

$$\sum_{k=0}^{10} \frac{\text{Derivative}[k][f][0]}{k!}x^k$$

$$x - \frac{x^3}{3} + \frac{x^5}{5} - \frac{x^7}{7} + \frac{x^9}{9}$$

```
Series[f[x], {x, 0, 10}]
```

$$x - \frac{x^3}{3} + \frac{x^5}{5} - \frac{x^7}{7} + \frac{x^9}{9} + O[x]^{11}$$

8.21 Obtain a representation of x^5 in powers of $x - 2$.

SOLUTION

```
Series[x⁵, {x, 2, 5}] //Normal
```

$$32 + 80(-2 + x) + 80(-2 + x)^2 + 40(-2 + x)^3 + 10(-2 + x)^4 + (-2 + x)^5$$

```
%//TraditionalForm
```

$$(x - 2)^5 + 10(x - 2)^4 + 40(x - 2)^3 + 80(x - 2)^2 + 80(x - 2) + 32$$

8.22 Construct a Taylor polynomial of degree 5 about $a = 1$ for the function $\sqrt{x}$ and use it to approximate $\sqrt{3/2}$.

SOLUTION

```
p[x_] = Series[√x, {x,1,5}] //Normal
```

$$1 + \frac{1}{2}(-1 + x) - \frac{1}{8}(-1 + x)^2 + \frac{1}{16}(-1 + x)^3 - \frac{5}{128}(-1 + x)^4 + \frac{7}{256}(-1 + x)^5$$

```
approx = p[3/2] //N
```

```
1.22498
```

```
exact = √3/2 //N
```

```
1.22474
```

```
Abs[% - %%]
```

```
0.000230715     ←This is the absolute error of the approximation.
```

8.23 Let $f(x) = \sin x$ and compute the Maclaurin polynomials of degrees 7, 9, and 11. Then plot $f(x)$ and the three polynomials on one set of axes, $0 \le x \le 2\pi$, and observe their behavior.

SOLUTION

```
f[x_] = Sin[x];
p7[x_] = Series[f[x], {x, 0, 7}] //Normal;
p9[x_] = Series[f[x], {x, 0, 9}] //Normal;
p11[x_] = Series[f[x], {x, 0, 11}] //Normal;
```

```
Plot[{f[x], p7[x], p9[x], p11[x]}, {x, 0, 2π},
      PlotStyle → {Thickness[.01], Thickness[.001],
                   Thickness[.001], Thickness[.001]}]
```

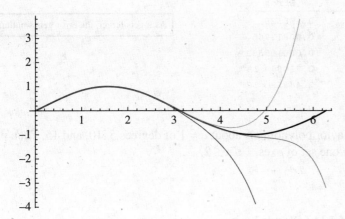

The higher the degree of the polynomial, the better the polynomial
approximates $f(x) = \sin x$.

8.24 Let $f(x) = \sin x$ and compute the Maclaurin polynomial of degree 11. Construct an error function and
compute its value from $x = 0$ to $x = 1$ in increments of 0.1. Place the results in the form of a table and
comment on the values of the error as x gets further from 0.

SOLUTION

```
f[x_] = Sin[x];

p11[x_] = Normal[Series[f[x], {x, 0, 11}]];

error[x_] = Abs[f[x] - p11[x]];

errorvalues = Table[{x, error[x]}, {x, 0, 6, 1.}];

TableForm[errorvalues, TableSpacing → {1,5},
          TableHeadings → {None, {"x"," error[x]"}}]
```

x	error[x]
0.	0.
1.	1.59828×10^{-10}
2.	1.29086×10^{-6}
3.	0.000245414
4.	0.0100021
5.	0.174693
6.	1.78084

As x gets further from 0, the error gets larger.

8.25 Let $f(x) = \sin x$. Construct the Maclaurin polynomials of degrees 1, 3, 5, 7, and 9 and compute their
value at $x = 1$. Determine the error in the approximations and express in a tabular form.

SOLUTION

```
f[x_] = Sin[x];

exactvalue = f[1];

value[n_] := Normal[Series[f[x], {x, 0, n}]] /. x → 1

data = Table[{n, N[value[n]], N[exactvalue],
              N[Abs[value[n] - exactvalue]]}, {n, 1, 9, 2}];
```

```
TableForm[data, TableSpacing → {1,5},
         TableHeadings → {None, {"n"," p(1)"," f(1)"," Error"}}]
```

n	p(1)	f(1)	Error
1	1.	0.841471	0.158529
3	0.833333	0.841471	0.00813765
5	0.841667	0.841471	0.000195682
7	0.841468	0.841471	2.73084×10^{-6}
9	0.841471	0.841471	2.48923×10^{-8}

As n gets larger, the error gets smaller.

8.26 Let $f(x) = \ln x$ and compute the Taylor polynomials about $a = 1$ of degrees 5, 10, and 15. Then plot $f(x)$ and the three polynomials on one set of axes, $1 \le x \le 2$.

SOLUTION

```
f[x_] = Log[x];
p5[x_] = Series[f[x], {x, 1, 5}]//Normal;
p10[x_] = Series[f[x], {x, 1, 10}]//Normal;
p15[x_] = Series[f[x], {x, 1, 15}]//Normal;
Plot[{f[x], p5[x], p10[x], p15[x]}, {x, 1, 2},
     PlotStyle → {Thickness[.01], Thickness[.001],
                  Thickness[.001], Thickness[.001]}]
```

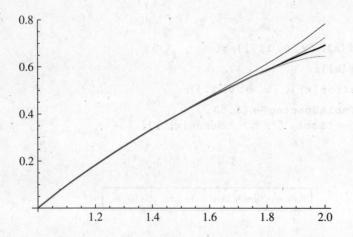

8.27 Let $f(x) = \ln x$ and construct the Taylor polynomial of degree 5 about $a = 1$. Construct an error function and compute its value from $x = 1$ to $x = 2$ in increments of 0.1. Place the results in the form of a table and comment on the values of the error as x gets further from 1.

SOLUTION

```
f[x_] = Log[x];
p5[x_] = Normal[Series[f[x], {x, 1, 5}]];
error[x_] = Abs[f[x] - p5[x]];
errorvalues = Table[{x, error[x]}, {x, 1, 2, .1}];
```

```
TableForm[errorvalues, TableSpacing → {1,5},
        TableHeadings → {None, {" x"," error[x]"}}]
```

x	error[x]
1.	0.
1.1	1.53529×10^{-7}
1.2	9.10987×10^{-6}
1.3	0.0000967355
1.4	0.000509097
1.5	0.00182656
1.6	0.00514837
1.7	0.0122941
1.8	0.026016
1.9	0.0502191
2.	0.0901862

As x gets further from 1, the error gets larger.

8.28 Let $f(x) = \ln x$. Construct the Taylor polynomials of degrees 1, 2, 3, . . . , 10 about $a = 1$ and compute their value at $x = 1.5$. Determine the error in the approximations and express in a tabular form.

SOLUTION

```
f[x_] = Log[x];

exactvalue = f[1.5];

value[n_] := Normal[Series[f[x], {x, 1, n}]] /. x → 1.5

data = Table[{n, N[value[n]], exactvalue,
            N[Abs[value[n] - exactvalue]]}, {n, 1, 10}];

TableForm[data, TableSpacing → {1,5},
        TableHeadings → {None, {"n"," pₙ(1.5)"," f(1.5)"," Error"}}]
```

n	$P_n(1.5)$	f(1.5)	Error
1	0.5	0.405465	0.0945349
2	0.375	0.405465	0.0304651
3	0.416667	0.405465	0.0112016
4	0.401042	0.405465	0.00442344
5	0.407292	0.405465	0.00182656
6	0.404688	0.405465	0.000777608
7	0.405804	0.405465	0.000338463
8	0.405315	0.405465	0.000149818
9	0.405532	0.405465	0.000067196
10	0.405435	0.405465	0.0000304603

8.29 What is the coefficient of the x^{20} term of the Maclaurin series for $\sin(x^2 + 1)$?

SOLUTION

```
s = Series[Sin[x² + 1], {x, 0, 20}];

SeriesCoefficient[s, 20]
```

$$-\frac{\sin[1]}{3\,628\,800}$$

CHAPTER 9

Integral Calculus

9.1 Antiderivatives

An antiderivative of a function f is another function F such that $F'(x) = f(x)$. In *Mathematica*, the **Integrate** command computes antiderivatives. You will notice, however, that the constant of integration, C, is omitted from the answer.

- **Integrate[f[x], x]** computes the antiderivative (indefinite integral) $\int f(x)\,dx$. The symbol $\int \square\, d\,\square$ from the Basic Math Input palette may also be used.

Mathematica can compute antiderivatives of elementary integrals found in standard tables, but if unable to evaluate an antiderivative in terms of elementary functions, the software will try to express the antiderivative in terms of special functions. If this is not possible, *Mathematica* returns the antiderivative unevaluated.

EXAMPLE 1

$\int x^2\, \text{Exp}[x]\, \text{Sin}[x]\, dx$ or **Integrate[x^2 Exp[x] Sin[x], x]**

$\frac{1}{2}\, e^x\, (-(-1+x)^2\, \text{Cos}[x] + (-1+x^2)\, \text{Sin}[x])$

EXAMPLE 2

$\int \text{Sin}[x^2]\, dx$ or **Integrate[Sin[x^2], x]**

$\sqrt{\frac{\pi}{2}}\, \text{Fresnel}\left[\sqrt{\frac{2}{\pi}}\, x\right]$

> This integral has no simple antiderivative, so *Mathematica* expresses it as a Fresnel sine integral: $\text{FresnelS}(x) = \int_0^x \sin\left(\pi \frac{t^2}{2}\right) dt$

EXAMPLE 3

$\int \text{Sin}[\text{Sin}[x]]\, dx$ or **Integrate[Sin[Sin[x]], x]**

$\int \text{Sin}[\text{Sin}[x]]\, dx$ ← *Mathematica* cannot evaluate this antiderivative.

Care must be taken when *general* antiderivatives involving parameters are requested.

EXAMPLE 4

$\int x^n\, dx$

$\frac{x^{1+n}}{1+n}$

Of course, this result is valid only if n ≠ −1, but if the value of n is specified, *Mathematica* knows what to do.

```
n = -1;
∫xⁿ dx
Log[x]
```

SOLVED PROBLEMS

9.1 Compute $\int \sqrt{x}\, dx$.

SOLUTION

$$\int \sqrt{x}\, dx \quad \text{or} \quad \text{Integrate}\left[\sqrt{x}, x\right]$$

$$\frac{2x^{3/2}}{3}$$

9.2 Compute $\int \sqrt{a^2 + x^2}\, dx$

SOLUTION

$$\int \sqrt{a^2+x^2}\, dx \quad \text{or} \quad \text{Integrate}\left[\sqrt{a^2+x^2}, x\right]$$

$$\frac{1}{2}x\sqrt{a^2+x^2} + \frac{1}{2}a^2 \text{Log}\left[2\left(x+\sqrt{a^2+x^2}\right)\right]$$

9.3 Compute $\int \dfrac{1}{\sqrt{u^2-a^2}}\, du$

SOLUTION

$$\int \frac{1}{\sqrt{u^2-a^2}}\, du \quad \text{or} \quad \text{Integrate[1/Sqrt[u}^2 - a^2\text{], u]}$$

$$\text{Log}\left[2\left(u+\sqrt{-a^2+u^2}\right)\right]$$

9.4 Compute $\int \tanh x\, dx$.

SOLUTION

$$\int \text{Tanh[x]dx} \quad \text{or} \quad \text{Integrate[Tanh[x], x]}$$
Log[Cosh[x]]

9.5 Evaluate (a) $\int f'(x)dx$ and (b) $\int g'(f(x))f'(x)dx$.

SOLUTION

(a) $\int$f'[x] dx   (b) $\int$g'[f[x]]f'[x]dx
    f[x]   g[f[x]]

9.6 Construct a table of integrals for $\int \sin^n x\, dx \quad n = 1, 2, 3, \ldots, 10$.

SOLUTION

```
anti[n_]:= ∫ Sin[x]ⁿ dx
tablevalues = Table[{n, Together[anti[n]]}, {n, 1, 10}];
TableForm[tablevalues, TableSpacing → {1, 5},
        TableHeadings → {None, {"n", "∫Sinⁿx dx"}}]
```

n	$\int \mathrm{Sin}^n[x]\,dx$
1	$-\mathrm{Cos}[x]$
2	$\frac{1}{4}(2x - \mathrm{Sin}[2x])$
3	$\frac{1}{12}(-9\,\mathrm{Cos}[x] + \mathrm{Cos}[3x])$
4	$\frac{1}{32}(12x - 8\,\mathrm{Sin}[2x] + \mathrm{Sin}[4x])$
5	$\frac{1}{240}(-150\,\mathrm{Cos}[x] + 25\,\mathrm{Cos}[3x] - 3\,\mathrm{Cos}[5x])$
6	$\frac{1}{192}(60x - 45\,\mathrm{Sin}[2x] + 9\,\mathrm{Sin}[4x] - \mathrm{Sin}[6x])$
7	$\frac{-1225\,\mathrm{Cos}[x] + 245\,\mathrm{Cos}[3x] - 49\,\mathrm{Cos}[5x] + 5\,\mathrm{Cos}[7x]}{2240}$
8	$\frac{840x - 672\,\mathrm{Sin}[2x] + 168\,\mathrm{Sin}[4x] - 32\,\mathrm{Sin}[6x] + 3\,\mathrm{Sin}[8x]}{3072}$
9	$\frac{-39\,690\,\mathrm{Cos}[x] + 8820\,\mathrm{Cos}[3x] - 2268\,\mathrm{Cos}[5x] + 405\,\mathrm{Cos}[7x] - 35\,\mathrm{Cos}[9x]}{80\,640}$
10	$\frac{2520x - 2100\,\mathrm{Sin}[2x] + 600\,\mathrm{Sin}[4x] - 150\,\mathrm{Sin}[6x] + 25\,\mathrm{Sin}[8x] - 2\,\mathrm{Sin}[10x]}{10\,240}$

9.7 Use **Manipulate** to evaluate $\int \sin^n x\,dx$ for $1 \le n \le 10$.

SOLUTION

```
Manipulate[ ∫Sin[x]ⁿ dx //Together, {n, 1, 10, 1}, ControlType → RadioButton]
```

```
n  ○1 ○2 ○3 ○4 ○5 ◉6 ○7 ○8 ○9 ○10

 1
---  (60 x - 45 Sin[2 x] + 9 Sin[4 x] - Sin[6 x])
192
```

9.2 Definite Integrals

A definite integral can be computed one of two ways: exactly, using the Fundamental Theorem of Calculus, or approximately, using numerical methods. You can instruct *Mathematica* which method you wish to use by choosing from two commands.

- **Integrate[f[x], {x, a, b}]** computes, whenever possible, the exact value of $\int_a^b f(x)\,dx$. The symbol $\int_\square^\square \square\,d\square$ on the Basic Math Input palette may be used as well.
- **NIntegrate[f[x], {x, a, b}]** computes an approximation to the value of $\int_a^b f(x)\,dx$ using strictly numerical methods.

NIntegrate evaluates the integral using an adaptive algorithm, subdividing the interval of integration until a desired degree of accuracy is achieved. The interval is divided recursively until the value of **AccuracyGoal** or **PrecisionGoal** is achieved.

- **AccuracyGoal** is an option that specifies how many digits to the right of the decimal point should be sought in the final result. **AccuracyGoal** effectively specifies the absolute error. The default for **NIntegrate** is **AccuracyGoal → Infinity**, which specifies that accuracy should *not* be used as the criterion for terminating the numerical procedure.
- **WorkingPrecision** is an option that specifies how many digits of precision should be maintained in internal computations. The default value is approximately 16.
- **PrecisionGoal** is an option that effectively specifies the relative error. The default setting, **PrecisionGoal → Automatic**, sets **PrecisionGoal** to half the value of **WorkingPrecision**. If defaults are not used, you should set **PrecisionGoal** to be less than the value of **WorkingPrecision**.

Other options, which control more precisely how the algorithm should be implemented, are available, but will not be discussed here. These options are useful for integrals involving "pathological" functions such as $\int_{.0001}^{1} \sin\left(\frac{1}{x}\right) dx$ or $\int_{-1000}^{1000} e^{-x^2} dx$. The interested reader should consult the *Mathematica* Documentation Center for details.

The sequence $\mathtt{N[Integrate[f[x], \{x, a, b\}]]}$ or $\int_a^b \mathtt{f[x]dx}$ $\mathtt{//N}$ evaluates the integral, whenever possible, by first finding the antiderivative and then using the Fundamental Theorem of Calculus. If this is impossible, $\mathtt{NIntegrate[f[x], \{x, a, b\}]}$ is called automatically.

EXAMPLE 5   To evaluate $\int_0^1 x e^x \sin x \, dx$ we input

$\mathtt{Integrate[x\,Exp[x]\,Sin[x], \{x, 0, 1\}]}$

$\frac{1}{2}(-1 + e\,\mathtt{Sin[1]})$

As an alternate representation, we can use the Basic Math Input palette.

$\int_0^1 \mathtt{x\,e^x\,Sin[x]\,dx}$

$\frac{1}{2}(-1 + e\,\mathtt{Sin[1]})$

If a numerical approximation is desired, we can type

$\int_0^1 \mathtt{x\,e^x\,Sin[x]\,dx}$ $\mathtt{//N}$ or $\mathtt{Integrate[x\,Exp[x]\,Sin[x], \{x, 0, 1\}]}$ $\mathtt{//N}$

0.643678

Here, the antiderivative of the function $x e^x \sin x$ was computed and then evaluated from 0 to 1. If a strictly numerical procedure is preferred, we can use $\mathtt{Nintegrate}$.

$\mathtt{NIntegrate[x\,Exp[x]\,Sin[x], \{x, 0, 1\}]}$

0.643678

EXAMPLE 6   Obtain an approximation to $\int_0^1 \sin(\sin x) \, dx$ accurate to (a) 6 significant digits and (b) 20 significant digits.

(a) $\int_0^1 \mathtt{Sin[Sin[x]]dx}$ $\mathtt{//N}$
0.430606

(b) $\mathtt{N}\left[\int_0^1 \mathtt{Sin[Sin[x]]dx}, 20\right]$
0.43060610312069060491

> *Mathematica* automatically adjusts $\mathtt{WorkingPrecision}$ and $\mathtt{PrecisionGoal}$ to achieve the desired result.

Mathematica can handle certain *improper* integrals. An improper integral of type I is an integral with one or two infinite limits of integration. We define $\int_a^\infty f(x) \, dx = \lim_{t \to \infty} \int_a^t f(x) \, dx$ and $\int_{-\infty}^b f(x) \, dx = \lim_{t \to -\infty} \int_t^b f(x) \, dx$ provided the limits exist. Such an integral is said to be *convergent*. If both $\int_{-\infty}^a f(x) \, dx$ and $\int_a^\infty f(x) \, dx$ converge, we define $\int_{-\infty}^\infty f(x) \, dx = \int_{-\infty}^a f(x) \, dx + \int_a^\infty f(x) \, dx$.

EXAMPLE 7

$\int_0^\infty \mathtt{e^{-x}dx}$
1

EXAMPLE 8

$\int_0^\infty \mathtt{x\,dx}$ ← This integral is divergent.

$\mathtt{Integrate::idiv : Integral\ of\ x\ does\ not\ converge\ on\ \{0, \infty\}. \gg}$

$\int_0^\infty \mathtt{x\,dx}$

EXAMPLE 9

$$\int_{-\infty}^{\infty} \frac{1}{1+x^2}\, dx$$

π

The value of a type I improper integral may depend upon the values of parameters within the integrand. The option **Assumptions** allows the specification of conditions to be imposed upon these parameters.

- **Assumptions** $\rightarrow$ *conditions* specifies *conditions* to be applied to parameters within the integral.

EXAMPLE 10 $\int_{1}^{\infty} x^n\, dx$ converges if $n < -1$ and diverges otherwise.

Integrate[x^n, {x, 1, ∞}, Assumptions $\rightarrow$ n < -1]

$-\dfrac{1}{1+n}$

Integrate[x^n, {x, 1, ∞}, Assumptions $\rightarrow$ n $\geq$ -1]

Integrate::idiv : Integral of x^n does not converge on {1, ∞}. »

Integrate[x^n, {x, 1, ∞}, Assumptions n $\geq$ -1]

An improper integral of type II is an integral whose integrand is discontinuous on the interval of integration. If f is continuous on $[a, b)$ but not at b, we define $\int_{a}^{b} f(x)\, dx = \lim_{t \to b^-} \int_{a}^{t} f(x)\, dx$, and if f is continuous on $(a, b]$ but not at a we define $\int_{a}^{b} f(x)\, dx = \lim_{t \to a^+} \int_{t}^{b} f(x)\, dx$. If the limit exists, we say the integral is *convergent*. If f has a discontinuity at $c \in (a, b)$ and both $\int_{a}^{c} f(x)\, dx$ and $\int_{c}^{b} f(x)\, dx$ are convergent, then $\int_{a}^{b} f(x)\, dx = \int_{a}^{c} f(x)\, dx + \int_{c}^{b} f(x)\, dx$.

EXAMPLE 11

$$\int_{0}^{1} \text{Log}[x]\, dx$$

-1

EXAMPLE 12

$\int_{-2}^{3} \frac{1}{x}\, dx$ or   **Integrate[1/x, {x, -2, 3}]**

Integrate::idiv : Integral of $\frac{1}{x}$ does not converge on {-2, 3}. »

$\int_{-2}^{3} \frac{1}{x}\, dx$

Because of the discontinuity at 0, the integral of Example 12 is improper. If we break up the integral into the sum of two integrals, $\int_{-2}^{0} \frac{dx}{x} + \int_{0}^{3} \frac{dx}{x}$, each integral, evaluated separately, diverges. However, if we consider the limits simultaneously,

$$\lim_{t \to 0^+} \left[\int_{-2}^{-t} \frac{dx}{x} + \int_{t}^{3} \frac{dx}{x} \right] = \lim_{t \to 0^+} \left[\ln |x| \Big|_{-2}^{-t} + \ln x \Big|_{t}^{3} \right]$$

$$= \lim_{t \to 0^+} \left[\ln t - \ln 2 + \ln 3 - \ln t \right]$$

$$= \ln 3 - \ln 2$$

$$= \ln \frac{3}{2}$$

This number is called the *Cauchy Principal Value*. The option **PrincipalValue** instructs **Integrate** to compute the Cauchy Principal Value of an integral.

- **PrincipalValue** $\rightarrow$ **True** specifies that the Cauchy Principal Value of an integral is to be determined.

EXAMPLE 13

```
Integrate[1/x, {x, -2, 3}, PrincipalValue→True]
```
$\text{Log}\left[\dfrac{3}{2}\right]$ ← Compare with the result of Example 12.

SOLVED PROBLEMS

9.8 Compute the area bounded by the curves $f(x) = 1 - x^2$ and $g(x) = x^4 - 3x^2$.

SOLUTION

```
f[x_] = 1 - x²;
g[x_] = x⁴ - 3x²;
Plot[{f[x], g[x]}, {x, -2, 2}]
```

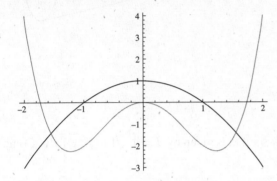

First we must find the points of intersection of the two curves.

```
intersectionpoints = Solve[f[x] == g[x]]
```
$$\left\{\left\{x \to -i\,\sqrt{-1+\sqrt{2}}\right\}, \left\{x \to i\,\sqrt{-1+\sqrt{2}}\right\}, \left\{x \to -\sqrt{1+\sqrt{2}}\right\}, \left\{x \to \sqrt{1+\sqrt{2}}\right\}\right\}$$

```
{a, b, c, d} = x /. intersectionpoints
```
$$\left\{-i\,\sqrt{-1+\sqrt{2}},\ i\,\sqrt{-1+\sqrt{2}},\ -\sqrt{1+\sqrt{2}},\ \sqrt{1+\sqrt{2}}\right\}$$

The points of intersection correspond to the real solutions of this equation c and d.

$\displaystyle\int_{c}^{d}(f[x] - g[x])\,dx$ //Simplify

$\dfrac{8}{15}\,\sqrt{1+\sqrt{2}}\,(4+\sqrt{2})$

```
% //N
```
```
4.48665
```

9.9 The volume of the solid of revolution obtained by rotating about the *x*-axis the area bounded by the curve $y = f(x)$, the *x*-axis, and the lines $x = a$ and $x = b$ is $\pi \displaystyle\int_{a}^{b}[f(x)]^2\,dx$. Compute the volume of the sphere obtained if the semicircle $y = \sqrt{r^2 - x^2}$, $-r \le x \le r$, is rotated about the *x*-axis.

SOLUTION

$y = \sqrt{r^2 - x^2}$

$\pi \displaystyle\int_{-r}^{r} y^2\,dx$

$\dfrac{4\,\pi\,r^3}{3}$

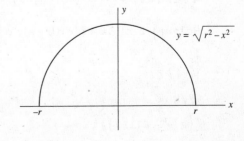

9.10 Compute the volume of a frustum of a cone with height h and radii r and R, and use this to derive the formula for the volume of a cone of radius R and height h.

SOLUTION

Position the frustum as shown in the diagram. The frustum is generated by rotating about the x-axis the region bounded by the line segment connecting $(0, r)$ and (h, R), the x-axis, and the vertical lines $x = r$ and $x = R$. The equation of the line segment is $y = \dfrac{R-r}{h}x + r$. The volume is $\pi \int_0^h y^2 dx$.

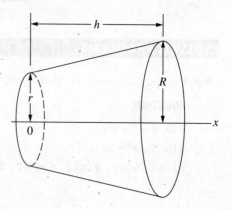

```
y = R - r
    ----- x + r
      h
```

```
    h
π  ∫  y² dx
    0
```

$$\tfrac{1}{3} h \pi (r^2 + r R + R^2)$$

```
% /. r → 0                    ← Let r = 0 for a cone.
```

$$\tfrac{1}{3} h \pi R^2$$

9.11 The arc length of a curve represented by $f(x)$, $a \le x \le b$, is given by $L = \int_a^b \sqrt{1 + [f'(x)]^2}\, dx$. Compute the length of arc of one "arch" of a sine curve.

SOLUTION

One arch of the curve is generated for $0 \le x \le \pi$.

```
f[x_] = Sin[x];
```

```
Integrate[Sqrt[1 + f'[x]^2], {x, 0, Pi}]    or    ∫₀^π √(1 + f'[x]²) dx
```

$$2 \sqrt{2} \; \text{EllipticE}\left[\tfrac{1}{2}\right]$$

```
% //N
```

```
3.8202
```

> *Mathematica* returns the value of the integral as a complete elliptic integral of the second kind, represented by `EllipticE[x]`. We easily obtain a numerical approximation to the arc length.

9.12 The Mean Value Theorem for integrals says that if f is continuous on a closed bounded interval $[a, b]$, there exists a number, c, between a and b, such that $\int_a^b f(x)\, dx = f(c)(b - a)$. Find the value of c that satisfies the mean value theorem for $f(x) = \ln x$ on the interval $[1, 2]$.

SOLUTION

```
f[x_] = Log[x];
```

```
a = 1; b = 2;
```

```
Solve[∫ₐ^b f[x] dx == f[c](b - a), c] //Simplify
```

$$\left\{\left\{c \to \tfrac{4}{e}\right\}\right\}$$

```
% //N
```

```
{{c → 1.47152}}
```

To get a visualization of the Mean Value Theorem for integrals, consider the following plot. Observe that the area below the curve, above the *x*-axis, is equal to the area enclosed by the rectangle determined by c.

```
c = 4/E
g1 = Plot[{f[x],f[c]}, {x, a, b},
      Ticks → {{1, 1.2, 1.4, 1.6, 1.8, 2.0,{c, "c"}}, Automatic}]
g2 = Graphics[Line[{{2, 0}, {2, f[2]}}]];
g3 = Graphics[{Dashed, Line[{{c, 0}, {c, f[c]}}]}];
Show[g1, g2, g3]
```

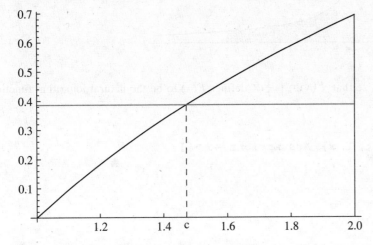

The area below the curve, above the *x*-axis, is equal to the area enclosed by the rectangle.

9.13 The work done in moving an object from *a* to *b* by a variable force, *f*(*x*), is $\int_a^b f(x)\,dx$. According to Hooke's law, the force required to hold a spring stretched beyond its natural length is directly proportional to the displaced distance. If the natural length of a spring is 10 cm, and the force that is required to hold the spring 5 cm beyond this length is 40 Newtons, how much work is done in stretching the spring from 10 to 15 cm?

SOLUTION

Hooke's law states that $f(x) = kx$ where *x* represents the distance beyond the spring's natural length. Since a force of 40 Newtons is required to hold the spring 5 cm (0.05 m) beyond its natural length, $40 = 0.05k$.

```
k = 40/0.05;
f[x_] = k x;
work = ∫₀^.05 f[x] dx
1.                              ← The work done is 1 Joule.
```

9.3 Functions Defined by Integrals

If *f* is continuous on [*a*, *b*], we can define a new function:

$$F(x) = \int_a^x f(t)\,dt$$

Intuitively, if $f(t) \geq 0$, $F(x)$ represents the area bounded by $f(t)$ and the *t*-axis from *a* to *x*, if $x \geq a$, and the negative of this area if $x < a$. The (second) Fundamental Theorem of Calculus tells us that *F* is differentiable on (*a*, *b*) and $F'(x) = f(x)$ for all $x \, \varepsilon \, (a, b)$.

EXAMPLE 14   Let $f(x) = 1/x$, $x > 0$. The shaded area in the diagram represents $F(x)$, assuming $x \geq 1$.

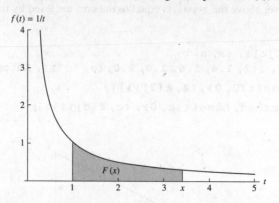

Students of calculus will recognize that $F(x) = \int_1^x \frac{1}{t}\,dt$ defines $F(x)$ to be the natural logarithm function. *Mathematica* knows this also.

```
f[x_] = 1/x;
F[x_] = Integrate[f[t],{t, 1, x}, Assumptions →x>0];
F[2]
Log[2]
F[1/2]
-Log[2]
```

EXAMPLE 15   The continuous function $f(x) = x^x$ has an antiderivative, but it cannot be put into "closed form" in terms of elementary functions. However, *Mathematica* can deal with it as a function defined by an integral. Since all antiderivatives of $f(x)$ differ by a constant, we define $F(x)$ to be the antiderivative for which $F(0) = 0$. Let us plot this antiderivative for $0 \leq x \leq 4$.

```
f[x_] = x^x;
F[x_] = ∫₀ˣ f[t]dt;          ← By making the lower limit 0, we force F(0) = 0.
Plot[F[x], {x, 0, 4}]
```

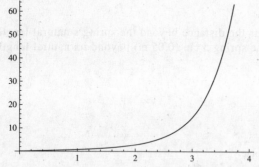

SOLVED PROBLEMS

9.14   Let $F(x) = \int_1^x e^{\sin t}\,dt$. Find $F'(x)$.

SOLUTION

```
F[x_] = ∫₁ˣ Exp[Sin[t]]dt;
F'[x]
e^Sin[x]
```

> This is in accordance with the Second Fundamental Theorem of Calculus.

9.15 Sketch, on one set of axes, the graphs of the *three* antiderivatives of $f(x) = e^{\sin x}$, $0 \le x \le 2\pi$, for which $F(0) = 0$, $F(1) = 0$, and $F(2) = 0$.

SOLUTION

Because of the complicated nature of $f(x)$, it is faster to use **NIntegrate**.

```
f[x_] = Exp[Sin[x]];
F1[x_] := NIntegrate[f[t], {t, 0, x}]
F2[x_] := NIntegrate[f[t], {t, 1, x}]
F3[x_] := NIntegrate[f[t], {t, 2, x}]
Plot[{F1[x], F2[x], F3[x]}, {x, 0, 2π}]
```

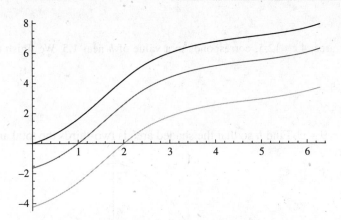

9.16 Consider the semicircle $x^2 + y^2 = 16$, $y \ge 0$ shown in the figure. Find the height, h, so that the shaded area is half the area of the semicircle.

SOLUTION

We solve for x as a function of y:

```
Solve[x² + y² == 16, x]
```

$$\left\{\left\{x \to -\sqrt{16 - y^2}\right\}, \left\{x \to \sqrt{16 - y^2}\right\}\right\}$$

By subdividing the y-axis and taking advantage of symmetry, we obtain the following representation for $A(h)$, the shaded area:

$$A(h) = 2\int_0^h x(y)\,dy \quad \text{where} \quad x(y) = \sqrt{16 - y^2}$$

```
x[y_] = √(16 - y²);

A[h_] = 2∫₀ʰ x[y] dy;
```

Compute the total area inside the semicircle. (Since we know a formula for the area of a circle, this is a good check for errors.)

```
A[4]
```

8π

To approximate the solution, draw a graph of $A(h)$:

```
Plot[A[h], {h, 0, 4}]
```

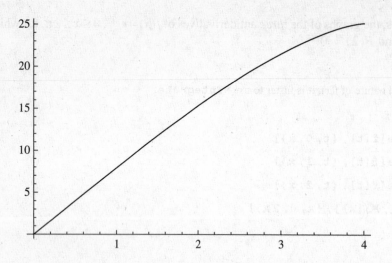

It appears that half the semicircular area, $4\pi \approx 12.5$, corresponds to a value of h near 1.5. We finish the job with **FindRoot**.

```
FindRoot[A[h] == 4π, {h, 1.5}]
```
$\{h \to 1.61589\}$

9.17 The curve shown is the parabola $y = 9 - x^2$. Find h so that the shaded area is two-thirds the total area bounded by the curve and the x-axis.

SOLUTION

```
Solve[y == 9 - x², x]
```
$\left\{\left\{x \to -\sqrt{9-y}\right\}, \left\{x \to \sqrt{9-y}\right\}\right\}$

```
x[y_] = √(9 - y);
```

$$A[h\_] = 2\int_0^h x[y]\,dy;$$

```
totalarea = A[9]
```
36

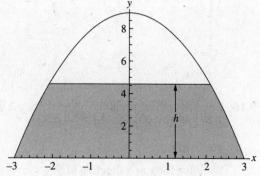

```
Plot[A[h], {h, 0, 9}, AxesLabel → {"h", "A(h)"}]
```

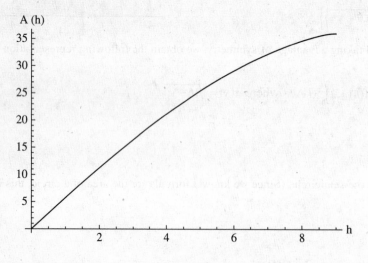

Two-thirds of the total area of 36 is 24 and appears to correspond to a value of h near 5.

```
FindRoot[A[h] == (2/3) totalarea, {h, 5}]
```
$\{h \to 4.67325\}$

9.18 Find a point on the parabola $y = x^2$ which is five units away from the origin along the curve.

SOLUTION

The length of arc of a function, $f(x)$, from $x = a$ to $x = b$ is $L = \int_a^b \sqrt{1 + [f'(x)]^2}\, dx$. Obviously there are two points. We shall find the point that lies in the first quadrant.

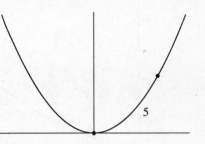

```
f[x_] = x²;

s[x_] = ∫₀ˣ √(1+f'[t]²) dt;

Plot[s[x], {x, 0, 3}, AxesLabel → {"x", "s(x)"}]
```

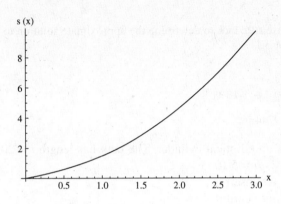

The graph shows $s(2) \approx 5$.

```
solution = FindRoot[s[x] == 5, {x, 2}]

{x→2.08401}

x = x /. solution;

{x, f[x]}

{2.08401, 4.34308}
```

9.19 A mixing bowl is a hemisphere of radius 5 in. Determine the height of 100 cubic inches of liquid.

SOLUTION

The equation of the hemisphere in three dimensions is $x^2 + y^2 + z^2 = 25$, $z \le 0$. Its intersection with the plane $z = z_0$ is the circle $x^2 + y^2 = 25 - z_0^2$, whose radius $r = \sqrt{25 - z_0^2}$ and whose area $\pi r^2 = \pi (25 - z_0^2)$. Integrating with respect to z, the volume of the shaded region is $V(h) = \pi \int_{-5}^{-5+h} (25 - z^2)\, dz$. ($z_0$ has been replaced by z for convenience.)

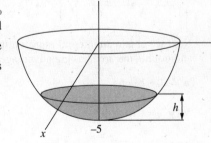

```
v[h_] = π ∫₋₅⁻⁵⁺ʰ (25 - z²) dz

(5h² - h³/3) π
```

As a check, $v[5]$ should give the volume of the hemisphere. The volume of the hemisphere is $\frac{2}{3}\pi r^3 = \frac{2}{3}\pi(5^3) = \frac{250\pi}{3}$.

```
v[5]

250π
───
 3
```

Plot v as a function of h.

```
Plot[v[h], {h, 0, 5}, AxesLabel → {"h", "v[h]"}]
```

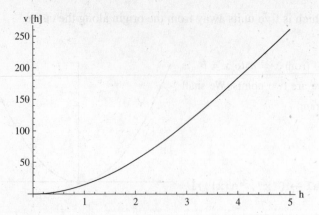

Since v[h] is a polynomial function, we can use **NSolve** to determine the approximate solution to the problem. (From the graph, it looks like *h* is near 3.)

```
NSolve[v[h] == 100]
```

$\{\{h \to -2.34629\}, \{h \to 2.79744\}, \{h \to 14.5489\}\}$

Obviously, the only realistic solution is *h* = 2.79744 in.

9.20 An underground fuel tank is in the shape of an elliptical cylinder. The tank has length *l* = 20 ft, semi-major axis *a* = 10 ft and semi-minor axis *b* = 5 ft. To measure the amount of fuel in the tank, we insert a stick vertically through the center of the cylinder until it touches the bottom of the tank and measure how high the fuel level is on the stick. How far from the end of the stick should a mark be placed to indicate that only 500 cubic feet of fuel remain?

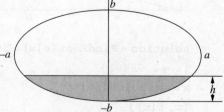

Cross-section of fuel tank.

SOLUTION

The equation of the ellipse is $\frac{x^2}{a^2} + \frac{y^2}{b^2} = 1$. We first want to define *x* as a function of *y*.

```
Solve[x²/a² + y²/b² == 1, x]
```

$\left\{\left\{x \to -\sqrt{a^2 - \frac{a^2 y^2}{b^2}}\right\}, \left\{x \to \sqrt{a^2 - \frac{a^2 y^2}{b^2}}\right\}\right\}$

Next we obtain an integral representing the cross-sectional area of the tank. We take the positive solution and double the area, taking advantage of symmetry.

```
a = 10; b = 5;
```

$$x[y\_] = \sqrt{a^2 - \frac{a^2 y^2}{b^2}};$$

$$area[h\_] = 2 \int_{-b}^{-b+h} x[y] \, dy;$$

As a check, we can compute area[0], area[b], and area[2b]. The area enclosed by the ellipse $\frac{x^2}{a^2} + \frac{y^2}{b^2} = 1$ is πab.

```
area[0]
```
0

```
area[b]
```
25π

```
area[2b]
```
50π

Since the tank has a uniform cross-section, its volume = length × cross-sectional area.

```
length = 20;
volume[h_] = length * area[h];
```

To approximate the location on the stick that corresponds to 500 cubic feet, we draw the graph of volume [h]. Then we use **FindRoot** to obtain a more accurate value. (We cannot use **NSolve**, as in the previous problem, because volume [h] is a non-algebraic function.)

```
Plot[volume[h], {h, 0, 2b}, AxesLabel → {"h", "volume"}]
```

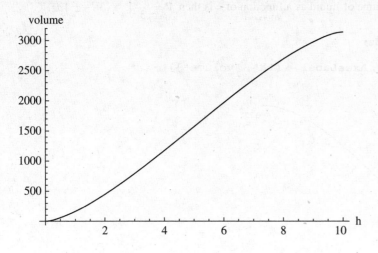

From the graph we observe that volume = 500 when h is near 2.

```
FindRoot[volume[h] == 500, {h, 2}]
```

$\{h \rightarrow 2.1623\}$

9.21 An underground fuel tank is in the shape of an ellipsoid with semi-axes 6, 10, and 6 ft. (This problem, although more difficult than the previous problem, is somewhat more realistic.) To measure the amount of fuel in the tank, we insert a stick vertically through the center of the ellipsoid until it touches the bottom of the tank and measure how high the fuel level is on the stick. How far from the end of the stick should a mark be placed to indicate that only 500 cubic feet of fuel remain?

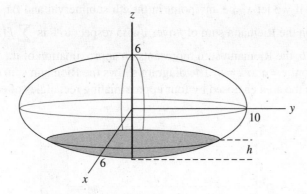

SOLUTION

The equation of this ellipsoid is $\frac{x^2}{a^2} + \frac{y^2}{b^2} + \frac{z^2}{c^2} = 1$ with $a = 6$, $b = 10$, and $c = 6$. The intersection of the ellipsoid with the plane $z = z_0$ is the ellipse $\frac{x^2}{a^2} + \frac{y^2}{b^2} = 1 - \frac{z_0^2}{c^2}$ whose area can be computed as a function of z_0. We then integrate with respect to z to obtain the volume.

To determine the area of the ellipse, we take advantage of the fact that the area enclosed by an ellipse is π times the product of its semi-major and semi-minor axes. If we re-write $\dfrac{x^2}{a^2}+\dfrac{y^2}{b^2}=1-\dfrac{z_0^2}{c^2}$ in the equivalent

form $\dfrac{x^2}{a^2\left(1-\dfrac{z_0^2}{c^2}\right)}+\dfrac{y^2}{b^2\left(1-\dfrac{z_0^2}{c^2}\right)}=1$, we see that the semi-axes are $a\sqrt{1-\dfrac{z_0^2}{c^2}}$ and $b\sqrt{1-\dfrac{z_0^2}{c^2}}$. The elliptical

area is then $\pi ab\left(1-\dfrac{z_0^2}{c^2}\right)$. For our values of a, b, and c, this becomes $60\,\pi\left(1-\dfrac{z_0^2}{6^2}\right)=\dfrac{5\pi}{3}\left(36-z_0^2\right)$. The

integral representing the volume of liquid as a function of h is then $V=\dfrac{5\pi}{3}\displaystyle\int_{-6}^{-6+h}\left(36-z^2\right)dz$.

```
v[h_]= 5π/3 ∫_{-6}^{-6+h} (36-z²) dz;

Plot [v[h], {h, 0, 12}, AxesLabel → {"h", "volume"}]
```

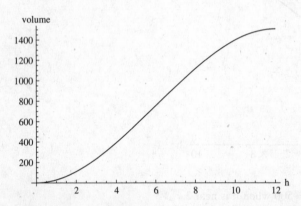

```
NSolve[v[h] == 500]
{{h → -3.63858}, {h → 4.62871}, {h → 17.0099}}
```

The mark should be placed approximately 4.62871 ft from the end of the stick. The other solutions are extraneous.

9.4   Riemann Sums

A partition, P, of the interval I = [a, b] is a collection of subintervals,

$$[x_0, x_1], [x_1, x_2], \ldots, [x_{n-1}, x_n]$$

where $x_0 = a$ and $x_n = b$. If we let x_i^* be any point in the ith subinterval and $\Delta x_i = x_i - x_{i-1}$ be the length of the ith subinterval, then the Riemann sum of f over I with respect to P is $\displaystyle\sum_{i=1}^{n} f\left(x_i^*\right)\Delta x_i$.

If $f(x) \geq 0$ for $a \leq x \leq b$, the Riemann sum represents an approximation of the area under the graph of $f(x)$, above the x-axis, from $x = a$ to $x = b$. The diagram shows the Riemann sum of the function $f(x) = x^2$ over the interval [1, 2] as the area enclosed by four approximating rectangles of equal width.

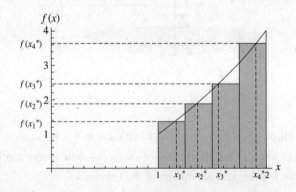

The Riemann sum, represented by the gray area enclosed by the rectangles, offers only an approximation to the area under the curve. However, as the width of each rectangle shrinks, the approximation gets better and the *exact* area under the curve is approached as a limit.

The definite integral of $f(x)$ over $[a, b]$ is defined in many calculus texts by

$$\int_a^b f(x)dx = \lim_{\|P\| \to 0} \sum_{i=1}^n f(x_i^*)\Delta x_i \quad \text{where} \quad \|P\| = \max_{1 \le i \le n} \Delta x_i$$

The condition $\|P\| \to 0$ guarantees that the lengths of all subintervals shrink toward 0 as we take more and more subintervals. If all subintervals are of equal length, this condition is equivalent to $n \to \infty$. For convenience we shall only consider subintervals of equal length. However, in theory, this need not be the case.

EXAMPLE 16 We will consider the function $f(x) = \sin x$ on the interval $[0, \pi/2]$. Because $\int_0^{\pi/2} \sin x \, dx = 1$, this is a good example for comparative purposes.

(a) We use 100 subintervals and choose x_i^* to be the *left* endpoint of each subinterval.

```
f[x_] = Sin[x];
a = 0; b = π/2; n = 100;
Δx = (b - a)/n;        ← Since each Δx has the same value, subscripts are not necessary.
xstar[i_] = a + (i - 1) Δx;
```
$$\sum_{i=1}^n f[\text{xstar}[i]]\,\Delta x \,//N$$
```
0.992125
```

(b) We choose the value of x_i^* to be the *right* endpoint of each subinterval. (This time we expect an overapproximation.)

```
f[x_] = Sin[x];
a = 0; b = π/2; n = 100;
Δx = (b - a)/n;
xstar[i_] = a + i Δx;
```
$$\sum_{i=1}^n f[\text{xstar}[i]]\,\Delta x \,//N$$
```
1.00783
```

To improve the accuracy of the approximation offered in Example 16, we can choose the value of x_i^* to be the midpoint of each subinterval. This leads to an approximation method called the *midpoint rule*.

EXAMPLE 17

```
f[x_] = Sin[x];
a = 0; b = π/2; n = 100;
Δx = (b - a)/n;.
xstar[i_] = a + (i - .5) Δx;
```
$$\sum_{i=1}^n f[\text{xstar}[i]]\,\Delta x \,//N$$
```
1.00001
```

As expected, the accuracy of the approximation improves.

Another simple approximation method, called the *trapezoidal rule*, improves accuracy by connecting the points on the curve corresponding to the points of subdivision with line segments, forming trapezoidal approximations of the area in place of rectangular approximations.

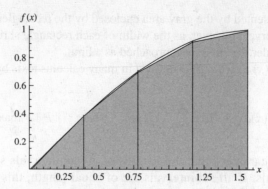

Trapezoidal approximation to $\int_0^{\pi/2} \sin x \, dx$ using four trapezoids.

The area enclosed by a trapezoid with base Δx and sides A and B is $\frac{\Delta x}{2}(A+B)$. Thus, the area enclosed by the trapezoid constructed in the ith interval, $[x_{i-1}, x_i]$, is $\frac{\Delta x_i}{2}\left[f(x_{i-1}) + f(x_i)\right]$.

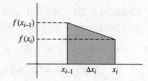

The total trapezoidal area, obtained by adding the individual areas, is

$$\frac{\Delta x_1}{2}[f(x_0) + f(x_1)] + \frac{\Delta x_2}{2}[f(x_1) + f(x_2)] + \frac{\Delta x_3}{2}[f(x_2) + f(x_3)] + \cdots + \frac{\Delta x_n}{2}[f(x_{n-1}) + f(x_n)]$$

If all intervals have the same length, Δx, this reduces to

$$\frac{\Delta x}{2}\Big[[f(x_0) + f(x_1)] + [f(x_1) + f(x_2)] + [f(x_2) + f(x_3)] + \cdots + [f(x_{n-1}) + f(x_n)]\Big]$$

or

$$\frac{\Delta x}{2}\Big[f(x_0) + 2f(x_1) + 2f(x_2) + 2f(x_3) + \cdots + 2f(x_{n-1}) + f(x_n)\Big]$$

EXAMPLE 18   Approximate $\int_0^{\pi/2} \sin x \, dx$ using the trapezoidal rule.

```
f[x_] = Sin[x];
a = 0; b = π/2; n = 100;
Δx = (b - a)/n;
x[i_] = a + i * Δx;
```

$$\text{approximation} = \frac{\Delta x}{2}\left(f[a] + 2\sum_{i=1}^{n-1} f[x[i]] + f[b]\right) //N$$

```
0.999979
```

SOLVED PROBLEMS

9.22 Compute the Riemann sums of $f(x) = x\, e^x \sqrt{x}$ over the interval $[0, 2]$ using
(a) the left endpoint of each subinterval.
(b) the right endpoint of each subinterval.
(c) the midpoint of each subinterval.
Compare with *Mathematica*'s approximation to the integral $\int_0^2 f(x)dx$.

SOLUTION

```
f[x_] = xe^x √x;
a = 0; b = 2;
∫_a^b f[x]dx //N
```

```
10.2406          ← Mathematica's approximation.
```

```
n = 100;
Δx = (b - a) / n;
xstar[i_] = a + (i - 1) Δx;
```

$$\sum_{i=1}^{n} f[xstar[i]] \, \Delta x \, //N$$

10.0328   ← Left endpoint approximation.

```
xstar[i_] = a + i Δx;
```

$$\sum_{i=1}^{n} f[xstar[i]] \, \Delta x \, //N$$

10.4508   ← Right endpoint approximation.

```
xstar[i_] = a + (i - .5) Δx;
```

$$\sum_{i=1}^{n} f[xstar[i]] \, \Delta x \, //N$$

10.24   ← Midpoint approximation.

9.23 Approximate $\int_{1}^{2} x \ln x \, dx$ using the trapezoidal rule with $n = 100$ and compare the result with *Mathematica*'s approximation.

SOLUTION

```
f[x_] = x Log[x];
a = 1; b = 2;
n = 100;
Δx = (b - a) / n;
x[i_] = a + i Δx;
```

$$\text{approximation} = \frac{\Delta x}{2}\left(f[a] + 2\sum_{i=1}^{n-1} f[x[i]] + f[b] \right) //N$$

0.6363

$$\int_{a}^{b} f[x] \, dx \, //N$$

| The error of 0.000006 is less than 0.001%. |

0.636294

9.24 Compute the lower and upper Riemann sums for the function $f(x) = x^2$ on the interval $[0, 1]$ for $n = 2, 4, 8, 16, \ldots, 2^{20}$ subintervals. Explain the behavior of the approximations in terms of the integral $\int_{0}^{1} x^2 \, dx$.

SOLUTION

```
f[x_] = x²;
a = 0; b = 1;
n = 2^m;
Δx = (b - a) / n;
nn = PaddedForm[n, 10];
```

$$\text{temp1} = \text{PaddedForm}\left[N\left[\sum_{i=1}^{n} f[a + (i - 1) \Delta x] \, \Delta x\right], \{8, 6\} \right];$$

$$\text{temp2} = \text{PaddedForm}\left[N\left[\sum_{i=1}^{n} f[a + i \, \Delta x] \, \Delta x\right], \{8, 6\} \right];$$

```
list = Table[{nn, temp1, temp2}, {m, 1, 20}];
TableForm[list, TableSpacing → {1, 5},
          TableHeadings → {None, {"          n", "     Lower", "     Upper"}}]
```

n	Lower	Upper
2	0.125000	0.625000
4	0.218750	0.468750
8	0.273438	0.398438
16	0.302734	0.365234
32	0.317871	0.349121
64	0.325562	0.341187
128	0.329437	0.337250
256	0.331383	0.335289
512	0.332357	0.334311
1024	0.332845	0.333822
2048	0.333089	0.333578
4096	0.333211	0.333455
8192	0.333272	0.333394
16384	0.333303	0.333364
32768	0.333318	0.333349
65536	0.333326	0.333341
131072	0.333330	0.333337
262144	0.333331	0.333335
524288	0.333332	0.333334
1048576	0.333333	0.333334

As n gets larger, the lower sums increase, approaching a limit of $\frac{1}{3}$, and the upper sums decrease, also approaching $\frac{1}{3}$.

$$\int_0^1 x^2\, dx$$

$\frac{1}{3}$

9.25 Compute an approximation of $\int_0^1 e^{x^2}\, dx$ using the trapezoidal rule with 10, 50, and 100 subintervals. Compare with *Mathematica*'s approximation.

SOLUTION

```
f[x_] = Exp[x²];
a = 0; b = 1;
∫ₐᵇ f[x] dx //N
```

1.46265   ← This is *Mathematica*'s approximation.

```
Δx = (b - a)/n;
x[i_] = a + i * Δx;
n = 10
```

$$\text{approximation} = \frac{\Delta x}{2}\left(f[a] + 2\sum_{i=1}^{n-1} f[x[i]] + f[b]\right) //N$$

1.46717   ← Error = 0.00452.

```
n = 50
```

$$\text{approximation} = \frac{\Delta x}{2}\left(f[a] + 2\sum_{i=1}^{n-1} f[x[i]] + f[b]\right) //N$$

1.46283   ← Error = 0.00018.

```
n = 100
```

$$\text{approximation} = \frac{\Delta x}{2}\left(f[a] + 2\sum_{i=1}^{n-1} f[x[i]] + f[b]\right) //N$$

1.4627   ← Error = 0.00005.

CHAPTER 10

Multivariate Calculus

10.1 Partial Derivatives

The commands **D**, ∂, and **Derivative** discussed in Chapter 8 are actually commands for computing *partial derivatives*. Of course, if there is only one variable present in a function, the partial derivative becomes an ordinary derivative. If two or more variables are present, however, all variables other than the one specified are treated as constants.

In the following descriptions, **f** stands for a function of several variables.

- **D[f, x]** or $\partial_x f$ (on the **Basic Math Input** palette) returns $\partial f / \partial x$, the partial derivative of **f** with respect to x.
- **D[f, (x, n)]** or $\partial_{\{x, n\}} f$ returns $\partial^n f / \partial x^n$, the nth order partial derivative of **f** with respect to x.
- **D[f, x₁, x₂, ..., xₖ]** or $\partial_{x_1, x_2, \ldots, x_k} f$ returns the "mixed" partial derivative $\dfrac{\partial^k f}{\partial x_1 \, \partial x_2 \, \ldots \, \partial x_k}$
- **D[f, {x₁, n₁}, {x₂, n₂}, ..., {xₖ, nₖ}]** or $\partial_{\{x_1, n_1\}, \{x_2, n_2\}, \ldots, \{x_k, n_k\}} f$ returns the partial derivative $\dfrac{\partial^n f [x_1, x_2, \ldots, x_k]}{\partial_{x_1}^{n_1} \, \partial_{x_2}^{n_2} \, \ldots \, \partial_{x_k}^{n_k}}$ where $n_1 + n_2 + \cdots + n_k = n$.

For convenience, an *invisible* comma may be used to separate variables in the partial derivative symbol. An invisible comma is entered by the three-key sequence [ESC] [.] [ESC]. An invisible comma works like an ordinary comma, but is hidden from the display.

EXAMPLE 1

D[x² y³ z⁴, x]

$2\,x\,y^3\,z^4$

∂_y **(x² y³ z⁴)**   ← The parentheses are important here. Why?

$3\,x^2\,y^2\,z^4$

D[x² y³ z⁴, {z, 2}]

$12\,x^2\,y^3\,z^2$

$\partial_{x, y}$ **(x² y³ z⁴)**

$6\,x\,y^2\,z^4$

EXAMPLE 2

Compute $\dfrac{\partial^7}{\partial x^3 \partial y^4} x^5 y^7$.

f[x_, y_] = x⁵ y⁷;
D[f[x, y], {x, 3}, {y, 4}]

$50\,400\,x^2\,y^3$

$\partial_{(x, 3), (y, 4)} f[x, y]$

$50400 x^2 y^3$

The **Derivative** command can also be used to construct partial derivatives. Suppose *f* is a function of *k* variables, $x_1, x_2, \ldots, x_k$.

- **Derivative[n₁, n₂, ..., nₖ][f]** gives the partial derivative $\dfrac{\partial^n f}{\partial_{x_1}^{n_1} \partial_{x_2}^{n_2} \ldots \partial_{x_k}^{n_k}}$ where

  $n_1 + n_2 + \ldots + n_k = n$. It returns a pure function (see the appendix) that may then be evaluated at $[x_1, x_2, \ldots, x_k]$.

EXAMPLE 3 (Continuation of Example 2)

```
f[x_, y_] = x⁵ Y⁷;
g = Derivative[3, 4][f]
50400 #1² #2³ &
g[x, y]
50400 x² y³
```

Although the command **D** can be used to evaluate partial derivatives at a given point, **Derivative** is perhaps a bit more convenient.

EXAMPLE 4 Let $f(x, y) = x^3 \sin y$. Evaluate f_{xy} at the point $(2, \pi)$.

```
f[x_, y_] = x³ Sin[y];
D[f[x, y], x, y] /. {x→2, y→π}
-12
Derivative[1, 1][f][2, π]
-12
```

SOLVED PROBLEMS

10.1 Compute the first- and second-order partial derivatives of $f(x, y) = xe^{xy}$.

SOLUTION

```
f[x_, y_] = x Exp[x y];
D[f[x, y], x]
eˣʸ + eˣʸ xy
D[f[x, y], y]
eˣʸ x²
D[f[x, y], {x, 2}]
2 eˣʸ y + eˣʸ x y²
D[f[x, y], {y, 2}]
eˣʸ x³
D[f[x, y], x, y]
2 eˣʸ x + eˣʸ x² y
```

10.2 The partial derivatives of $f(x, y)$ are defined by the following limits:

$$f_x(x, y) = \lim_{h \to 0} \frac{f(x+h, y) - f(x, y)}{h}$$

$$f_y(x, y) = \lim_{h \to 0} \frac{f(x, y+h) - f(x, y)}{h}$$

Compute the derivatives of $f(x, y) = \ln(x^2 + y^3)$ using the definition and verify using the *Mathematica* **D** command.

SOLUTION

```
f[x_, y_] = Log[x² + y³];
```

$$\text{Limit}\left[\frac{f[x + h, y] - f[x, y]}{h}, h \to 0\right]$$

$$\frac{2x}{x^2 + y^3}$$

```
D[f[x, y], x]
```

$$\frac{2x}{x^2 + y^3}$$

$$\text{Limit}\left[\frac{f[x, y + h] - f[x, y]}{h}, h \to 0\right]$$

$$\frac{3y^2}{x^2 + y^3}$$

```
D[f[x, y], y]
```

$$\frac{3y^2}{x^2 + y^3}$$

10.3 Let $z = e^{xy}$. Compute $\dfrac{\partial^3 z}{\partial^2 x \, \partial y}$.

SOLUTION

```
z = Exp[x y];
D[z, {x, 2}, y]   or   ∂{x, 2}, y z
```
$2e^{xy} y + e^{xy} x y^2$

10.4 Verify that $u = e^{-a^2 k^2 t} \sin kx$ is a solution of the heat equation: $\dfrac{\partial u}{\partial t} = a^2 \dfrac{\partial^2 u}{\partial x^2}$.

SOLUTION

```
u[x_, t_] = Exp[-a² k² t] Sin[k x];
lhs = D[u[x, t], t]
```
$-a^2 e^{-a^2 k^2 t} k^2 \sin[kx]$

```
rhs = a² D[u[x, t], {x, 2}]
```
$-a^2 e^{-a^2 k^2 t} k^2 \sin[kx]$

```
lhs == rhs
True
```

10.5 A function of three variables, $f(x, y, z)$, is said to be harmonic if it satisfies Laplace's equation: $\dfrac{\partial^2 f}{\partial x^2} + \dfrac{\partial^2 f}{\partial y^2} + \dfrac{\partial^2 f}{\partial z^2} = 0$. Let $f(x, y, z) = \dfrac{1}{\sqrt{x^2 + y^2 + z^2}}$. Compute f_{xx}, f_{yy}, and f_{zz} and show that f is harmonic.

SOLUTION

```
f[x_, y_, z_] = 
```
$$\frac{1}{\sqrt{x^2 + y^2 + z^2}};$$

```
∂{x, 2} f[x, y, z] //Together
```

$$\frac{2x^2 - y^2 - z^2}{(x^2 + y^2 + z^2)^{5/2}}$$

```
∂{y,2}f[x, y, z] //Together
```

$$\frac{-x^2 + 2y^2 - z^2}{(x^2 + y^2 + z^2)^{5/2}}$$

```
∂{z,2}f[x, y, z] //Together
```

$$\frac{-x^2 - y^2 + 2z^2}{(x^2 + y^2 + z^2)^{5/2}}$$

```
%%% + %% + % //Together
```

```
0
```

10.6 The plane tangent to the surface defined by $z = f(x, y)$ at the point (x_0, y_0, z_0) is

$$z = z_0 + f_x(x_0, y_0)(x - x_0) + f_y(x_0, y_0)(y - y_0)$$

Determine the equation of the plane tangent to the paraboloid $z = 10 - x^2 - 2y^2$ at the point where $x = 1$ and $y = 2$. Sketch the paraboloid and its tangent plane.

SOLUTION

```
f[x_, y_] = 10 - x² - 2 y²;
z = f[1, 2] + Derivative[1, 0][f][1, 2](x - 1) +
          Derivative[0, 1][f][1, 2](y - 2)//Expand
```

```
19 - 2x - 8y
```

| The tangent plane has equation $z = 19 - 2x - 8y$. |

```
g1 = Plot3D[f[x, y], {x, -5, 5}, {y, -5, 5}];
g2 = Plot3D[z, {x, -5, 5}, {y, -5, 5}]
Show[g1, g2, PlotRange→All, ViewPoint→{2.330, -2.223, 1.040}]
```

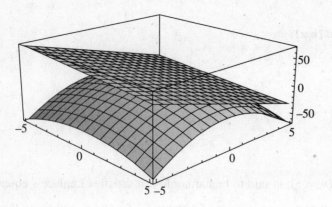

10.7 The plane tangent to the surface $f(x, y, z) = 0$ at the point (x_0, y_0, z_0) is

$$f_x(x_0, y_0, z_0)(x - x_0) + f_y(x_0, y_0, z_0)(y - y_0) + f_z(x_0, y_0, z_0)(z - z_0) = 0$$

Sketch the sphere $x^2 + y^2 + z^2 = 14$ and its tangent plane at the point $(1, 2, 3)$.

SOLUTION

The sphere is centered at the origin and has a radius of $\sqrt{14}$. Its equation is rewritten as $x^2 + y^2 + z^2 - 14 = 0$. We can use the graphics primitive **Sphere** to construct its graph. (See Chapter 5.)

```
f[x_, y_, z_] = x² + y² + z² - 14;

g1 = Graphics3D[Sphere[{0, 0, 0}, √14]];

a = Derivative[1, 0, 0][f][1, 2, 3];

b = Derivative[0, 1, 0][f][1, 2, 3];

c = Derivative[0, 0, 1][f][1, 2, 3];

Solve[a (x - 1) + b (y - 2) + c (z - 3) == 0, z]
```

$$\left\{\left\{z \rightarrow \tfrac{1}{3}(14 - x - 2y)\right\}\right\}$$

```
g2 = Plot3D[⅓ (14 - x - 2 y), {x, -5, 5}, {y, -5, 5}];

Show[g1, g2]
```

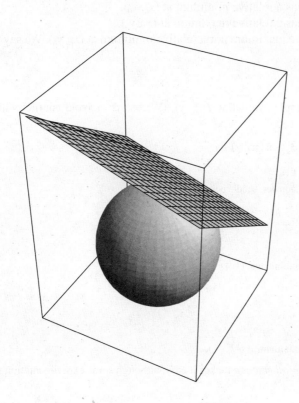

10.2   Maximum and Minimum Values

A function, f, has a relative (or local) maximum at (x_0, y_0) if there exists an open disk centered at (x_0, y_0) such that $f(x, y) \le f(x_0, y_0)$ for all (x, y) in the disk. A similar definition (with the inequality reversed) holds for a relative minimum. If f has either a relative maximum or relative minimum at (x_0, y_0), we say that f has a relative *extremum* at (x_0, y_0).

If f is differentiable, a necessary condition for $f(x, y)$ to have a relative extremum at the point (x_0, y_0) is $f_x(x_0, y_0) = f_y(x_0, y_0) = 0$. The point (x_0, y_0) is called a *critical point* of f.

EXAMPLE 5   To find the critical point(s) for the function $f(x, y) = x^4 + y^4 - 4xy$, we compute the first-order partial derivatives, set them both equal to 0, and solve the resulting equations.

```
f[x_, y_] = x⁴ + y⁴ - 4 x y
pdx = D[f[x, y], x]
4 x³ - 4 y
pdy = D[f[x, y], y]
-4 x + 4 y³
Solve[{pdx == 0, pdy == 0}, {x, y}]
{{x → -1, y → -1}, {x → 0, y → 0}, {x → -i, y → i}, {x → i, y → -i},
  {x → 1, y → 1}, {x → -(-1)¹ᐟ⁴, y → -(-1)³ᐟ⁴}, {x → (-1)¹ᐟ⁴, y → (-1)³ᐟ⁴},
  {x → -(-1)³ᐟ⁴, y → -(-1)¹ᐟ⁴}, {x → (-1)³ᐟ⁴, y → (-1)¹ᐟ⁴}}
```

The only *real* critical points are $(-1, -1)$, $(0, 0)$, and $(1, 1)$.

Not all critical points turn out to be relative extrema. To determine whether a function has a relative extremum at a critical point, and if so, whether it is a maximum or minimum, we use the *Second Partial Derivatives Test:*

Let $D(x, y) = f_{xx}(x, y) f_{yy}(x, y) - [f_{xy}(x, y)]^2$ and let (x_0, y_0) be a critical point of f.

1. If $D(x_0, y_0) > 0$ and $f_{xx}(x_0, y_0) > 0$, then f has a relative minimum at (x_0, y_0).
2. If $D(x_0, y_0) > 0$ and $f_{xx}(x_0, y_0) < 0$, then f has a relative maximum at (x_0, y_0).
3. If $D(x_0, y_0) < 0$, then f has neither a relative maximum nor a relative minimum at (x_0, y_0). We say that f has a *saddle point* at (x_0, y_0).

If $D(x_0, y_0) = 0$, the test is inconclusive.

EXAMPLE 6 Continuing with the previous example, we define $D(x, y)$. (We use **d** to avoid conflict with **D**, *Mathematica*'s derivative operator.)

```
d[x_, y_] = ∂_{x,2} f[x, y] ∂_{y,2} f[x, y] - (∂_{x,y} f[x, y])²;
d[0, 0]
-16                              ← Negative number; saddle point at (0, 0).
d[1, 1]
128
∂_{x,2} f[x, y] /. {x → 1, y → 1}
12                               ← Relative minimum at (1, 1).
d[-1, -1]
128
∂_{x,2} f[x, y] /. {x → -1, y → -1}
12                               ← Relative minimum at (-1, -1).
```

It is certainly worthwhile plotting this function. *Mathematica* makes it easy, although some experimentation with the options is necessary to show the details clearly.

```
Plot3D[f[x, y], {x, -2, 2}, {y, -2, 2}, PlotRange → {-2, 5},
    ViewPoint → {1.761, -2.816, 0.647}]
```

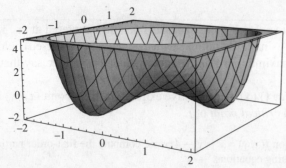

To find the maximum and minimum values of a function $f(x, y)$ subject to the constraint $g(x, y) = 0$, the method of *Lagrange multipliers* can be used. Geometrically, it can be shown that the maximum (minimum) value of f will occur where the level curves of f and the level curves of g share a common tangent line. At this point the gradient of f[1] and the gradient of g will be parallel and $\nabla f(x, y) = \lambda \nabla g(x, y)$. It follows that

$$f_x(x, y) = \lambda g_x(x, y)$$
$$f_y(x, y) = \lambda g_y(x, y)$$

Using these equations, together with $g(x, y) = 0$, λ can be eliminated and the values of x and y corresponding to the maximum and minimum values of f can be determined. The next example illustrates the procedure.

EXAMPLE 7 Suppose we wish to find the maximum and minimum values of $f(x, y) = 2x^2 + 3y^2$ subject to the constraint $x^2 + y^2 = 4$. We define $g(x, y) = x^2 + y^2 - 4$ and eliminate λ.

> **Eliminate**[*equations*, λ] eliminates λ between a set of simultaneous equations. See Chapter 6.

```
f[x_, y_] = 2 x^2 + 3 y^2;

g[x_, y_] = x^2 + y^2 - 4;

conditions = Eliminate[{∂_x f[x, y] == λ ∂_x g[x, y], ∂_y f[x, y] == λ ∂_y g[x, y], g[x, y] == 0}, λ]
```
$$x^2 == 4 - y^2 \;\&\&\; x\,y == 0 \;\&\&\; -4y + y^3 == 0$$

```
points = Solve[conditions]
```
$$\{\{x \to -2, y \to 0\}, \{x \to 0, y \to -2\}, \{x \to 0, y \to 2\}, \{x \to 2, y \to 0\}\}$$

To determine the maximum and minimum values of f, we compute its values at these points.

```
functionvalues = f[x, y] /. points
```
$$\{8, 12, 12, 8\}$$
```
Max[functionvalues]
```
12
```
Min[functionvalues]
```
8

The method of Lagrange multipliers can be extended to functions of three (or more) variables.

EXAMPLE 8 To find the maximum and minimum values of $f(x, y, z) = xyz$, subject to the constraint $x^2 + 2y^2 + 3z^2 = 6$, we define $g(x, y, z) = x^2 + 2y^2 + 3z^2 - 6$.

```
f[x_, y_, z_] = x y z;

g[x_, y_, z_] = x^2 + 2 y^2 + 3 z^2 - 6;

conditions = Eliminate[{∂_x f[x, y, z] == λ ∂_x g[x, y, z],
    ∂_y f[x, y, z] == λ ∂_y g[x, y, z], ∂_z f[x, y, z] == λ ∂_z g[x, y, z], g[x, y, z] == 0}, λ]
```
$$x^2 == 6 - 2y^2 - 3z^2 \;\&\&\; 4y^2 z == z(6 - 3z^2) \;\&\&\; x(2y^2 - 3z^2) == 0 \;\&\&\; x\,y(-2 + 3z^2) == 0 \;\&\&$$
$$x\,z(-2 + 3z^2) == 0 \;\&\&\; y\,z(-2 + 3z^2) == 0 \;\&\&\; 4z - 8z^3 + 3z^5 == 0 \;\&\&\; y^3 + y(-3 + 3z^2) == 0$$

```
points = Solve[conditions]
```
$$\{\{x \to 0, y \to 0, z \to -\sqrt{2}\}, \{x \to 0, y \to 0, z \to -\sqrt{2}\},$$
$$\{x \to 0, y \to 0, z \to \sqrt{2}\}, \{x \to 0, y \to 0, z \to \sqrt{2}\}, \{x \to 0, y \to -\sqrt{3}, z \to 0\},$$

---

[1]The gradient of $f(x, y)$ is the vector function $\nabla f(x, y) = f_x(x, y)\,\mathbf{i} + f_y(x, y)\,\mathbf{j}$.
 The gradient of $f(x, y, z)$ is $\nabla f(x, y, z) = f_x(x, y, z)\,\mathbf{i} + f_y(x, y, z)\,\mathbf{j} + f_z(x, y, z)\,\mathbf{k}$.

$\{x \rightarrow 0, \, y \rightarrow -\sqrt{3}, \, z \rightarrow 0\}, \{x \rightarrow 0, \, y \rightarrow \sqrt{3}, \, z \rightarrow 0\}, \{x \rightarrow 0, \, y \rightarrow \sqrt{3}, \, z \rightarrow 0\},$

$\{x \rightarrow -\sqrt{2}, \, y \rightarrow -1, \, z \rightarrow -\sqrt{\dfrac{2}{3}}\}, \{x \rightarrow -\sqrt{2}, \, y \rightarrow -1, \, z \rightarrow \sqrt{\dfrac{2}{3}}\},$

$\{x \rightarrow -\sqrt{2}, \, y \rightarrow 1, \, z \rightarrow -\sqrt{\dfrac{2}{3}}\}, \{x \rightarrow -\sqrt{2}, \, y \rightarrow 1, \, z \rightarrow \sqrt{\dfrac{2}{3}}\},$

$\{x \rightarrow \sqrt{2}, \, y \rightarrow -1, \, z \rightarrow -\sqrt{\dfrac{2}{3}}\}, \{x \rightarrow \sqrt{2}, \, y \rightarrow -1, \, z \rightarrow \sqrt{\dfrac{2}{3}}\},$

$\{x \rightarrow \sqrt{2}, \, y \rightarrow 1, \, z \rightarrow -\sqrt{\dfrac{2}{3}}\}, \{x \rightarrow \sqrt{2}, \, y \rightarrow 1, \, z \rightarrow \sqrt{\dfrac{2}{3}}\}, \{x \rightarrow -\sqrt{6}, \, y \rightarrow 0, \, z \rightarrow 0\},$

$\{x \rightarrow -\sqrt{6}, \, y \rightarrow 0, \, z \rightarrow 0\}, \{x \rightarrow \sqrt{6}, \, y \rightarrow 0, \, z \rightarrow 0\}, \{x \rightarrow \sqrt{6}, \, y \rightarrow 0, \, z \rightarrow 0\}\}$

```
functionvalues = f[x, y, z] /. points
```

$\left\{0, 0, 0, 0, 0, 0, 0, 0, -\dfrac{2}{\sqrt{3}}, \dfrac{2}{\sqrt{3}}, \dfrac{2}{\sqrt{3}}, -\dfrac{2}{\sqrt{3}}, \dfrac{2}{\sqrt{3}}, -\dfrac{2}{\sqrt{3}}, -\dfrac{2}{\sqrt{3}}, \dfrac{2}{\sqrt{3}}, 0, 0, 0, 0\right\}$

```
Max[functionvalues]
```

$\dfrac{2}{\sqrt{3}}$

```
Min[functionvalues]
```

$-\dfrac{2}{\sqrt{3}}$

SOLVED PROBLEMS

10.8 Find all relative extrema of the function $f(x, y) = x^2 - y^2$. Sketch the surface.

SOLUTION

```
f[x_, y_] = x² - y²;
Solve[{∂ₓf[x, y] == 0, ∂_y f[x, y] == 0}, {x, y}]
{{x → 0, y → 0}}
d[x_, y_] = ∂_{x, 2} f[x, y] ∂_{y, 2} f[x, y] - (∂_{x, y} f[x, y])²;
d[0, 0]
-4                                    ← Negative number; saddle point at (0, 0).
Plot3D[f[x, y], {x, -5, 5}, {y, -5, 5}, BoxRatios → {1, 1, 1}]
```

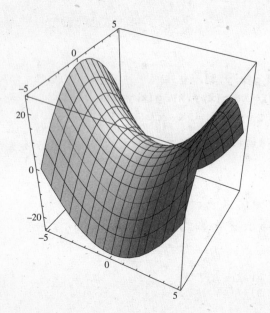

10.9 Find all relative extrema of the function $f(x,y) = xye^{-x^2-y^2}$. Sketch the surface.

SOLUTION

```
f[x_, y_] = x y Exp[-x² - y²]
pdx = ∂ₓ f[x, y] //Factor
```
$-e^{-x^2-y^2}(-1+2x^2)y$

```
pdy = ∂_y f[x, y] //Factor
```
$-e^{-x^2-y^2}x(-1+2y^2)$

If we try to use **Solve** to find where the partial derivatives are 0, we will get an error message due to the presence of the (non-algebraic) exponential. However, since $-e^{-x^2-y^2}$ cannot equal zero, we can ignore its presence.

```
Solve[{-1+2x²) y == 0, x (-1+2y²) == 0, {x, y}]
```

$$\left\{\{x \to 0,\ y \to 0\}, \left\{x \to -\frac{1}{\sqrt{2}},\ y \to -\frac{1}{\sqrt{2}}\right\}, \left\{x \to -\frac{1}{\sqrt{2}},\ y \to \frac{1}{\sqrt{2}}\right\},\right.$$

$$\left.\left\{x \to \frac{1}{\sqrt{2}},\ y \to -\frac{1}{\sqrt{2}}\right\}, \left\{x \to \frac{1}{\sqrt{2}},\ y \to -\frac{1}{\sqrt{2}}\right\}\right\}$$

```
d[x_, y_] = ∂_{x,2} f[x, y] ∂_{y,2} f[x, y] - (∂_{x,y} f[x, y])²;
d[0, 0]
```
-1 ← Negative number; no relative extremum.

```
d[-1/√2, -1/√2]
```
$\dfrac{4}{e^2}$

```
∂_{x,2} f[x, y]/.{x → -1/√2, y → -1/√2}
```
$-\dfrac{2}{e}$ ← Relative maximum at $\left(\dfrac{-1}{\sqrt{2}}, \dfrac{-1}{\sqrt{2}}\right)$.

```
d[-1/√2, 1/√2]
```
$\dfrac{4}{e^2}$

```
∂_{x,2} f[x, y]/.{x → -1/√2, y → 1/√2}
```
$\dfrac{2}{e}$ ← Relative minimum at $\left(\dfrac{-1}{\sqrt{2}}, \dfrac{1}{\sqrt{2}}\right)$.

```
d[1/√2, -1/√2]
```
$\dfrac{4}{e^2}$

```
∂_{x,2} f[x, y]/.{x → 1/√2, y → -1/√2}
```
$\dfrac{2}{e}$ ← Relative minimum at $\left(\dfrac{1}{\sqrt{2}}, \dfrac{-1}{\sqrt{2}}\right)$.

```
d[1/√2, 1/√2]
```
$\dfrac{4}{e^2}$

```
∂_{x,2} f[x, y]/.{x → 1/√2, y → 1/√2}
```
$-\dfrac{2}{e}$ ← Relative maximum at $\left(\dfrac{1}{\sqrt{2}}, \dfrac{1}{\sqrt{2}}\right)$.

We sketch the surface showing two views.

```
Plot3D[f[x, y], {x, -3, 3}, {y, -3, 3}, PlotPoints → 30,
       ViewPoint → {1.391, -3.001, 0.713}, PlotRange → All]

Plot3D[f[x, y], {x, -3, 3}, {y, -3, 3}, PlotPoints → 30,
       ViewPoint → {0.617, -3.318, 0.245}, PlotRange → All]
```

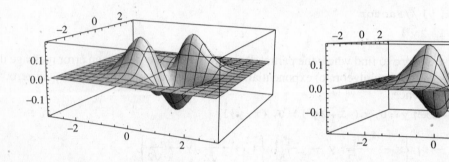

10.10 Use Lagrange multipliers to find the points on the circle $x^2 + y^2 - 2x - 4y = 0$ closest to and farthest from P(4, 4).

SOLUTION

```
circle = ContourPlot[x² + y² - 2x - 4y == 0, {x, -5, 5}, {y, -1, 5}];
point = Graphics[{PointSize[.01], Point[{4, 4}]}];
Show[circle, point, Axes → True, Frame → False, AspectRatio → Automatic]
```

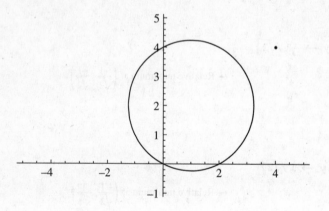

```
f[x_, y_] = (x - 4)² + (y - 4)²;          ← We minimize the square of the distance from P.
g[x_, y_] = x² + y² - 2x - 4y;

conditions = Eliminate[{∂ₓf[x, y] == λ∂ₓg[x, y],
    ∂ᵧf[x, y] == λ∂ᵧg[x, y], g[x, y] == 0}, λ]
```

$$2x == -4 + 3y \ \&\& \ -52y + 13y^2 == -32$$

```
points = Solve[conditions]
```

$$\left\{\left\{x \to \tfrac{1}{13}\left(13 - 3\sqrt{65}\right), \ y \to \tfrac{2}{13}\left(13 - \sqrt{65}\right)\right\}, \left\{x \to \tfrac{1}{13}\left(13 + 3\sqrt{65}\right), \ y \to \tfrac{2}{13}\left(13 + \sqrt{65}\right)\right\}\right\}$$

```
√f[x, y] /. points //N
{5.84162, 1.36948}
```

> Based upon the computed distances, the first point is farthest from P and the second point is closest.

10.11 Find the points on the sphere $x^2 + y^2 + z^2 = 1$ that are closest to and farthest from $(1, 2, 3)$.

SOLUTION

```
f[x_, y_, z_] = (x - 1)² + (y - 2)² + (z - 3)²;
g[x_, y_, z_] = x² + y² + z² - 1;

conditions = Eliminate[{∂ₓf[x, y, z] == λ∂ₓg[x, y, z],
    ∂_y f[x, y, z] == λ∂_y g[x, y, z], ∂_z f[x, y, z] == λ∂_z g[x, y, z], g[x, y, z] == 0}, λ]

3 x == z && 3 y == 2 z && 14 z² == 9
Points = Solve[conditions, {x, y, z}]
```

$$\left\{\left\{x \to -\frac{1}{\sqrt{14}},\ y \to -\sqrt{\frac{2}{7}},\ z \to -\frac{3}{\sqrt{14}}\right\},\right.$$ ←Farthest point.

$$\left.\left\{x \to \frac{1}{\sqrt{14}},\ y \to \sqrt{\frac{2}{7}},\ z \to \frac{3}{\sqrt{14}}\right\}\right\}$$ ←Closest point.

```
√f[x, y, z] /.Points //N

{4.74166, 2.64166}
```

10.3 The Total Differential

The command **D**, discussed in Section 10.1, gives the partial derivative of a function of several variables. All variables other than the variable of differentiation are considered as constants. If f is a function, say, of two variables, x and y, but y is a function of x, **D** will compute an incorrect derivative.

Dt gives the *total differential* of a function.

- **Dt[f[x, y]]** returns the total differential of f[x, y].
- **Dt[f[x, y], x]** returns the total derivative of f[x, y] with respect to x.

Of course, f may be a function of more than two variables and the independent variable, listed as x in the above description, can be any of the variables defining f.

D[f[x, y], x] returns $\dfrac{\partial f}{\partial x}$ but **Dt[f[x, y], x]** returns $\dfrac{df}{dx}$.

EXAMPLE 9

	Calculus equivalent
`f[x_, y_] = x² y³;`	$f(x, y) = x^2 y^3$
`D[f[x, y], x]`	
`2 x y³`	$\dfrac{\partial f}{\partial x} = 2xy^3$
`D[f[x, y], y]`	
`3 x² y²`	$\dfrac{\partial f}{\partial y} = 3x^2 y^2$
`Dt[f[x, y]]`	
`2 x y³ Dt[x] + 3 x² y² Dt[y]`	$df = 2xy^3 dx + 3x^2 y^2 dy$
`Dt[f[x, y], x]`	
`2 x y³ + 3 x² y² Dt[y, x]`	$\dfrac{df}{dx} = 2xy^3 + 3x^2 y^2 \dfrac{dy}{dx}$
`Dt[f[x, y], y]`	
`3 x² y² + 2 x y³ Dt[x, y]`	$\dfrac{df}{dy} = 3x^2 y^2 + 2xy^3 \dfrac{dx}{dy}$

EXAMPLE 10   Suppose $f(x, y) = x^4 y^5$ where $y = x^3$. The following sequence gives an *incorrect* result for the derivative dz/dx.

```
z = x⁴ y⁵;
D[z, x] /. y → x³        ← The partial derivative is computed, then y is replaced by x³.
4 x¹⁸
```

In reality $z = x^{19}$, so $\dfrac{dz}{dx}$ should be $19 x^{18}$.

```
Dt[z, x] /. y → x³        ← The total derivative is computed, then y is replaced by x³.
19 x¹⁸
```

In some expressions, constants represented by letters might cause confusion. The option **Constants** can be used to instruct *Mathematica* to treat a particular symbol as a constant.

- **Constants → {** *objectlist* **}** causes all symbols in *objectlist* to be treated as constants.

EXAMPLE 11

```
Dt[xⁿ, x] //Expand
n x⁻¹⁺ⁿ + xⁿ Dt[n, x] Log[x]

Dt[xⁿ, x, Constants → {n}]
n x⁻¹⁺ⁿ
```

SOLVED PROBLEMS

10.12   Let $z = \sin xy$. Let $x = 1$, $y = 2$, $dx = \Delta x = 0.03$, $dy = \Delta y = 0.02$. Compute dz and compare it with the value of Δz.

SOLUTION

```
z = f[x_, y_] = Sin[x y];
Δz = f[x + Δx, y + Δy] – f[x, y];
Dt[z] /. {x → 1, y → 2, Dt[x] → 0.03, Dt[y] → 0.02}
 -0.0332917
Δz /. {x → 1, y → 2, Δx → 0.03, Δy → 0.02}
 -0.0364571
```

10.13   Use differentials to approximate $e^{0.1}\sqrt{4.01}$ and determine the percentage error of the estimate.

SOLUTION

We take advantage of the fact that 0.1 is near 0 and 4.01 is near 4.

```
f[x_, y_] = Exp[x] Sqrt[y];
approximation = f[0, 4] + Dt[f[x, y]] /. {x → 0, y → 4, Dt[x] → 0.1, Dt[y] → 0.01}
2.2025
exactvalue = f[0.1, 4.01]
2.2131
percenterror = Abs[approximation – exactvalue]/exactvalue ∗ 100;
Print["The error is ", percenterror, "%"]
The error is 0.479103 %
```

10.14   Use differentials to approximate the amount of metal in a tin can with height 30 cm and radius 10 cm if the thickness of the metal in the wall of the cylinder is 0.05 cm and the top and bottom are each 0.03 cm thick.

SOLUTION

$v = \pi r^2 h;$

$Dt[v] /. \{h \rightarrow 30, r \rightarrow 10, Dt[r] \rightarrow 0.05, Dt[h] \rightarrow 0.06\}$

113.097

> The change in height is the sum of the thicknesses of the top and bottom.

The amount of metal is approximately 113 cm^3.

10.15 If three resistors of resistance, R_1, R_2, and R_3 ohms are connected in parallel, their effective resistance is $\frac{1}{\frac{1}{R_1} + \frac{1}{R_2} + \frac{1}{R_3}}$ ohms. If a 20-ohm, 30-ohm, and 50-ohm resistor, each with maximum error of 1%, are connected in parallel, what range of resistance is possible from this combination?

SOLUTION

$f[R1\_, R2\_, R3\_] = \frac{1}{\frac{1}{R1} + \frac{1}{R2} + \frac{1}{R3}};$

$f[20, 30, 50] //N$

9.67742

$Dt[f[R1, R2, R3]] /. \{R1 \rightarrow 20, R2 \rightarrow 30, R3 \rightarrow 50, Dt[R1] \rightarrow 0.2,$
    $Dt[R2] \rightarrow 0.3, Dt[R3] \rightarrow 0.5\}$

0.0967742

The combined resistance is 9.67742 ± 0.0967742 ohms.

10.4   Multiple Integrals

Multiple integrals, or more precisely iterated integrals, are invoked by the **Integrate** command and are an extension of the command for a function of one variable.

- **Integrate[f[x, y], {x, xmin, xmax}, {y, ymin, ymax}]** evaluates the double integral $\int_{xmin}^{xmax} \int_{ymin}^{ymax} f(x,y) \, dy \, dx$.

- **Integrate[f[x, y, z], {x, xmin, xmax}, {y, ymin, ymax}, {z, zmin, zmax}]** evaluates the triple integral $\int_{xmin}^{xmax} \int_{ymin}^{ymax} \int_{zmin}^{zmax} f(x,y,z) \, dz \, dy \, dx$.

Higher order iterated integrals are evaluated in a similar manner. Note that the *rightmost* variable of integration in the **Integrate** command is the variable that is evaluated first.

As an alternative, the integral symbol from the Basic Math Input palette may be used repeatedly for the evaluation of multiple integrals.

EXAMPLE 12   To evaluate $\int_1^2 \int_1^x (x + y) \, dy \, dx$, we would type

Integrate[x + y, {x, 1, 2}, {y, 1, x}]

$\frac{3}{2}$

or

$\int_1^2 \int_1^x (x + y) \, dy \, dx$

$\frac{3}{2}$

EXAMPLE 13   To evaluate the triple integral $\int_0^2 \int_0^x \int_0^{xy} xyz \, dz \, dy \, dx$, we can type either

```
Integrate[x y z, {x, 0, 2}, {y, 0, x}, {z, 0, x y}]
```

4

or

$$\int_0^2 \int_0^x \int_0^{xy} x\,y\,z\,dz\,dy\,dx$$

4

If the integral is such that it's *exact* value cannot be evaluated, numerical integration can be used instead.

- **NIntegrate[f[x,y], {x, xmin, xmax}, {y, ymin, ymax}]** returns a numerical approximation of the value of the double integral $\int_{xmin}^{xmax} \int_{ymin}^{ymax} f(x,y) \, dy \, dx$.

- **NIntegrate[f[x, y, z], {x, xmin, xmax}, {y, ymin, ymax}, {z, zmin, zmax}]** returns a numerical approximation of the value of the triple integral $\int_{xmin}^{xmax} \int_{ymin}^{ymax} \int_{zmin}^{zmax} f(x,y,z) \, dz \, dy \, dx$.

Higher-order iterated integrals are approximated in a similar manner. If the **Basic Math Input** palette is used, the **N** command (or **//N** to the right of the integral) may be used. All of the options for **Nintegrate** as applied to single integrals apply to multiple integrals.

EXAMPLE 14

```
NIntegrate[Exp[x^2 y^2], {x, 0, 1}, {y, 0, 1}]   or   ∫₀¹∫₀¹ e^{x²y²} dy dx  //N
```

1.1351

SOLVED PROBLEMS

10.16   Use a double integral to compute the area bounded by the parabola $y = x^2 - 2x + 2$ and the line $y = x + 1$.

SOLUTION

```
f[x_] = x² - 2x + 2;
g[x_] = x + 1;
Plot[{f[x], g[x]}, {x, -1, 3}]
```

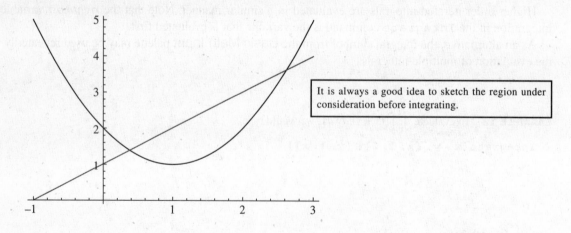

> It is always a good idea to sketch the region under consideration before integrating.

```
intersections = Solve[f[x] == g[x]]
```

$$\left\{\left\{x \to \tfrac{1}{2}(3 - \sqrt{5})\right\}, \left\{x \to \tfrac{1}{2}(3 + \sqrt{5})\right\}\right\}$$

```
xvalues = x /.intersections
```

$$\left\{\frac{1}{2}(3-\sqrt{5}),\ \frac{1}{2}(3+\sqrt{5})\right\}$$

```
a = xvalues[[1]]; b = xvalues[[2]];
```

$$\int_a^b \int_{f[x]}^{g[x]} dy\, dx$$

$$\frac{5\sqrt{5}}{6}$$

10.17 Find the center of mass of the lamina bounded by the parabola $y = 9 - x^2$ and the x-axis if the density at each point is proportional to its distance from the x-axis.

SOLUTION

Let R be the region bounded by $y = 9 - x^2$ and the x-axis.

```
Plot[9 - x², {x, -3, 3}]
```

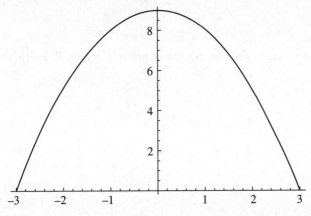

The graph intersects the x-axis at -3 and 3. The coordinates of the center of mass are $\left(\dfrac{M_y}{M}, \dfrac{M_x}{M}\right)$ where

$$M_y = \text{moment about the } y\text{-axis} = \iint_R x\rho(x,y)dA$$

$$M_x = \text{moment about the } x\text{-axis} = \iint_R y\rho(x,y)dA$$

$$M = \text{mass of lamina} = \iint_R \rho(x,y)dA$$

The density function $\rho(x,y) = ky$.

```
ρ[x_, y_]=k y;
```

$$my = \int_{-3}^{3}\int_0^{9-x^2} x\,\rho[x, y]\,dy\,dx$$

0

$$mx = \int_{-3}^{3}\int_0^{9-x^2} y\,\rho[x, y]\,dy\,dx$$

$$\frac{23\,328\,k}{35}$$

$$m = \int_{-3}^{3}\int_0^{9-x^2} \rho[x, y]\,dy\,dx$$

$$\frac{648\,k}{5}$$

$$\left\{\frac{my}{m}, \frac{mx}{m}\right\}$$

$$\left\{0, \frac{36}{7}\right\}$$

260   CHAPTER 10   *Multivariate Calculus*

10.18 Compute the shaded area. The curve shown is the Spiral of Archimedes and has polar equation $r = \theta$. It is shown for $0 \le \theta \le 6\pi$.

SOLUTION

The area inside a polar region, R, is $\iint\limits_{R} r\, dr\, d\theta$. The smaller arc of the shaded region is described by $r = \theta$, $2\pi \le \theta \le 5\pi/2$ and the larger arc may be represented by $r = \theta + 2\pi$, $2\pi \le \theta \le 5\pi/2$. The enclosed area can be expressed as $\int_{2\pi}^{5\pi/2} \int_{\theta}^{\theta + 2\pi} r\, dr\, d\theta$.

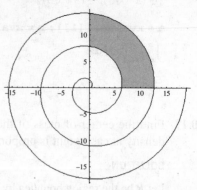

$$\int_{2\pi}^{5\pi/2} \int_{\theta}^{\theta+2\pi} \mathbf{r\, dr\, d\theta}$$

$$\frac{13\pi^3}{4}$$

10.19 Compute the volume under the paraboloid $z = x^2 + y^2$, above the region bounded by $y = x^2$ and $y = \sqrt{x+1}$.

SOLUTION

The volume bounded by a surface $z = f(x, y)$ and the x-y plane, above a region R, is $\iint\limits_{R} f(x,y)\, dA$. First let us look at R.

```
Plot[{x², √(x+1)},{x,-1,2}]
```

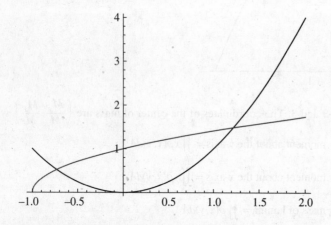

Next we find the points of intersection. Because of the complicated nature of the solution, we will obtain a numerical approximation.

```
NSolve[x² == √(x+1)]
```

$\{\{x \to 1.22074\}, \{x \to -0.724492\}$

Now we can express the volume as a double integral. Two solutions are shown.

$$\int_{-.724492}^{1.22074} \int_{x^2}^{\sqrt{x+1}} (x^2 + y^2)\, dy\, dx$$

1.11738

```
NIntegrate[x^2 + y^2, {x, -0.724492, 1.22074}, {y, x^2, Sqrt[x + 1]}]
```

1.11738

10.20 Find the volume under the hemisphere $z = 4 - x^2 - y^2$ above the region in the *x-y* plane bounded by the cardioid $r = 1 - \cos\theta$.

SOLUTION

We will translate the problem into cylindrical coordinates. Since $r^2 = x^2 + y^2$, the equation of the hemisphere becomes $z = 4 - r^2$. The region of integration, *R*, is the cardioid shown.

```
PolarPlot[1-Cos[θ], {θ, 0, 2π}]
```

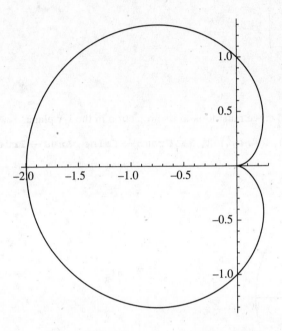

The volume, $V = \iint\limits_R (4 - r^2)\, dA = \int_0^{2\pi} \int_0^{1-\cos\theta} (4 - r^2)\, r\, dr\, d\theta$

$$\int_0^{2\pi} \int_0^{1-\text{Cos}[\theta]} (4 - r^2)\, r\, dr\, d\theta$$

$$\frac{61\pi}{16}$$

10.21 Find the volume of the solid that lies under the paraboloid $z = x^2 + y^2$, above the *x-y* plane, and inside the cylinder $(x-1)^2 + y^2 = 1$.

SOLUTION

The cylinder $(x-1)^2 + y^2 = 1$ is a cylinder of radius 1 whose axis is translated from the *z*-axis by the vector $(1, 0, 0)$.

```
s1 = Graphics3D[Cylinder[{{1, 0, 0}, {1, 0, 8}}]];

s2 = Plot3D[x² + y², {x, -2, 2}, {y, -2, 2}];
```

```
Show[s1, s2, PlotRange → {0,4}, ViewPoint → {1.217, -3.125, 0.447}]
```

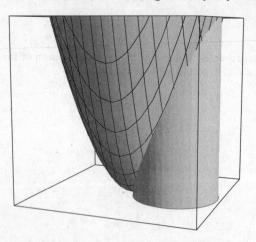

Now that we know what the region looks like, we must look at its projection in the *x-y* plane.

```
ContourPlot[(x - 1)² + y² == 1, {x, 0, 2}, {y, -1, 1}, Frame → False, Axes → True]
```

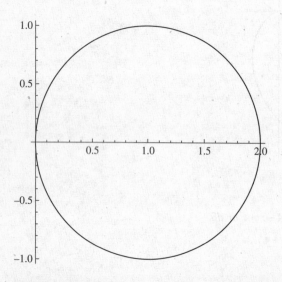

Although the problem can be solved in rectangular coordinates, it is easier to solve it using cylindrical coordinates. The equation of the circle can be expanded to $x^2 + y^2 = 2x$, which is equivalent, in polar coordinates, to $r = 2\cos\theta$, and the paraboloid $z = x^2 + y^2$ becomes $z = r^2$. The complete circle is generated as θ varies from $-\pi/2$ to $\pi/2$. The required volume may be expressed as a double integral: $\int_{-\pi/2}^{\pi/2}\int_0^{2\cos\theta}(r^2)\,r\,dr\,d\theta$

$$\int_{-\pi/2}^{\pi/2}\int_0^{2\cos[\theta]} r^3 \, dr \, d\theta$$

$$\frac{3\pi}{2}$$

10.22   The area of the surface $z = f(x, y)$ above the region R in the *x-y* plane is $\iint_R \sqrt{[f_x(x,y)]^2 + [f_y(x,y)]^2 + 1}\ dA$. Compute the surface area of a sphere of radius a.

SOLUTION

We compute the surface area of the portion of the sphere in the first octant and, by symmetry, multiply by 8. The equation of the sphere is $x^2 + y^2 + z^2 = a^2$. Solving for z we get $z = f(x,y) = \sqrt{a^2 - x^2 - y^2}$ as the function representing the upper hemisphere. Since the projection of the hemisphere onto the *x-y* plane is a circle of radius a centered at the origin, it is most convenient to use cylindrical coordinates.

```
f[x_, y_] = √(a² - x² - y²);
1 + (∂_x f[x, y])² + (∂_y f[x, y])² // Together
```

$$- \frac{a^2}{-a^2 + x^2 + y^2}$$

```
% /. x² + y² → r² // Simplify                    ← Replace x² + y² by r².
```

$$\frac{a^2}{a^2 - r^2}$$

$$8 \int_0^{\pi/2} \int_0^a \sqrt{\frac{a^2}{a^2 - r^2}} \; r \, dr \, d\theta$$

$$4\sqrt{a^2} \, \pi \, \text{Abs}[a]$$

$$\boxed{8 \iint_R \sqrt{1 + (\partial_x f[x, y])^2 + (\partial_y f[x, y])^2} \, dA = \\ 8 \int_0^{\pi/2} \int_0^a \sqrt{\frac{a^2}{a^2 - r^2}} \, r \, dr \, d\theta}$$

```
Simplify[%, Assumptions → a > 0]
```

$$4 \, a^2 \, \pi$$

10.23 Find the volume of the "ice cream cone" bounded by the cone $z = 3\sqrt{x^2 + y^2}$ and the sphere $x^2 + y^2 + (z - 9)^2 = 9$.

SOLUTION

The required volume is represented by the triple integral $\iiint_G dV$. Because of the nature of the bounding surfaces, this problem is done most conveniently using cylindrical coordinates. First, rewrite the equation of the sphere, solving for z in terms of x and y.

```
Solve[x² + y² + (z - 9)² == 9, z]
```

$$\left\{ \left\{ z \to 9 - \sqrt{9 - x^2 - y^2} \right\}, \left\{ z \to 9 + \sqrt{9 - x^2 - y^2} \right\} \right\}$$

Using the second solution (corresponding to the upper hemisphere), and replacing $x^2 + y^2$ by r^2, the equation of the sphere becomes $z = 9 + \sqrt{9 - r^2}$. The equation of the cone, $z = 3\sqrt{x^2 + y^2}$, becomes $z = 3r$. Now we can sketch the surfaces that form our region.

```
cone = RevolutionPlot3D[3 r, {r, 0, 3}, {θ, 0, 2π}];
hemisphere = RevolutionPlot3D[9 + √(9 - r²), {r, 0, 3}, {θ, 0, 2π}];

Show[cone, hemisphere, PlotRange → All, BoxRatios → {1, 1, 2},
     Axes → False, Boxed → False]
```

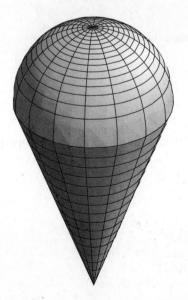

To compute the volume, we observe that the projection of the region onto the x-y plane is a circle. To determine its radius, we find the intersection of the cone and the hemisphere.

```
Solve[3 r == 9 + √(9 - r²)]
```

$\{\{r \to 3\}\}$

The projection onto the x-y plane is a circle of radius 3 centered at the origin. The required volume is

$$\int_0^{2\pi} \int_0^3 \int_{3r}^{9+\sqrt{9-r^2}} r \, dz \, dr \, d\theta$$

45π

10.24 A "silo" is formed above the x-y plane by the intersection of a right circular cylinder of radius 3 and a sphere of radius 5. Compute its volume.

SOLUTION

It is easiest to work in cylindrical coordinates. The equation of the spherical cap is $z = \sqrt{25 - r^2}$. It will intersect the cylinder when $r = 3$. The height of the cylinder will be 4.

```
cylinder = Graphics3D[Cylinder[{{0, 0, 0}, {0, 0, 4}}, 3]];

cap = RevolutionPlot3D[√(25 - r²), {r, 0, 3}, {θ, 0, 2π}];

Show[cylinder, cap, Boxed → False, PlotRange → {0, 5}]
```

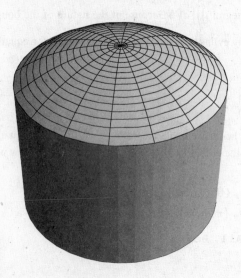

The projection of the solid is a circle of radius 3, centered at the origin.

$$\text{volume} = \int_0^{2\pi} \int_0^3 \int_0^{\sqrt{25-r^2}} r \, dz \, dr \, d\theta$$

$\dfrac{122\pi}{3}$

10.25 Find the center of mass of a solid hemisphere of radius a if its density at each point is proportional to its distance above the x-y plane.

SOLUTION

The center of mass has coordinates $(\bar{x}, \bar{y}, \bar{z})$ where

$$\bar{x} = \frac{\iiint\limits_G x \, \sigma(x,y,z) \, dV}{\iiint\limits_G \sigma(x,y,z) \, dV}, \quad \bar{y} = \frac{\iiint\limits_G y \, \sigma(x,y,z) \, dV}{\iiint\limits_G \sigma(x,y,z) \, dV}, \quad \bar{z} = \frac{\iiint\limits_G z \, \sigma(x,y,z) \, dV}{\iiint\limits_G \sigma(x,y,z) \, dV}.$$

The density function, $\sigma(x,y,z) = kz$, where k is the constant of proportionality. The problem is most conveniently solved by using spherical coordinates:

$$x = \rho\sin\phi\cos\theta, \quad y = \rho\sin\phi\sin\theta, \quad z = \rho\cos\phi$$

```
x = ρ Sin[φ] Cos[θ];
y = ρ Sin[φ] Sin[θ];
z = ρ Cos[φ];
σ = k z;
mass = ∫₀²π ∫₀^π/2 ∫₀ᵃ σ ρ² Sin[φ]dρ dφ dθ
```

$$\frac{1}{4}a^4 k\pi$$

$$\mathbf{centerofmass} = \left\{ \frac{\int_0^{2\pi}\int_0^{\pi/2}\int_0^a x\,\sigma\,\rho^2\,\mathbf{Sin[\phi]}d\rho\,d\phi\,d\theta}{\mathbf{mass}}, \frac{\int_0^{2\pi}\int_0^{\pi/2}\int_0^a y\,\sigma\,\rho^2\,\mathbf{Sin[\phi]}\,d\rho\,d\phi\,d\theta}{\mathbf{mass}}, \right.$$

$$\left. \frac{\int_0^{2\pi}\int_0^{\pi/2}\int_0^a z\,\sigma\,\rho^2\,\mathbf{Sin[\phi]}d\rho\,d\phi\,d\theta}{\mathbf{mass}} \right\}$$

$$\left\{0, 0, \frac{8a}{15}\right\}$$

10.26 Find the moment of inertia of a solid hemisphere of radius a about its axis if its density is equal to the distance from the center of its base.

SOLUTION

The moment of inertia about the z-axis is $\iiint\limits_G \left(\delta(x, y, z)\right)^2 \sigma(x, y, z)\, dV$ where $\delta(x, y, z)$ is the distance from the point (x, y, z) to the z-axis and $\sigma(x, y, z)$ is the density at the point (x, y, z). In this problem we should use a spherical coordinate system:

$$x = \rho\sin\phi\cos\theta, \quad y = \rho\sin\phi\,\sin\theta, \quad z = \rho\cos\phi$$

$$\delta(x, y, z) = \sqrt{x^2 + y^2} = \sqrt{(\rho\sin\phi\cos\theta)^2 + (\rho\sin\phi\sin\theta)^2} = \rho\sin\phi$$

$$\sigma(x, y, z) = \sqrt{x^2 + y^2 + z^2} = \rho$$

```
δ = ρ Sin[φ];
σ = ρ;
∫₀²π ∫₀^π/2 ∫₀ᵃ δ² σ ρ² Sin[φ] dρ dφ dθ
```

$$\frac{2a^6\pi}{9}$$

CHAPTER 11

Ordinary Differential Equations

11.1 Analytical Solutions

Simply put, a differential equation is an equation expressing a relationship between a function and one or more of its derivatives. A function that satisfies a differential equation is called a *solution*.

The *Mathematica* command **DSolve** is used to solve differential equations. As with algebraic or transcendental equations, a double equal sign, **==**, is used to separate the two sides of the equation.

- **DSolve [*equation*, y[x], x]** gives the general solution, y[x], of the differential equation, *equation*, whose independent variable is x.
- **DSolve [*equation*, y, x]** gives the general solution, y, of the differential equation expressed as a "pure" function (see appendix) within a list. **ReplaceAll (/.)** may then be used to evaluate the solution. Alternatively, one may use **Part** or **[[ ]]** to extract the solution from the list.

EXAMPLE 1   To solve the first-order differential equation $\frac{dy}{dx} = x + y$, we simply type

```
DSolve [y' [x] == x + y [x] , y [x] , x]
```
$\{\{y [x] \rightarrow -1 - x + e^x C [1] \}\}$

EXAMPLE 2   To obtain the solution of $\frac{dy}{dx} = x + y$ as a pure function (see appendix, Section A.1), we enter

```
solution = DSolve [y' [x] == x + y [x] , y, x]
```
$\{\{y \rightarrow Function [\{x\}, -1 - x + e^x C [1]] \}\}$

If we wish to evaluate the solution, we can type

```
y [x] /. solution
```
$\{\{y [x] \rightarrow -1 - x + e^x C [1] \}\}$

Using the pure function, we can evaluate the derivatives of the solution. This would be clumsy using the solution of Example 1.

```
y' [x] /. solution
```
$\{-1 + e^x C [1] \}$
```
y'' [x] + y' [x] /. solution
```
$\{-1 + 2 e^x C [1] \}$

We can define a function, f, representing the solution:

```
f = solution [[1, 1, 2]]
```
Function [{x}, -1 - x + e^x C [1]]

We can then directly evaluate f or any of its derivatives.

f[x]

$-1 - x + e^x C[1]$

f'[x]

$-1 + e^x C[1]$

f''[x]

$e^x C[1]$

It is *extremely important* that the unknown function be represented $y[x]$, not y, within the differential equation. Similarly, its derivatives must be represented $y'[x]$, $y''[x]$, etc. The next example illustrates some common errors.

EXAMPLE 3

DSolve[y'[x] == x + y, y[x], x]

DSolve::dvnoarg : The function y appears with no arguments. »

DSolve[y'[x] == x + y, y[x], x]

> ┌───┐
> ⋮ The function and its derivative must be specified as $y[x]$ and ⋮
> ⋮ $y'[x]$, respectively. ⋮
> └───┘

DSolve[y' == x + y[x], y[x], x]

DSolve::dvnoarg : The function y' appears with no arguments. »

DSolve[y' == x + y, y[x], x]

The solution of a first-order differential equation *without* initial conditions involves an arbitrary constant, labeled, by default, $C[1]$. Additional constants (for higher-order equations) are labeled $C[2]$, $C[3]$, ... If a different labeling is desired, the option **GeneratedParameters** may be used.

- **GeneratedParameters → constantlabel** specifies that the constants should be labeled constantlabel[1], constantlabel[2], etc.

EXAMPLE 4

DSolve[y'[x] == x + y[x], y[x], x, GeneratedParameters → mylabel]

$\{\{y[x] \rightarrow -1 - x + e^x \, mylabel[1]\}\}$

Higher order differential equations are solved in a similar manner. The derivatives are represented as $y'[x], y''[x], y'''[x]$, ... Alternatively, **D**, ∂, or **Derivative** may be used.

EXAMPLE 5

DSolve[y''[x] + y[x] == 0, y[x], x]

$\{\{y[x] \rightarrow C[1] \cos[x] + C[2] \sin[x]\}\}$

DSolve[D[y[x], {x, 2}] + y[x] == 0, y[x], x]

$\{\{y[x] \rightarrow C[1] \cos[x] + C[2] \sin[x]\}\}$

DSolve[$\partial_{\{x, 2\}}$y[x] + y[x] == 0, y[x], x]

$\{\{y[x] \rightarrow C[1] \cos[x] + C[2] \sin[x]\}\}$

DSolve[Derivative[2][y][x] + y[x] == 0, y[x], x]

$\{\{y[x] \rightarrow C[1] \cos[x] + C[2] \sin[x]\}\}$

More complicated differential equations are solved, if possible, using special functions. If *Mathematica* cannot solve the equation, it will return the equation either unsolved or in terms of unevaluated integrals. In such cases a numerical solution (see Section 11.2) may be more appropriate.

EXAMPLE 6 $x^2\dfrac{d^2y}{dx^2}+x\dfrac{dy}{dx}+(x^2-4)y=0$ is a special case of Bessel's equation. The solution is expressed in terms of Bessel functions of the first (BesselJ) and second (BesselY) kind.

```
DSolve[x²y''[x]+xy'[x]+(x²-4)y[x] == 0, y[x], x]
{{y[x] → BesselJ[2,x] C[1] + BesselY[2,x] C[2]}}
```

EXAMPLE 7 $\dfrac{d^2y}{dx^2}+\dfrac{dy}{dx}+y^2=0$ is a nonlinear differential equation that *Mathematica* cannot solve.

```
DSolve[y''[x] +y'[x] +y[x]² == 0, y[x], x]
DSolve[[y[x]² +y'[x] +y''[x] == 0, y[x], x]
```

If values of y, and perhaps one or more of its derivatives, are specified along with the differential equation, the task of finding y is known as an *initial value problem*. The differential equation and the initial conditions are specified as a list within the **DSolve** command. A unique solution is returned, provided an appropriate number of initial conditions are supplied.

EXAMPLE 8   Solve the equation $\dfrac{dy}{dx}=x+y$ with initial condition $y(0)=2$. Then plot the solution.

```
solution = DSolve[{y'[x] == x+y[x], y[0] == 2}, y[x], x]
{{y[x] → -1 + 3 eˣ - x}}
Plot[y[x] /. solution, {x, -5, 2}, AxesOrigin → {0, 0}]
```

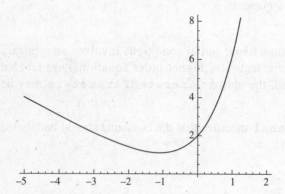

Here is another way the solution can be plotted:

```
solution = DSolve[{y'[x] == x + y[x], y[0] == 2}, y, x]
{{y → Function[{x}, -1 + 3 eˣ - x]}}
f = solution[[1, 1, 2]];
Plot[f[x], {x, -5, 2}, AxesOrigin → {0, 0}]
```

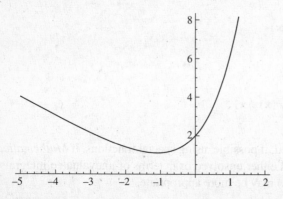

A useful way of visualizing the solution of a first-order differential equation is to introduce the concept of a vector field. A vector field on $\mathbb{R}^2$ is a vector function that assigns to each point (x, y) a two-dimensional vector $\mathbf{F}(x, y)$. By drawing the vectors $\mathbf{F}(x, y)$ for a (finite) subset of $\mathbb{R}^2$, one obtains a geometric interpretation of the behavior of $\mathbf{F}$.

- `VectorPlot[{Fx, Fy}, {x, xmin, xmax}, {y, ymin, ymax}]` produces a vector field plot of the two-dimensional vector function $\mathbf{F}$, whose components are `Fx` and `Fy`. The direction of the arrow is the direction of the vector field at the point (x, y). The size of the arrow is proportional to the magnitude of the vector field.
  - `Axes → Automatic` may be used if axes are desired. By default, no axes are drawn.
  - `Frame → False` may be used if a frame around the plot is not desired. The default is `Frame → True`.
  - `VectorScale` is an option that determines the length and arrowhead size of the field vectors that are drawn. The default is `ScaleFactor → Automatic`. Options include `Tiny`, `Small`, `Medium` and `Large`.

Note: Starting with version 7, `VectorPlot` can be found in the *Mathematica* kernel. If you are using version 6, you must use `VectorFieldPlot`, located in the package `VectorFieldPlots`` which must be loaded prior to use. See the Documentation Center for appropriate usage.

EXAMPLE 9   Plot the vector field $\mathbf{F}(x, y) = -y\,\mathbf{i} + x\,\mathbf{j}$.

```
VectorPlot [{-y, x}, {x, -5, 5}, {y, -5, 5}]
```

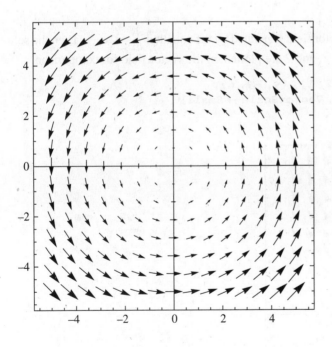

Any first-order differential equation can be used to define a vector field. Indeed, the vector field $\mathbf{i} + f(x, y)\,\mathbf{j}$, corresponding to the equation $\dfrac{dy}{dx} = f(x, y)$, generates a field whose vectors are tangent to the solution. The next example, although simple, illustrates this nicely.

EXAMPLE 10   Plot the vector field of the solution of the equation $\dfrac{dy}{dx} = 2x$. The solutions to this equation, parabolas $y = x^2 + c$, can be seen quite vividly.

```
VectorPlot[{1, 2 x}, {x, -1, 1}, {y, -1, 1}, Axes → Automatic]
```

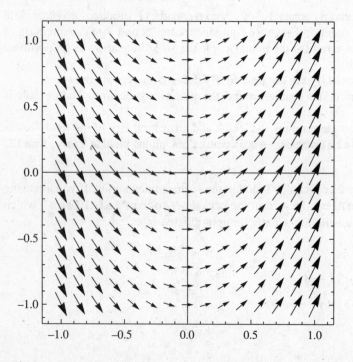

EXAMPLE 11   In this example we plot the vector field generated by the equation $\dfrac{dy}{dx} = 2x + y$. Then the solutions with initial conditions $y(0) = -2, -1, 0, 1$, and 2 are plotted on the vector field for comparison.

```
vf = VectorPlot[{1, 2x + y}, {x, -2, 1}, {y, -4, 6},
                Axes → Automatic, VectorScale → Small]
```

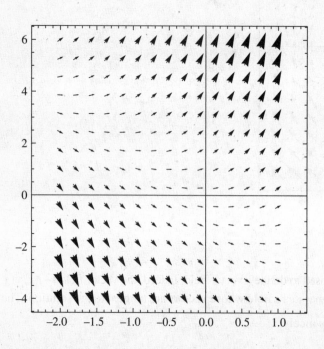

```
solutions = Table[DSolve[{y'[x] == 2x + y[x], y[0] == k}, y[x], x], {k, -2, 2}];
Do[g[k] = Plot[solutions[[k, 1, 1, 2]], {x, -2, 1}, PlotRange → All,
        Frame → True, PlotStyle → Thickness[.005]], {k, 1, 5}]
Show[g[1], g[2], g[3], g[4], g[5], vf, AspectRatio → 1, PlotRange → {-4,6}]
```

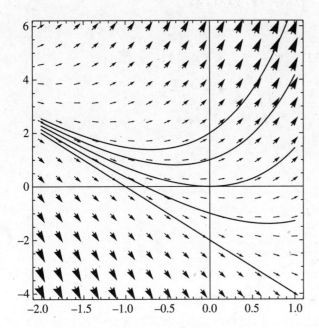

A *system* of differential equations consists of *n* differential equations involving *n* + 1 variables. Solving a system of differential equations with *Mathematica* is similar to solving a single equation.

EXAMPLE 12 This example illustrates how to solve the system $\frac{dx}{dt} = t^2$, $\frac{dy}{dt} = t^3$ with initial conditions $x(0) = 2$, $y(0) = 3$. The equation and its initial conditions are contained within a list.

```
solution = DSolve[{x'[t] == t², y'[t] == t³, x[0] == 2, y[0] == 3}, {x[t], y[t]}, t]
```

$$\left\{\left\{x[t] \rightarrow \tfrac{1}{3}\,(6 + t^3),\; y[t] \rightarrow \tfrac{1}{4}\,(12 + t^4)\right\}\right\}$$

Instead of specifying the values of *f* and its derivatives at a single point, values at two distinct points may be given. The problem of solving the differential equation then becomes known as a *boundary value problem*. However, unlike initial value problems, which can be shown to have unique solutions for a wide variety of cases, a boundary value problem may have no solution even for the simplest of equations.

EXAMPLE 13 The equation $\frac{d^2y}{dx^2} + y = 0$ with boundary conditions $y(0) = 0$, $y(\pi) = 1$ has no solution.

```
DSolve[{y''[x] + y[x] == 0, y[0] == 0, y[π] == 1}, y[x], x]
```

DSolve::bvnul : For some branches of the general solution,
   the given boundary conditions lead to an empty solution. >>

{}

The same equation with $y(0) = 0$, $y(\pi/2) = 1$ has a unique solution.

```
DSolve[{y''[x] + y[x] == 0, y[0] == 0, y[π/2] == 1}, y[x], x]
```

{{y[x] → Sin[x]}}

SOLVED PROBLEMS

11.1 Solve the differential equation $\dfrac{dy}{dx} = xy$ with initial condition $y(1) = 2$ and graph the solution for $-2 \le x \le 2$.

SOLUTION

```
solution = DSolve[{y'[x] == x y[x], y[1] == 2}, y[x], x]
```

$$\left\{ \left\{ y[x] \rightarrow 2\, e^{-\frac{1}{2} + \frac{x^2}{2}} \right\} \right\}$$

```
Plot[y[x] /.solution, {x, -2, 2}]
```

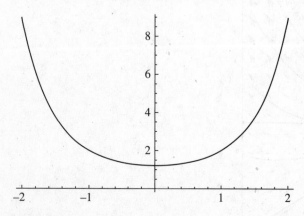

11.2 Plot the vector field for the differential equation of Problem 11.1.

SOLUTION

```
VectorPlot[{1, x y}, {x, -2, 2}, {y, -10, 10}, VectorScale → Tiny,
        Axes → Automatic]
```

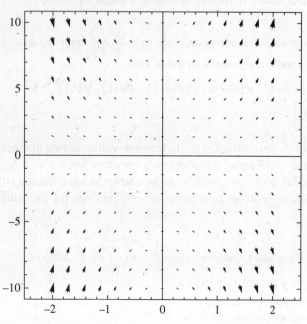

11.3 Plot the vector field for the equation $\dfrac{dy}{dx} = x^2 + y$ together with its solutions for $y(0) = 0, 1, 2, 3,$ and 4.

SOLUTION

```
vf = VectorPlot[{1, x^2 + y}, {x, 0, 1}, {y, 0, 12}, Axes → Automatic,
        VectorScale → Tiny];
```

```
solutions = Table[DSolve[{y'[x] == x² + y[x], y[0] == k}, y[x], x], {k, 0, 4}];
Do[g[k] = Plot[solutions[[k, 1, 1, 2]], {x, 0, 1},
        PlotStyle → Thickness[.007], PlotRange → All], {k, 1, 5}]
Show[g[1], g[2], g[3], g[4], g[5], vf, Frame → True, AspectRatio → 1]
```

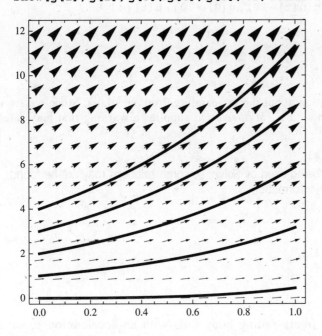

11.4 The *escape velocity* is the minimum velocity with which an object must be propelled in order to escape the gravitational field of a celestial body. Compute the escape velocity for the planet Earth.

SOLUTION

We shall assume that the initial velocity is in a radial direction away from Earth's center. According to Newton's laws of motion, the acceleration of a particle is inversely proportional to the square of the distance of the particle from the center of Earth. If r represents that distance, R the radius of the earth (approximately 3,960 miles), v the velocity of the particle, and a its acceleration, then $a = \dfrac{dv}{dt} = \dfrac{k}{r^2}$. At Earth's surface $(r = R)$, $a = -g$, where $g = 32.16$ ft/sec^2 = .00609 mi/sec^2. It follows that $k = -gR^2$, so $a = -\dfrac{gR^2}{r^2}$. Since $a = \dfrac{dv}{dt}$ and $v = \dfrac{dr}{dt}$, by the chain rule we have $a = \dfrac{dv}{dt} = \dfrac{dv}{dr}\dfrac{dr}{dt} = v\dfrac{dv}{dr}$. If v_0 represents the escape velocity, we are led to the differential equation $v\dfrac{dv}{dr} = -\dfrac{gR^2}{r^2}$ with initial condition $v = v_0$ when $r = R$.

```
DSolve[{v[r]v'[r] == -gR²/r², v[R] == v0}, v[r], r]
```

$$\left\{\left\{v[r] \to -\sqrt{\frac{-2grR + 2gR^2 + rv0^2}{r}}\right\}, \left\{v[r] \to \sqrt{\frac{-2grR + 2gR^2 + rv0^2}{r}}\right\}\right\}$$

Since the velocity is positive at the surface of Earth $(r = R)$, and must remain positive for the duration of the flight, we reject the first solution. Furthermore, $v(r)$ will remain positive if and only if $-2gR + v_0^2 \geq 0$, so $v_0 \geq \sqrt{2gR}$.

```
√(2gR) /.{g → .00609, R → 3960}
```

```
6.94498
```

The escape velocity is 6.94498 mi/sec.

11.5 According to Newton's law of cooling, the temperature of an object changes at a rate proportional to the difference in temperature between the object and the outside medium. If an object whose temperature is 70°F is placed in a medium whose temperature is 20°, and is found to be 40° after 3 minutes, what will its temperature be after 6 minutes?

SOLUTION

If $u(t)$ represents the temperature of the object at time t, $\frac{du}{dt} = k(u - 20)$. The initial condition is $u(0) = 70$.

```
solution1 = DSolve[{u'[t] == k(u[t] - 20), u[0] == 70}, u[t], t]
```

$$\left\{\left\{ u[t] \to 10\left(2 + 5 e^{kt}\right)\right\}\right\}$$

```
u[t_] = solution1[[1, 1, 2]]
```

$10(2 + 5\, e^{kt})$

We determine k using the information about the temperature 3 minutes later. Since we are using **Solve** for the transcendental function e^x, *Mathematica* supplies a warning that may safely be ignored.

```
solution2 = Solve[u[3] == 40, k]
```

Solve::ifun : Inverse functions are being used by Solve, so some solutions may not be found; use Reduce for complete solution information. »

$$\left\{\left\{ k \to -\frac{1}{3} \mathrm{Log}\left[\frac{5}{2}\right]\right\}\right\}$$

```
u[6] /. k → solution2[[1, 1, 2]]
```

28

The temperature 6 min later is 28°F.

11.6 If air resistance is neglected, a freely falling body falls with an acceleration g, which is approximately 32.16 ft/sec². If air resistance is considered, its motion is changed dramatically. If an object whose mass is 5 slugs is dropped from a height of 1000 ft, determine how long it will take to hit the ground (a) neglecting air resistance and (b) assuming that the force of air resistance is equal to the velocity of the object. Draw a graph of the height functions with and without air resistance.

SOLUTION

Let $h(t)$ represent the height of the object at time t, $v(t)$ its velocity, and $a(t)$ its acceleration. Recall that $v(t) = h'(t)$ and $a(t) = v'(t) = h''(t)$ and, by Newton's law, the sum of the external forces acting upon the object is equal to its mass times its acceleration: $ma(t) = \sum F$.

(a) If air resistance is neglected, the only force acting on the object is gravity, so $ma(t) = -mg$. We can divide by m and solve the differential equation $h''(t) = -g$ with initial conditions $h'(0) = 0$, $h(0) = 1000$. (*Note:* We take "up" to be the positive direction.)

```
g = 32.16;
solution = DSolve[{h''[t] == -g, h'[0] == 0, h[0] == 1000}, h[t], t];
height1[t_] = solution[[1, 1, 2]]
```

$1000 - 16.08t^2$

When the object reaches the ground its height will be 0.

```
Solve[height1[t] == 0, t]
```

$\{\{t \to -7.886\}, \{t \to 7.886\}\}$

It takes 7.886 sec to reach the ground.

(b) If air resistance is taken into account, there is an external force acting upon the object in addition to gravity, equal to $v(t)$. The differential equation becomes

$$ma(t) = -mg - v(t)$$

or

$$mh''(t) = -mg - h'(t)$$

with initial conditions as in (a).

```
m = 5; g = 32.16;
solution = DSolve[{mh''[t] == -mg - h'[t], h'[0] == 0, h[0] == 1000}, h[t], t];
height2[t_] = solution[[1, 1, 2]]
```
$$e^{-0.2t}(-804. + 1804. \, e^{0.2t} - 160.8 \, e^{0.2t} \, t)$$
```
FindRoot[height2[t] == 0, {t, 10}]
```
$$\{t \rightarrow 10.6213\}$$

It now takes 10.6213 sec to reach the ground.

```
<< PlotLegends`
Plot[{height1[t], height2[t]}, {t, 0, 11},
  PlotStyle → {Thick, Thin}, PlotRange → {0, 1000},
  AxesLabel → {"Time", "Height"},
  PlotLegend → {"Without air resistance", "Air resistance included"},
    LegendSize → {1, .5}]
```

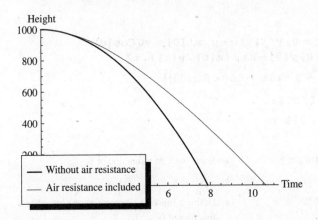

11.7 A baseball is hit with a velocity of 100 ft/sec at an angle of 30° with the horizontal. The height of the bat is 3 ft above the ground. Neglecting air and wind resistance, will it clear a 35-ft-high fence located 200 ft from home plate? (Assume $g = 32.16$ ft/sec².)

SOLUTION

```
g = 32.16; h = 3; θ = 30 Degree; v0 = 100;
solution = DSolve[{x''[t] == 0, y''[t] == - g, x'[0] == v0 Cos[θ], y'[0] == v0 Sin[θ],
  x[0] == 0, y[0] == h}, {x[t], y[t]}, t];
xsolution[t_] = solution[[1, 1, 2]]
```
$$50 \sqrt{3} \, t$$
```
ysolution[t_] = solution[[1, 2, 2]]
```
$$3 + 50 \, t - 16.08 \, t^2$$
```
ParametricPlot[{xsolution[t], ysolution[t]}, {t, 0, 3.2},
            AxesLabel → {"x", "y"}]
```

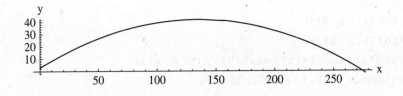

From the graph it is questionable whether $y \geq 35$ when $x = 200$, so we compute precisely when the ball reaches the fence and then calculate its height at that instant.

```
Solve[xsolution[t] == 200]
```

$$\left\{\left\{t \rightarrow \frac{4}{\sqrt{3}}\right\}\right\}$$

```
ysolution[t] /.%
```

$\{32.7101\}$

Since the height of the ball is less than 35 ft, the ball will *not* clear the fence.

11.8   At what angle should the ball in the previous problem be hit so that it goes over the fence?

SOLUTION

First we want to get a relationship between y and θ.

```
Clear[θ]

g = 32; h = 3; v0 = 100;

solution = DSolve[{x''[t] == 0, y''[t] == -g, x'[0] == v0 Cos[θ],
   y'[0] == v0 Sin[θ], x[0] == 0, y[0] == h}, {x[t], y[t]}, t]
```

$$\left\{\left\{x[t] \rightarrow 100\, t\, \text{Cos}[\theta]\, , y[t] \rightarrow 3 - 16\, t^2 + 100\, t\, \text{Sin}[\theta]\right\}\right\}$$

```
horiz[t_] = solution[[1, 1, 2]];

vert[t_] = solution[[1, 2, 2]]
```

$3 - 16\, t^2 + 100\, t\, \text{Sin}[\theta]$

```
temp = Solve[horiz[t] == 200, t]        ← Solve for t as a function of θ.
```

$\{\{t \rightarrow 2\, \text{Sec}[\theta]\}\}$

```
height[θ_] = vert[t] /. temp            ← Define a function representing the height as a
```
                                                       function of θ.

$\{3 - 64\, \text{Sec}[\theta]^2 + 200\, \text{Tan}[\theta]\}$

```
NSolve[height[θ Degree] == 35, θ]       ← Find θ which gives a height of 35 ft. The value of
```
                                                       θ is expressed in degrees.

Solve::ifun : Inverse functions are being used by Solve, so some solutions may not be found; use Reduce for complete solution information. >>

                                                     This warning may be safely disregarded.

$\{\{\theta \rightarrow -149.364\}, \{\theta \rightarrow -111.545\}, \{\theta \rightarrow 30.6357\}, \{\theta \rightarrow 68.4546\}\}$

The negative values of θ may be disregarded. The ball will go over the fence if θ lies between $30.6357°$ and $68.4546°$. We conclude by sketching these two trajectories. The vertical line represents the 35-ft fence located 200 ft from home plate.

```
θ = 30.6357 Degree;

horiz[t_] = solution[[1, 1, 2]];

vert[t_] = solution[[1, 2, 2]];

graph1 = ParametricPlot[{horiz[t], vert[t]}, {t, 0, 6}];


θ = 68.4546 Degree;

horiz[t_] = solution[[1, 1, 2]];

vert[t_] = solution[[1, 2, 2]];

graph2 = ParametricPlot[{horiz[t], vert[t]}, {t, 0, 6}];

graph3 = Graphics[Line[{{200, 0}, {200, 35}}]];
```

```
Show[graph1, graph2, graph3, PlotRange → {-50, 150}, AxesLabel → {"x", "y"}]
```

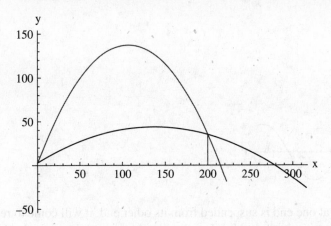

11.9 A culture of microorganisms grows at a rate proportional to the amount present at any given time. If there are 500 bacteria present after one day and 1200 after two days, how many bacteria will be present after four days?

SOLUTION

The differential equation described by this situation is $\frac{dN}{dt} = kN$, where N is the number of bacteria present in the culture and k is a constant to be determined by the given information. The initial condition is $N = 500$ when $t = 1$.

```
solution = DSolve[{n'[t] == k n[t], n[1] == 500}, n[t], t]
```

$$\left\{\left\{n[t] \rightarrow 500\, e^{-k+kt}\right\}\right\}$$

```
population[t_] = solution[[1, 1, 2]];
Solve[population[2] == 1200, k]
```

Solve::ifun : Inverse functions are being used by Solve, so some solutions may not be found; use Reduce for complete solution information. ≫

$$\left\{\left\{k \rightarrow \text{Log}\left[\frac{12}{5}\right]\right\}\right\}$$

```
population[4] /. k → Log[12/5]
```

```
6912
```

11.10 The equation governing the amount of current, I, flowing through a simple resistance–inductance circuit when an EMF (voltage) E is applied is $L\frac{dI}{dt} + RI = E$. The units for E, I, and L are, respectively, volts, amperes, and henries. If $R = 10$ ohms, $L = 1$ henry, the EMF source is an alternating voltage whose equation is $E(t) = 10 \sin 5t$, and the current is initially 4 amperes, find an expression for the current at time t and plot the graph of the current for the first 3 seconds.

SOLUTION

Note: Care must be taken not to use the conventional symbols **E** or **I** to represent voltage and current.

```
r = 10; l = 1; e[t_] = 10 Sin[5t];
solution = DSolve[{l i'[t] + r i[t] == e[t], i[0] == 4}, i[t], t]
```

$$\left\{\left\{i[t] \rightarrow -\frac{2}{5}\, e^{-10t}\left(-11 + e^{10t}\text{Cos}[5t] - 2\, e^{10t}\text{Sin}[5t]\right)\right\}\right\}$$

```
i[t_] = solution[[1, 1, 2]]
```

```
Plot[i[t], {t, 0, 3}, AxesLabel → {"t", "Current"}]
```

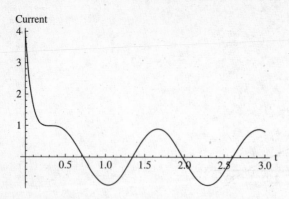

11.11 If a spring with mass m attached at one end is suspended from its other end, it will come to rest in an equilibrium position. If the system is then perturbed by releasing the mass with an initial velocity of v_0 at a distance y_0 below its equilibrium position, its motion satisfies the differential equation $m\dfrac{d^2y}{dt^2} + a\dfrac{dy}{dt} + ky = 0$, $y'(0) = v_0$, $y(0) = y_0$. a is a damping constant (determined experimentally) due to friction and air resistance, and k is the spring constant given in Hooke's law.

A mass of ¼ slug is attached to a spring with a spring constant, k, of 6 lb/ft. The mass is pulled downward from its equilibrium position 1 ft and then released. Assuming a damping constant, a, of ½, determine the motion of the mass and sketch its graph for the first 5 seconds.

SOLUTION

```
m = 1/4; y0 = -1; v0 = 0; a = 1/2; k = 6;
solution = DSolve[{m y''[t] + a y'[t] + k y[t] == 0, y'[0] == v0, y[0] == y0}, y[t], t]
```

$$\left\{\left\{y[t] \to -\frac{1}{23}\, e^{-t}\left(23\,\text{Cos}\left[\sqrt{23}\,t\right] + \sqrt{23}\,\text{Sin}\left[\sqrt{23}\,t\right]\right)\right\}\right\}$$

```
height[t_] = solution[[1, 1, 2]];
Plot[height[t], {t, 0, 5}, AxesLabel → {"t", "Height"}]
```

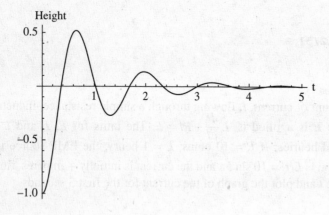

11.12 If a cable of uniform cross-section is suspended between two supports, the cable will sag forming a curve called a catenary. If we assume the lowest point on the curve to lie on the y-axis, a distance y_0 above the origin, the differential equation governing its shape can be shown to be $\dfrac{d^2y}{dx^2} = \dfrac{1}{a}\sqrt{1 + \left(\dfrac{dy}{dx}\right)^2}$ $y(0) = y_0$, $y'(0) = 0$, where a is a positive constant dependent upon the physical properties of the cable. Find an equation of the catenary and sketch its graph.

SOLUTION

```
solution = DSolve[{y''[x] == (1/a)√(1 + y'[x]²), y'[0] == 0, y[0] == y0}, y[x], x]
```

Solve::ifun : Inverse functions are being used by Solve, so some solutions may not be found;
use Reduce for complete solution information. >>

$$\left\{\left\{y[x] \to a + y0 - a \, \text{Cosh}\left[\frac{x}{a}\right]\right\}, \left\{y[x] \to -a + y0 + a \, \text{Cosh}\left[\frac{x}{a}\right]\right\}\right\}$$

Since $= y_0 > 0$, the second solution applies. We take $y_0 = a = 1$ and plot its graph.

```
catenary[x_] = solution[[2, 1, 2]] /. {a→1, y0→1};
Plot[catenary[x], {x, -1, 1}, Ticks→{Automatic, {0, 1, 2}}, PlotRange→{0, 2}]
```

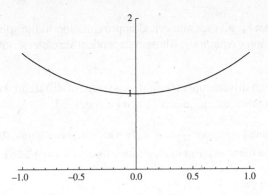

11.13 The logistic equation for population growth, $\frac{dp}{dt} = ap - bp^2$, was discovered in the mid-nineteenth century by the biologist Pierre Verhulst. The constant b is generally small in comparison to a so that for small population size p the quadratic term in p will be negligible and the population will grow approximately exponentially. For large p, however, the quadratic term serves to slow down the rate of growth of the population. Solve the logistic equation and sketch the solution for $a = 2$, $b = .05$, and an initial population $p_0 = 1$ (thousand). Then determine the limiting value of the population as $t \to \infty$.

SOLUTION

```
solution = DSolve[{p'[t] == a p[t] - b p[t]², p[0] == p0}, p[t], t]
```

Solve::ifun : Inverse functions are being used by Solve, so some solutions may not be found;
use Reduce for complete solution information. >>

$$\left\{\left\{p[t] \to \frac{a \, e^{at} \, p0}{a - b \, p0 + b \, e^{at} \, p0}\right\}\right\}$$

```
population[t_] = solution[[1, 1, 2]];
Plot[population[t] /. {p0→1, a→2, b→.05}, {t, 0, 5}, PlotRange→All]
```

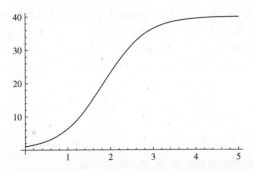

```
Limit[population[t] /. {p0→1, a→2, b→.05}, t→∞]
```
40.

11.14 Solve the boundary value problem $\dfrac{d^2y}{dx^2} + 4\pi^2 y = 0$, $y(0) = y(1) = 0$.

SOLUTION

```
DSolve[{y''[x] + 4π² y[x] == 0, y[0] == 0, y[1] == 0}, y[x], x]
   {{y[x] → C[2] Sin[2π x]}}
```

11.2   Numerical Solutions

Although certain types of differential equations can be solved analytically in terms of elementary functions, the vast majority of equations that arise in applications cannot. Even if unique solutions can be shown to exist, it may only be possible to obtain numerical approximations. The command **NDSolve** is designed specifically for this purpose.

- **NDSolve[*equations*, y, {x, xmin, xmax}]** gives a numerical approximation to the solution, y, of the differential equation with initial conditions, *equations*, whose independent variable, x, satisfies xmin ≤ x ≤ xmax.

Because **NDSolve** yields a numerical solution to a differential equation, or system of differential equations, an appropriate set of initial conditions that guarantees uniqueness must be specified.

EXAMPLE 14   In this example we consider the differential equation $\dfrac{dy}{dx} = x^2 + \sqrt{y}$ with initial condition $y(0) = 1$. Although this equation has a unique solution, it cannot be found in terms of elementary functions using **DSolve**.

```
DSolve[{y'[x] == x² + √(y[x]), y[0] == 1}, y[x], x]

DSolve[{y'[x] == x² + √(y[x]), y[0] == 1}, y[x], x]
```

We can only obtain a numerical approximation to the solution of this equation. Because numerical techniques construct approximations at only a finite number of points, *Mathematica* interpolates, i.e., constructs a smooth function passing through these points and returns the solution as an **InterpolatingFunction** object.

EXAMPLE 15

```
temp = NDSolve[{y'[x] == x² + √(y[x]), y[0] == 1}, y, {x, 0, 1}]

{{y → InterpolatingFunction[{{0., 1.}}, <>]}}
```

The actual interpolating function can now be extracted from this expression:

```
solution = temp[[1, 1, 2]]
InterpolatingFunction[{{0.; 1.}}, <>]
```

Only the *domain* of an **InterpolatingFunction** object is printed explicitly. The remaining elements are represented as < >. To see the data used in its construction, enter the command **FullForm[solution]**. Using the interpolated solution, solution, we can compute the solution at one or more points, and we can even plot it. One must be careful, however, to stay within the domain of the interpolating function or a warning will be generated.

```
solution[0.5]

1.60643

solution[1.5]

InterpolatingFunction::dmval : Input value {1.5} lies outside the range of data in the
   interpolating function. Extrapolation will be used. >>

4.32575          ← An extrapolated value is not as reliable as an interpolated value, in terms of accuracy.
```

```
list = Table[{x, solution[x]}, {x, 0, 1, .1}];
TableForm[list, TableSpacing → {1, 5}]
```

0	1.
0.1	1.10284
0.2	1.21273
0.3	1.33181
0.4	1.46228
0.5	1.60643
0.6	1.76656
0.7	1.94504
0.8	2.14429
0.9	2.36672
1.	2.61479

```
Plot[solution[x], {x, 0, 1}, AxesOrigin → {0, 0}]
```

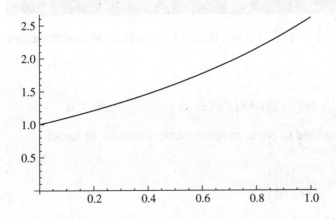

Although the default settings for **NDSolve** work nicely for most differential equations, *Mathematica* provides some options that can be used to set parameters to handle abnormal situations.

- **WorkingPrecision** is an option that specifies how many digits of precision should be maintained internally in computation. The default (on most computers) is **WorkingPrecision → 16**.
- **AccuracyGoal** is an option that specifies how many significant digits of accuracy are to be obtained. The default is **AccuracyGoal → Automatic**, which is half the value of **WorkingPrecision**. **AccuracyGoal** effectively specifies the absolute error allowed in a numerical procedure.
- **PrecisionGoal** is an option that specifies how many effective digits of precision should be sought in the final result. The default is **PrecisionGoal → Automatic**, which is half the value of **WorkingPrecision**. **PrecisionGoal** effectively specifies the relative error allowed in a numerical procedure.
- **MaxSteps** is the maximum number of steps to take in obtaining the solution. The default is **MaxSteps → Automatic**, which, for ordinary differential equations, is 10,000.
- **MaxStepSize** specifies the maximum size of each step in the iteration.
- **StartingStepSize** specifies the initial step size. The default is **StartingStepSize → Automatic**. (*Mathematica* automatically determines the best step size for the given equation.)

EXAMPLE 16   The differential equation $\dfrac{d^2y}{dx^2} + y = 0$ with initial conditions $y(0) = 0, y'(0) = 1$ has a unique solution $y = \sin x$. We attempt to solve it for $0 \le x \le 10{,}000$.

```
equation = NDSolve[{y''[x] + y[x] == 0, y[0] == 0, y'[0] == 1}, y, {x, 0, 10000}]
```

NDSolve::mxst : Maximum number of 10000 steps reached
    at the point x == 1422.780656413783`

```
{{y → InterpolatingFunction[{{0.,1422.78}}, <>]}}
```

Because of the wide interval, [0, 10000], over which the solution is to be obtained, more than 10,000 steps are necessary.

```
equation = NDSolve[{y''[x] + y[x] == 0, y[0] == 0, y'[0] == 1}, y,
             {x, 0, 10 000}, MaxSteps → 100 000]
```

```
InterpolatingFunction[{{0.,10 000.}}, <>]}}
```

Having obtained a solution, we check it for accuracy. The solution at $x = (4k+1)\dfrac{\pi}{2}$ should be 1.

```
f = equation[[1, 1, 2]];
f[633 π/2]
```

```
1.00002
```

SOLVED PROBLEMS

11.15   Solve the differential equation $\dfrac{dy}{dx} = 1 + \dfrac{1}{2}y^2 = 0, \;\; y(0) = 1, 0 \le x \le 1$, using **DSolve** and **NDSolve** and compare the results.

SOLUTION

$$\text{equation1} = \text{DSolve}\left[\left\{y'[x] == 1 + \frac{1}{2}y[x]^2, y[0] == 1\right\}, \; y[x], x\right]$$

Solve::ifun : Inverse functions are being used by Solve, so some solutions may not be found; use Reduce for complete solution information. ≫

$$\left\{\left\{y[x] \to \sqrt{2} \;\; \text{Tan}\left[\frac{1}{2}\left(\sqrt{2}x + 2\,\text{ArcTan}\left[\frac{1}{\sqrt{2}}\right]\right)\right]\right\}\right\}$$

```
solution1[x_] = equation1[[1, 1, 2]];
```

$$\text{equation2} = \text{NDSolve}\left[\left\{y'[x] == 1 + \frac{1}{2}y[x]^2, y[0] == 1\right\}, [y[x], \{x, 0, 1\}]\right]$$

```
{{y[x] → InterpolatingFunction[{{0., 1.}}, <>] [x]}}
solution2[x_] = equation2[[1, 1, 2]];
tabledata = Table[{x, solution1[x], solution2[x]}, {x,0,1,.1}];
TableForm[tabledata, TableSpacing → {1, 15},
         TableHeadings → {None, {"x","analytic", "numerical"}}]
```

x	analytic	numerical
0	1	1.
0.1	1.15817	1.15817
0.2	1.33582	1.33582
0.3	1.53895	1.53895
0.4	1.77601	1.77601
0.5	2.05935	2.05935
0.6	2.40786	2.40786
0.7	2.85196	2.85196
0.8	3.44406	3.44406
0.9	4.28301	4.28301
1.	5.58016	5.58016

11.16 Plot the solution to the differential equation $\dfrac{d^2y}{dt^2} + \left(\dfrac{dy}{dt} + 1\right)^2 \dfrac{dy}{dt} + y = 0$, $y(0) = 1$, $y'(0) = 0$ for $0 \le t \le 10$.

SOLUTION

```
solution = NDSolve[{y''[t] + (y'[t] + 1)² y'[t] + y[t] == 0, y[0] == 1, y'[0] == 0},
                    y[t], {t, 0, 10}]
{{y[t] → InterpolatingFunction[{{0., 10.}}, <>][t]}}
Plot[y[t] /. solution, {t, 0, 10}, PlotRange → All]
```

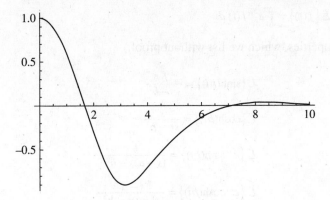

11.17 Plot the (five) solutions to $\dfrac{d^2y}{dx^2} + 0.3\dfrac{dy}{dx} + \sin y = 0$ for $0 \le x \le 30$ using initial conditions $y'(0) = 0$, $y(0) = -2, -1, 0, 1,$ and 2.

SOLUTION

```
Do[{solution = NDSolve[{y''[x] + 0.3 y'[x] + Sin[y[x]] == 0, y[0] == i, y'[0] == 1},
                        y[x], {x, 0, 30}];
    f[x_] = solution[[1, 1, 2]];
    graph[i] = Plot[f[x], {x, 0, 30}, PlotStyle → Hue[.2 i +.5],
                    PlotRange → All]}, {i, -2, 2}];
Show[graph[-2], graph[-1], graph[0], graph[1], graph[2]]
```

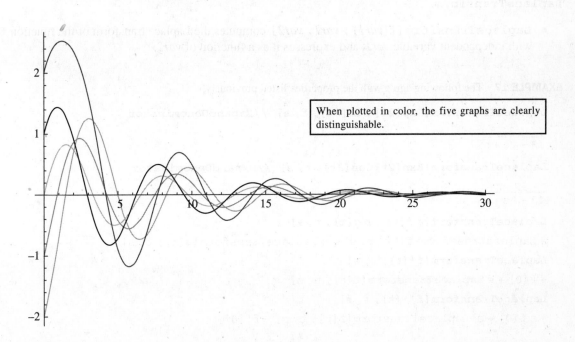

> When plotted in color, the five graphs are clearly distinguishable.

11.3 Laplace Transforms

In this section we describe an ingenious method for solving differential equations. Although the procedure can be used in a wide variety of problems, its real power lies in its ability to solve a differential equation whose "right hand side" is either discontinuous or zero except on a very short interval when its value is large. Because most of these types of problems arise within the context of time as the independent variable, we will express y and its derivatives as functions of t. We shall discuss Laplace transforms heuristically, and shall not concern ourselves with conditions sufficient for existence.

If f is defined on the interval $[0, \infty)$, the Laplace transform of $f(t)$ is defined

$$\mathcal{L}\{f(t)\} = \int_0^\infty e^{-st} f(t)\, dt$$

Its usefulness lies in the following properties, which we list without proof:

$$\mathcal{L}\{1\} = \frac{1}{s} \qquad\qquad \mathcal{L}\{\sinh(bt)\} = \frac{b}{s^2 - b^2}$$

$$\mathcal{L}\{t\} = \frac{1}{s^2} \qquad\qquad \mathcal{L}\{\cosh(bt)\} = \frac{s}{s^2 - b^2}$$

$$\mathcal{L}\{t^n\} = \frac{n!}{s^{n+1}} \quad \text{for positive integers } n \qquad \mathcal{L}\{e^{at}\sinh(bt)\} = \frac{b}{(s-a)^2 - b^2}$$

$$\mathcal{L}\{e^{at}\} = \frac{1}{s-a} \qquad\qquad \mathcal{L}\{e^{at}\cosh(bt)\} = \frac{s-a}{(s-a)^2 - b^2}$$

$$\mathcal{L}\{\sin(bt)\} = \frac{b}{s^2 + b^2} \qquad\qquad \mathcal{L}\{f'(t)\} = s\mathcal{L}\{f(t)\} - f(0)$$

$$\mathcal{L}\{\cos(bt)\} = \frac{s}{s^2 + b^2} \qquad\qquad \mathcal{L}\{f''(t)\} = s^2\mathcal{L}\{f(t)\} - sf(0) - f'(0)$$

$$\mathcal{L}\{e^{at}\sin(bt)\} = \frac{b}{(s-a)^2 + b^2} \qquad\qquad \mathcal{L}\{af(t) + bg(t)\} = a\,\mathcal{L}\{f(t)\} + b\,\mathcal{L}\{g(t)\}$$

$$\mathcal{L}\{e^{at}\cos(bt)\} = \frac{s-a}{(s-a)^2 + b^2} \qquad \text{If } F(s) = \mathcal{L}\{f(t)\}, \text{then } \mathcal{L}\{e^{at}f(t)\} = F(s-a)$$

Mathematica computes the Laplace transform of a function, f, by the invocation of the command `LaplaceTransform`.

- `LaplaceTransform[f[var1], var1, var2]` computes the Laplace transform of the function f, with independent variable *var1*, and expresses it as a function of *var2*.

EXAMPLE 17 The following agree with the properties listed previously.

`LaplaceTransform[Exp[2t] Sin[3t], t, s] //ExpandDenominator`

$$\frac{3}{13 - 4s + s^2}$$

`LaplaceTransform[Exp[2t] Cos[3t], t, s] //ExpandDenominator`

$$\frac{-2 + s}{13 - 4s + s^2}$$

`LaplaceTransform[a f[t] + b g[t], t, s]`

`a LaplaceTransform[f[t], t, s] + b LaplaceTransform[g[t], t, s]`

`LaplaceTransform[f'[t], t, s]`

`- f[0] + s LaplaceTransform[f[t], t, s]`

`LaplaceTransform[f''[t], t, s]`

`- s f[0] + s² LaplaceTransform[f[t], t, s] - f'[0]`

The power of the Laplace transform is derived from the fact that there is a one-to-one correspondence between $f(t)$ and $\mathcal{L}\{f(t)\}$. This means that if $\mathcal{L}\{f(t)\}$ is known, then $f(t)$ is uniquely determined. If $F(s) = \mathcal{L}\{f(t)\}$, then $f(t) = \mathcal{L}^{-1}\{F(s)\}$. $\mathcal{L}^{-1}$ is called the *inverse* Laplace transform.

- **InverseLaplaceTransform[F[***var1***]***,* ***var1****,* ***var2***]** computes the inverse Laplace transform of the function F, with independent variable *var1*, and expresses it as a function of *var2*.

EXAMPLE 18

```
InverseLaplaceTransform[ 1/(s - 3), s, t]
```

e^{3t}

```
InverseLaplaceTransform[ 1/(s³ - 8), s, t]
```

$\frac{1}{12} e^{-t} \left(e^{3t} - \text{Cos}[\sqrt{3}\, t] - \sqrt{3}\; \text{Sin}[\sqrt{3}\, t] \right)$

> Traditionally, one would factor the denominator, expand into partial fractions, and find the inverse transformation separately for each term. *Mathematica* does it all automatically.

The next example illustrates how the Laplace transform can be used to solve a simple differential equation.

EXAMPLE 19 Solve the differential equation $\dfrac{d^2y}{dt^2} - 3\dfrac{dy}{dt} + 2y = t^2$, $y'(0) = 1$, $y(0) = 2$. First we compute the Laplace transform of both sides of the equation. This can be done in one step.

```
equation = y''[t] - 3 y'[t] + 2 y[t] == t²;
temp = LaplaceTransform[equation, t, s]
```

$2\,\text{LaplaceTransform}[y[t], t, s] + s^2\,\text{LaplaceTransform}[y[t], t, s] -$

$\quad 3\,(s\,\text{LaplaceTransform}[y[t], t, s] - y[0]) - s\,y[0] - y'[0] == \dfrac{2}{s^3}$

Then we solve for the Laplace transform satisfying the given initial conditions.

```
temp2 = Solve[temp, LaplaceTransform[y[t], t, s]] /. {y'[0] → 1, y[0] → 2}
```

$\left\{ \left\{ \text{LaplaceTransform}[y[t], t, s] \to \dfrac{2 - 5\,s^3 + 2\,s^4}{s^3\,(2 - 3\,s + s^2)} \right\} \right\}$

Next we extract the transform as a function of s.

```
temp3 = temp2[[1, 1, 2]]
```

$\dfrac{2 - 5\,s^3 + 2\,s^4}{s^3\,(2 - 3\,s + s^2)}$

Finally, we compute the inverse Laplace transform to get the solution of the equation.

```
InverseLaplaceTransform[temp3, s, t]
```

$\dfrac{1}{4}\left(7 + 4\,e^{t} - 3\,e^{2t} + 6\,t + 2\,t^2 \right)$

As indicated at the beginning of this section, Laplace transforms are the ideal tool to use when dealing with discontinuous "right-hand sides." In this context we shall find it convenient to introduce the Heaviside theta function and the unit step function.

- **UnitStep[x]** returns a value of 0 if $x < 0$ and 1 if $x \geq 0$.
- **HeavisideTheta[x]** returns a value of 0 if $x < 0$ and 1 if $x > 0$.

The unit step function, which we represent as $u(t)$, offers a convenient way to define *piecewise defined* functions.

EXAMPLE 20 Plot the graph of $g(x) = \begin{cases} x & \text{if } x < 1 \\ x^3 & \text{if } x \geq 1 \end{cases}$ for $0 \leq x \leq 2$.

```
g[x_] = UnitStep[1 - x] x + UnitStep[x - 1] x³;
Plot[g[x], {x, 0, 2}]
```

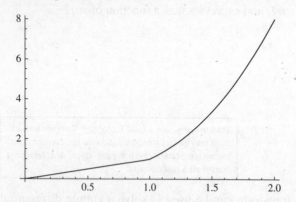

It is easily shown that $\mathcal{L}\{u(t-c)\} = \dfrac{e^{-cs}}{s}$ and, if $F(s) = \mathcal{L}\{f(t)\}$, then $\mathcal{L}\{u(t-c)f(t-c)\} = e^{-cs}F(s)$. These properties make it convenient to solve differential equations involving piecewise continuous functions.

EXAMPLE 21 Solve $\dfrac{d^2y}{dt^2} - 3\dfrac{dy}{dt} + 2y = g(t)$, $y(0) = y'(0) = 0$ where $g(t) = \begin{cases} 1 & \text{if } 0 \leq t < 1 \\ 0 & \text{if } t > 1 \end{cases}$

Plot the solution for $0 \leq t \leq 2$.

For $t \geq 0$, $g(t) = $ UnitStep[1 - t].

```
temp = LaplaceTransform[y''[t] - 3 y'[t] + 2 y[t] == UnitStep[1 - t], t, s]
```

2 LaplaceTransform[y[t], t, s] + s² LaplaceTransform[y[t], t, s] -

   3 (s LaplaceTransform[y[t], t, s] - y[0]) - s y[0] - y'[0] == $\dfrac{1 - \text{Cosh}[s] + \text{Sinh}[s]}{s}$

```
temp2=Solve[temp, LaplaceTransform[y[t], t, s]] /. {y'[0] →0, y[0] →0}
```

$\left\{\left\{\text{LaplaceTransform}[y[t], t, s \to \dfrac{1 - \text{Cosh}[s] + \text{Sinh}[s]}{s(2 - 3s + s^2)}\right\}\right\}$

```
temp3 = temp2[[1, 1, 2]]
```

$\dfrac{1 - \text{Cosh}[s] + \text{Sinh}[s] - 3 s y[0]}{s(2 - 3s + s^2)}$

```
f[t_] = InverseLaplaceTransform[temp3, s, t]
```

$\dfrac{1}{2}\left((-1 + e^t)^2 - (-1 + e^{-1+t})^2 \text{HeavisideTheta}[-1 + t]\right)$

```
Plot[f[t], {t, 0, 2}]
```

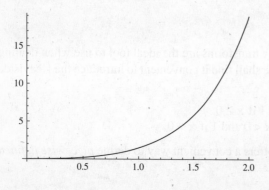

> Observe that the solution is continuous, even though the equation involves a discontinuous function.

In physical and biological applications, we are often led to differential equations whose right-hand side, $f(t)$, is a function of an *impulsive* nature, that is, $f(t)$ has zero value everywhere except over a short interval of time where its value is positive.

The Dirac delta function is an idealized impulse function. Although not a true function in the classical sense, its validity is justified by the *theory of distributions*, developed by Laurent Schwartz in the mid-twentieth century. It is defined by the following pair of conditions:

$$\delta(t - t_0) = 0 \ \text{ if } \ t \neq t_0$$

$$\int_{-\infty}^{\infty} \delta(t - t_0)dt = 1$$

An immediate consequence of the definition is the result that $\int_{-\infty}^{\infty} f(t)\delta(t - t_0)dt = f(t_0)$. It follows, therefore, that $\mathcal{L}\{\delta(t - t_0)\} = \int_0^{\infty} e^{-st}\delta(t - t_0)\,dt = e^{-st_0}$, provided that $t_0 \geq 0$. Otherwise its value is 0.

- **DiracDelta[t]** returns $\delta(t)$, the Dirac delta function, which satisfies $\delta(t) = 0$ if $t \neq 0$, $\int_{-\infty}^{\infty} \delta(t)dt = 1$.

EXAMPLE 22

```
∫∞−∞ DiracDelta[t] dt
```

1

```
∫∞−∞ f[t] DiracDelta[t - 5] dt
```

f[5]

EXAMPLE 23

```
LaplaceTransform[DiracDelta[t - a], t, s]
```
e^{-as} HeavisideTheta[a]
```
LaplaceTransform[DiracDelta[t - 3], t, s]
```
e^{-3s}
```
LaplaceTransform[DiracDelta[t + 3], t, s]
```
0

← Since *Mathematica* does not know whether *a* is negative or non-negative, HeavisideTheta[a] is included in the Laplace transform.

Since we know the Laplace transform of the Dirac delta function, we can solve differential equations involving impulses much the same way as described in Example 19. The following example illustrates the method.

EXAMPLE 24 Find the solution of the differential equation $\dfrac{d^2y}{dt^2} - 2\dfrac{dy}{dt} + y = \delta(t - 1)$, $y(0) = y'(0) = 0$.

```
equation = y''[t] - 2y'[t] + y[t] == DiracDelta[t - 1];
temp = LaplaceTransform[equation, t, s]
```

LaplaceTransform[y[t],t,s] + s² LaplaceTransform[y[t],t,s] −
   2 (s LaplaceTransform[y[t],t,s] − y[0]) − s y[0] − y'[0] == e⁻ˢ

```
temp2 = Solve[temp, LaplaceTransform[y[t], t, s]] /. {y'[0] → 0, y[0] → 0}
```

$$\left\{\left\{\text{LaplaceTransform}[y[t],t,s] \to \frac{e^{-s}}{1 - 2s + s^2}\right\}\right\}$$

```
temp3 = temp2[[1, 1, 2]]
```

$$\frac{e^{-s}}{1 - 2s + s^2}$$

```
solution[t_] = InverseLaplaceTransform[temp3, s, t]
```

e^{-1+t} (−1 + t) HeavisideTheta[−1 + t]

```
Plot[solution[t], {t, 0, 2}, PlotStyle → Thickness[.01]]
```

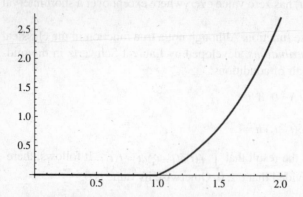

Laplace transforms can be used to solve systems of differential equations. The technique is similar to that of a single equation, except that a different transform is defined for each dependent variable. The next example illustrates the method for solving a system of two first-order equations. It generalizes in a natural way to larger and higher-order systems.

EXAMPLE 25   Solve the system $\begin{cases} \dfrac{dx}{dt} + y = t \\ 4x + \dfrac{dy}{dt} = 0 \end{cases}$ with initial condition $x(0) = 1$, $y(0) = -1$.

```
system = {x'[t] + y[t] == t, 4x[t] + y'[t] == 0};
temp = LaplaceTransform[system, t, s]
```

$\{s \, \text{LaplaceTransform}[x[t], t, s] + \text{LaplaceTransform}[y[t], t, s] - x[0] == \dfrac{1}{s^2},$

$4 \, \text{LaplaceTransform}[x[t], t, s] + s \, \text{LaplaceTransform}[y[t], t, s] - y[0] == 0\}$

```
temp2 = Solve[temp, {LaplaceTransform[x[t], t, s],
            LaplaceTransform[y[t], t, s]}] /. {x[0] → 1, y[0] → -1}
```

$\left\{\left\{\text{LaplaceTransform}[x[t], t, s] \rightarrow -\dfrac{-1 - s - s^2}{s(-4 + s^2)},\right.\right.$

$\left.\left.\text{LaplaceTransform}[y[t], t, s] \rightarrow -\dfrac{4 + 4s^2 + s^3}{s^2(-4 + s^2)}\right\}\right\}$

```
temp3a = temp2[[1, 1, 2]]
```

$\dfrac{-1 - s - s^2}{s(-4 + s^2)}$

```
temp3b = temp2[[1, 2, 2]]
```

$-\dfrac{4 + 4s^2 + s^3}{s^2(-4 + s^2)}$

```
InverseLaplaceTransform[temp3a, s, t]
```

$-\dfrac{1}{4} + \dfrac{3e^{-2t}}{8} + \dfrac{7e^{2t}}{8}$

```
InverseLaplaceTransform[temp3b, s, t]
```

$\dfrac{3e^{-2t}}{4} - \dfrac{7e^{2t}}{4} + t$

The solution to the system is

$$x = -\dfrac{1}{4} + \dfrac{3e^{-2t}}{8} + \dfrac{7e^{2t}}{8}, \quad y = \dfrac{3e^{-2t}}{4} - \dfrac{7e^{2t}}{4} + t$$

SOLVED PROBLEMS

11.18 Solve the equation $\dfrac{d^2y}{dt^2} + y = \sin t$ with initial conditions $y(0) = 0,\ y'(0) = 2$.

SOLUTION

```
equation = y''[t] + y[t] == Sin[t];
temp = LaplaceTransform[equation, t, s]
```

$$\text{LaplaceTransform}[y[t],t,s] +$$
$$s^2\,\text{LaplaceTransform}[y[t],t,s] - s\,y[0] - y'[0] == \frac{1}{1+s^2}$$

```
temp2 = Solve[temp, LaplaceTransform[y[t], t, s]] /. {y'[0] → 2, y[0] → 0}
```

$$\left\{\left\{\text{LaplaceTransform}[y[t],t,s] \to \frac{3+2\,s^2}{(1+s^2)^2}\right\}\right\}$$

```
temp3 = temp2[[1, 1, 2]];
InverseLaplaceTransform[temp3, s, t]
```

$$\frac{1}{2}\,(-t\,\text{Cos}[t] + 5\,\text{Sin}[t])$$

11.19 Solve $\dfrac{d^2y}{dt^2} + \dfrac{dy}{dt} + y = e^t,\ y(0) = 3,\ y'(0) = 2$.

SOLUTION

```
equation = y''[t] + y'[t] + y[t] == Exp[t];
temp = LaplaceTransform[equation, t, s]
```

$$\text{LaplaceTransform}[y[t],t,s] + s\,\text{LaplaceTransform}[y[t],t,s] +$$
$$s^2\,\text{LaplaceTransform}[y[t],t,s] - y[0] - s\,y[0] - y'[0] == \frac{1}{-1+s}$$

```
temp2 = Solve[temp, LaplaceTransform[y[t], t, s]] /. {y'[0] → 2, y[0] → 3}
```

$$\left\{\left\{\text{LaplaceTransform}[y[t],t,s] \to \frac{-4+2\,s+3\,s^2}{(-1+s)\,(1+s+s^2)}\right\}\right\}$$

```
temp3 = temp2[[1, 1, 2]];
InverseLaplaceTransform[temp3, s, t]
```

$$\frac{1}{3}\,e^{-t/2}\left(e^{3t/2} + 8\,\text{Cos}\left[\frac{\sqrt{3}\,t}{2}\right] + 6\,\sqrt{3}\,\text{Sin}\left[\frac{\sqrt{3}\,t}{2}\right]\right)$$

11.20 Solve the equation $\dfrac{d^2y}{dt^2} - 2\dfrac{dy}{dt} + y = g(t),\ y(0) = y'(0) = 0$, where $g(t) = \begin{cases} t & \text{if } 0 \le t \le 1 \\ t^2 & \text{if } t > 1 \end{cases}$ and plot the solution for $0 \le x \le 4$.

SOLUTION

```
equation = y''[t] - 2y'[t] + y[t] == t UnitStep[1 - t] + t² UnitStep[t - 1];
temp = LaplaceTransform[equation, t, s]
```

$$\text{LaplaceTransform}[y[t],t,s] + s^2\,\text{LaplaceTransform}[y[t],t,s] -$$
$$2\,(s\,\text{LaplaceTransform}[y[t],t,s] - y[0]) - s\,y[0] - y'[0] ==$$
$$\frac{e^{-s}\,(2+2\,s+s^2)}{s^3} + \frac{1 - \text{Cosh}[s] - s\,\text{Cosh}[s] + \text{Sinh}[s] + s\,\text{Sinh}[s]}{s^2}$$

```
temp2 = Solve[temp, LaplaceTransform[y[t], t, s]] /.{y'[0] → 0, y[0] → 0}
```

$$\{\{\text{LaplaceTransform}[y[t], t, s] \to$$

$$\frac{e^{s}(2 + 2\,s + e^{s}\,s + s^2 - e^{s}\,s\,\text{Cosh}[s] - e^{s}\,s^2\,\text{Cosh}[s] + e^{s}\,s\,\text{Sinh}[s] + e^{s}\,s^2\,\text{Sinh}[s])}{s^3(1-2\,s+s^2)}\}\}$$

```
temp3 = temp2[[1, 1, 2]]
```

$$\frac{e^{-s}(2 + 2s + e^s s + s^2 - e^s s \, Cosh[s] - e^s s^2 \, Cosh[s] + e^s s \, Sinh[s] + e^s s^2 \, Sinh[s])}{s^3(1 - 2s + s^2)}\}\}$$

```
f[t_] = InverseLaplaceTransform[temp3, s, t]
```

$$\frac{e(2 + e^t(-2 + t) + t) + (e^t(-11 + 3t) + e(4 + 3t + t^2)) \, HeavisideTheta[-1 + t]}{e}$$

```
Plot[f[t], {t, 0, 4}]
```

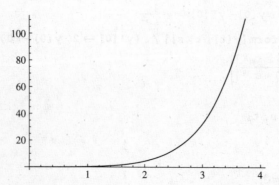

11.21   Solve the system

$$\begin{cases} \dfrac{dx}{dt} + y = t \sin t \\[2mm] x + \dfrac{dy}{dt} = t \cos t \end{cases} \qquad x(0) = y(0) = 0$$

SOLUTION

```
system = {x'[t] + y[t] == t Sin[t], x[t] + y'[t] == t Cos[t]};
temp = LaplaceTransform[system, t, s]
```

$$\{s \, LaplaceTransform[x[t], t, s] + LaplaceTransform[y[t], t, s] -$$
$$x[0] == \frac{2s}{(1 + s^2)^2}, \, LaplaceTransform[x[t], t, s] +$$
$$s \, LaplaceTransform[y[t], t, s] - y[0] == \frac{-1 + s^2}{(1 + s^2)^2}\}$$

```
temp2 = Solve[temp, {LaplaceTransform[x[t], t, s],
 LaplaceTransform[y[t], t, s]}] /. {x[0] → 0, y[0] → 0}
```

$$\{\{LaplaceTransform[x[t], t, s] \to \frac{1}{(-1 + s^2)(1 + s^2)},$$
$$LaplaceTransform[y[t], t, s] \to -\frac{3s - s^3}{(-1 + s^2)(1 + s^2)^2}\}\}$$

```
temp3a = temp2[[1, 1, 2]];
temp3b = temp2[[1, 2, 2]];
InverseLaplaceTransform[temp3a, s, t] //Simplify
```

$$\frac{1}{4}(-e^{-t} + e^t - 2 \, Sin[t])$$

```
InverseLaplaceTransform[temp3b, s, t] //Simplify
```

$$\frac{1}{4}(-e^{-t} - e^t + 2 \, Cos[t] + 4 \, t \, Sin[t])$$

The solution of the system is

$$\begin{cases} x = \dfrac{1}{4}(-e^{-t} + e^t - 2\sin t) \\[2mm] y = \dfrac{1}{4}(-e^{-t} - e^t + 2\cos t + 4t\sin t) \end{cases}$$

11.22 The equation governing the amount of current, *I*, flowing through a simple resistance-inductance circuit when an EMF (voltage) *E* is applied is $L\frac{dI}{dt} + RI = E$. The units for *E*, *I*, and *L* are, respectively, volts, amperes, and henries. Suppose $L = 1$ and $R = 10$. If 1 volt is applied at time $t = 0$ and removed 1 sec later, plot the current in the circuit during the first 2 seconds.

SOLUTION

```
e[t_] = UnitStep[1 - t];
Plot[e[t], {t, 0, 2}, PlotStyle → Thickness[.01]];
```

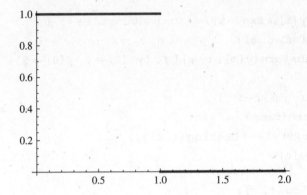

```
l = 1; r = 10;
equation = l i'[t] + r i[t] == e[t];
temp = LaplaceTransform[equation, t, s]
```

$-i[0] + 10\,\text{LaplaceTransform}[i[t], t, s] +$
$\quad s\,\text{LaplaceTransform}[i[t], t, s] == \dfrac{1 - \text{Cosh}[s] + \text{Sinh}[s]}{s}$

```
temp2 = Solve[temp, LaplaceTransform[i[t], t, s]] /. i[0] → 0
```

$\left\{\left\{\text{LaplaceTransform}[i[t], t, s] \rightarrow \dfrac{1 - \text{Cosh}[s] + \text{Sinh}[s]}{s(10 + s)}\right\}\right\}$

> The initial current is assumed to be 0.

```
temp3 = temp2[[1, 1, 2]]
```

$\dfrac{1 - \text{Cosh}[s] + \text{Sinh}[s]}{s(10 + s)}$

```
f[t_] = InverseLaplaceTransform[temp3, s, t]
```

$\frac{1}{10} e^{-10t} (-1 + e^{10t} + (e^{10} - e^{10t})\,\text{HeavisideTheta}[-1 + t])$

```
Plot[f[t], {t, 0, 2}, PlotStyle → Thickness[.01],
     AxesLabel → {"Time", "Current"}]
```

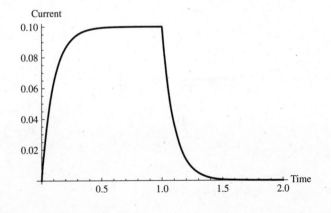

11.23 A particle of mass *m* is attached to one end of a spring and allowed to come to rest in an equilibrium position. If an external force, $f(t)$, is then applied to the particle, its motion is described by the equation $m\frac{d^2y}{dt^2} + a\frac{dy}{dt} + ky = f(t)$, $y'(0) = 0$, $y(0) = 0$ where *a* is a damping constant and *k* is the spring's stiffness constant. Assuming $m = 1$, $a = 2$, $k = 1$, and $f(t) = e^{-t}$, describe the motion of the spring. Then determine the motion of the spring if an impulse of 1 lb-sec is applied after 1 sec. Plot both graphs on one set of axes.

SOLUTION

```
m = 1; a = 2; k = 1;
equation = m y''[t] + a y'[t] + k y[t] == Exp[-t];  (* without impulse *)
temp = LaplaceTransform[equation, t, s];
temp2 = Solve[temp, LaplaceTransform[y[t], t, s]] /. {y'[0]→0, y[0]→0};
temp3 = temp2[[1, 1, 2]];
Print["Solution Without Impulse (dashed)"]
f1[t_] = InverseLaplaceTransform[temp3, s, t]
g1 = Plot[f1[t], {t, 0, 10}, PlotStyle → Dashing[{.01}]];
equation = m y''[t] + a y'[t] + k y[t] ==
  Exp[-t] + DiracDelta[t - 1]      (* with impulse *)
temp = LaplaceTransform[equation, t, s];
temp2 = Solve[temp, LaplaceTransform[y[t], t, s]] /. {y'[0] →0, y[0] →0};
temp3 = temp2[[1, 1, 2]];
Print["Solution With Impulse (solid)"]
f2[t_] = InverseLaplaceTransform[temp3, s, t]
g2 = Plot[f2[t], {t, 0, 10}];
```

Solution Without Impulse (dashed)

$\frac{1}{2} e^{-t} t^2$

Solution With Impulse (solid)

$\frac{1}{2} e^{-t} t^2 + e^{1-t}(-1 + t) \text{UnitStep}[-1 + t]$

```
Show[g1, g2 , PlotRange → All]
```

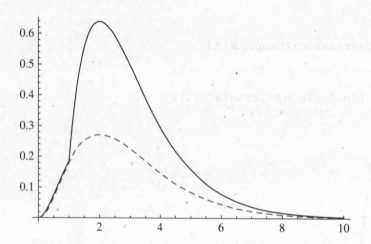

CHAPTER 12

Linear Algebra

12.1 Vectors and Matrices

Vectors and matrices are represented as lists (Chapter 3) in *Mathematica*. A vector is a simple list and a matrix is a list of vectors.

The elements of a vector or a matrix may be entered manually as a list or, more conveniently, by the use of built-in commands.

Vectors

- **Table[*expression*, {i, n}]** constructs an n-dimensional vector whose elements are the values of *expression* for i = 1, 2, 3, . . . , n.
- **Array[f, n]** generates an n-dimensional vector whose elements are f[1], f[2], . . . , f[n]. f is a function of one variable.

Matrices

- **Table[*expression*, {i, m}, {j, n}]** constructs an m × n matrix whose elements are the values of *expression* for (i, j) = (1, 1), . . . ,(m,n).
- **Array[f, {m, n}]** generates an m × n matrix whose elements are f[1,1], . . . , f[m,n]. f is a function of two variables.
- **DiagonalMatrix[*list*]** creates a diagonal matrix whose diagonal entries are the elements of the one-dimensional array *list*.
- **IdentityMatrix[n]** creates an n × n identity matrix.

Although matrices are represented as lists, they may be viewed as matrices by using the **MatrixForm** command.

- **MatrixForm[*list*]** prints the elements of the two-dimensional array *list* in a rectangular arrangement enclosed by brackets. If *list* is a simple (one-dimensional) array, **MatrixForm** prints it as a column vector, i.e., an *n* × 1 matrix.

Using **//MatrixForm** to the right of *list* is equivalent to **MatrixForm[*list*]** and is a bit more convenient. Care must be taken, however, not to use **//MatrixForm** in the definition of the matrix. This command is for display purposes only. (See Problem 12.3.)

EXAMPLE 1

```
m = {{1, 1}, {1, 2}};
m //MatrixForm
```
$$\begin{pmatrix} 1 & 1 \\ 1 & 2 \end{pmatrix}$$

Additionally, a matrix can be introduced via the menu Insert ⇒ Table/Matrix ⇒ New . . . (On a PC, a matrix can also be inserted by right-clicking the mouse and selecting Insert Table/Matrix . . .)

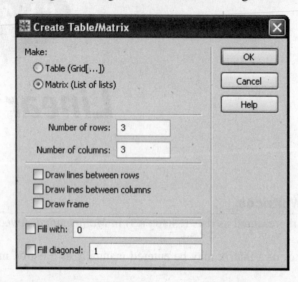

This produces a grid as shown. Once the grid has been set up, you can conveniently enter the numbers, using the [TAB] key to go from cell to cell. Options for filling with 0s and 1s are particularly convenient for large, sparse matrices.

$$\begin{pmatrix} \square & \square & \square \\ \square & \square & \square \\ \square & \square & \square \end{pmatrix}$$

EXAMPLE 2   A matrix is a list in *Mathematica*. Even if it is input via the Create Table/Matrix menu, it is represented internally as a list of depth 2.

$$m = \begin{pmatrix} 1 & 2 & 3 \\ 4 & 5 & 6 \\ 7 & 8 & 9 \end{pmatrix}$$ ← Matrix m is created using the Create Table/Matrix menu.

{{1, 2, 3}, {4, 5, 6}, {7, 8, 9}}

EXAMPLE 3   A vector can be represented as a simple list or as an $n \times 1$ matrix. Either way, **MatrixForm** will print it as a column vector.

v = {1, 2, 3, 4, 5};
v // MatrixForm

$$\begin{pmatrix} 1 \\ 2 \\ 3 \\ 4 \\ 5 \end{pmatrix}$$

$$v = \begin{pmatrix} 1 \\ 2 \\ 3 \\ 4 \\ 5 \end{pmatrix}$$ ← v is input using Create Table/Matrix.

{{1}, {2}, {3}, {4}, {5}}   ← v is output as a list.

v // MatrixForm

$$\begin{pmatrix} 1 \\ 2 \\ 3 \\ 4 \\ 5 \end{pmatrix}$$

EXAMPLE 4   To generate a vector whose entries are the squares of the first five consecutive integers, we could simply enter them by hand.

```
squares = {1, 4, 9, 16, 25}
```
{1, 4, 9, 16, 25}

More conveniently, however, we can use the **Table** or **Array** command.

```
squares = Table[i², {i, 5}]
```
{1, 4, 9, 16, 25}
```
f[i_] = i²;
```
```
squares = Array[f, 5]
```
{1, 4, 9, 16, 25}

To view as a vector,

```
MatrixForm[squares]   or   squares //MatrixForm
```

$$\begin{pmatrix} 1 \\ 4 \\ 9 \\ 16 \\ 25 \end{pmatrix}$$

EXAMPLE 5   We will construct a 5×7 matrix whose entries are the sum of its row and column positions. For example, $a_{2,3} = 5$. We could, of course, input the entries directly using the tool in Insert $\Rightarrow$ Table/Matrix $\Rightarrow$ New . . . , but it is certainly preferable to use one of the standard *Mathematica* commands. Here are two ways it can be done:

```
matrix = Table[i + j, {i, 5}, {j, 7}]
```
{{2, 3, 4, 5, 6, 7, 8}, {3, 4, 5, 6, 7, 8, 9}, {4, 5, 6, 7, 8, 9, 10}, {5, 6, 7, 8, 9, 10, 11}, {6, 7, 8, 9, 10, 11, 12}}

```
f[i_, j_] = i + j;
```
```
matrix = Array[f, {5, 7}]
```
{{2, 3, 4, 5, 6, 7, 8}, {3, 4, 5, 6, 7, 8, 9}, {4, 5, 6, 7, 8, 9, 10}, {5, 6, 7, 8, 9, 10, 11}, {6, 7, 8, 9, 10, 11, 12}}

Either way we can view the generated array as a matrix.

```
matrix //MatrixForm
```

$$\begin{pmatrix} 2 & 3 & 4 & 5 & 6 & 7 & 8 \\ 3 & 4 & 5 & 6 & 7 & 8 & 9 \\ 4 & 5 & 6 & 7 & 8 & 9 & 10 \\ 5 & 6 & 7 & 8 & 9 & 10 & 11 \\ 6 & 7 & 8 & 9 & 10 & 11 & 12 \end{pmatrix}$$

EXAMPLE 6   Submatrices can be constructed from a given matrix by careful implementation of [[ ]] (see the **Part** command, Chapter 3). First we construct a 5×5 matrix of consecutive integers.

```
matrix = Partition[Range[25],5];
```
```
matrix //MatrixForm
```

$$\begin{pmatrix} 1 & 2 & 3 & 4 & 5 \\ 6 & 7 & 8 & 9 & 10 \\ 11 & 12 & 13 & 14 & 15 \\ 16 & 17 & 18 & 19 & 20 \\ 21 & 22 & 23 & 24 & 25 \end{pmatrix}$$

We can obtain a particular element of the matrix, say the element in row 3, column 4.

```
matrix[[3, 4]]
```

14

If we want the entire fourth row, we extract the fourth sublist from `matrix`.

```
matrix[[4]]
```

{16, 17, 18, 19, 20}

The entire fourth column can be obtained using the **All** directive.

```
matrix[[All, 4]]
```

{4, 9, 14, 19, 24}

We can obtain the submatrix whose elements are in rows 1, 3, and 5 and columns 2 and 4.

```
matrix[[{1, 3, 5}, {2, 4}]] // MatrixForm
```

$$\begin{pmatrix} 2 & 4 \\ 12 & 14 \\ 22 & 24 \end{pmatrix}$$

Or we can obtain the 3 × 5 matrix consisting of **matrix** with rows 2 and 4 deleted.

```
matrix[[{1, 3, 5}, All]] // MatrixForm
```

$$\begin{pmatrix} 1 & 2 & 3 & 4 & 5 \\ 11 & 12 & 13 & 14 & 15 \\ 21 & 22 & 23 & 24 & 25 \end{pmatrix}$$

With careful use of the **Take** command (Chapter 3) we can even construct the submatrix of `matrix` consisting of those elements in rows 2 through 4 and columns 3 through 5.

```
Take[matrix, {2, 4}, {3, 5}] // MatrixForm
```

$$\begin{pmatrix} 8 & 9 & 10 \\ 13 & 14 & 15 \\ 18 & 19 & 20 \end{pmatrix}$$

Although many matrices can be created using **Table** or **Array**, *Mathematica* offers some commands for constructing certain specialized matrices.

- **ConstantArray[c, {m, n}]** generates an m × n array, each element of which is c.
- **HilbertMatrix[n]** creates an n × n Hilbert matrix
- **HilbertMatrix[m, n]** creates an m × n Hilbert matrix.
- **HankelMatrix[n]** creates a Hankel matrix whose first row (and column) is {1, 2, 3, . . . , n}
- **HankelMatrix[n, *list*]** creates a Hankel matrix whose first row (and column) is *list*.

EXAMPLE 7

```
ConstantArray[0, {3, 5}] //MatrixForm
```

$$\begin{pmatrix} 0 & 0 & 0 & 0 & 0 \\ 0 & 0 & 0 & 0 & 0 \\ 0 & 0 & 0 & 0 & 0 \end{pmatrix}$$

```
HilbertMatrix[5] //MatrixForm
```

$$\begin{pmatrix} 1 & \frac{1}{2} & \frac{1}{3} & \frac{1}{4} & \frac{1}{5} \\ \frac{1}{2} & \frac{1}{3} & \frac{1}{4} & \frac{1}{5} & \frac{1}{6} \\ \frac{1}{3} & \frac{1}{4} & \frac{1}{5} & \frac{1}{6} & \frac{1}{7} \\ \frac{1}{4} & \frac{1}{5} & \frac{1}{6} & \frac{1}{7} & \frac{1}{8} \\ \frac{1}{5} & \frac{1}{6} & \frac{1}{7} & \frac{1}{8} & \frac{1}{9} \end{pmatrix}$$

```
HankelMatrix[{a, b, c, d, e}] //MatrixForm
```

$$\begin{pmatrix} a & b & c & d & e \\ b & c & d & e & 0 \\ c & d & e & 0 & 0 \\ d & e & 0 & 0 & 0 \\ e & 0 & 0 & 0 & 0 \end{pmatrix}$$

SOLVED PROBLEMS

12.1 Construct a ten-dimensional vector of powers of 2.

SOLUTION

```
powersof2 = Table[2^k, {k, 1, 10}]
```

{2, 4, 8, 16, 32, 64, 128, 256, 512, 1024}

```
powersof2 // MatrixForm
```

$$\begin{pmatrix} 2 \\ 4 \\ 8 \\ 16 \\ 32 \\ 64 \\ 128 \\ 256 \\ 512 \\ 1024 \end{pmatrix}$$

12.2 Construct a 5×5 matrix of random digits.

SOLUTION

```
Table[RandomInteger[9], {i, 5}, {j, 5}] // MatrixForm
```

$$\begin{pmatrix} 5 & 7 & 9 & 9 & 4 \\ 2 & 8 & 6 & 9 & 7 \\ 1 & 0 & 8 & 2 & 2 \\ 7 & 8 & 8 & 8 & 1 \\ 9 & 2 & 2 & 3 & 4 \end{pmatrix}$$

12.3 What happens if **//MatrixForm** is included within the definition of a matrix?

SOLUTION

```
m1 = {{1, 1}, {1, 2}} //MatrixForm
m2 = {{2, 3}, {4, 5}} //MatrixForm
```

$$\begin{pmatrix} 1 & 1 \\ 1 & 2 \end{pmatrix}$$

$$\begin{pmatrix} 2 & 3 \\ 4 & 5 \end{pmatrix}$$

```
m1 + m2
```

$$\begin{pmatrix} 1 & 1 \\ 1 & 2 \end{pmatrix} + \begin{pmatrix} 2 & 3 \\ 4 & 5 \end{pmatrix}$$ ← We do *not* get the sum of the two matrices.

Mathematica cannot perform the indicated operation because m1 and m2 are not lists. Now we do it correctly.

```
m1 = {{1, 1}, {1, 2}}
m2 = {{2, 3}, {4, 5}}
{{1, 1}, {1, 2}}
{{2, 3}, {4, 5}}
m1 + m2 //MatrixForm
```

$$\begin{pmatrix} 3 & 4 \\ 5 & 7 \end{pmatrix}$$

12.4   Construct a 10×10 diagonal matrix whose diagonal entries are the first ten primes.

SOLUTION

```
primelist = Array[Prime, 10]
```
| `Prime` is a built-in *Mathematica* function. |

$\{2, 3, 5, 7, 11, 13, 17, 19, 23, 29\}$

```
DiagonalMatrix[primelist]  //MatrixForm
```

$$\begin{pmatrix} 2 & 0 & 0 & 0 & 0 & 0 & 0 & 0 & 0 & 0 \\ 0 & 3 & 0 & 0 & 0 & 0 & 0 & 0 & 0 & 0 \\ 0 & 0 & 5 & 0 & 0 & 0 & 0 & 0 & 0 & 0 \\ 0 & 0 & 0 & 7 & 0 & 0 & 0 & 0 & 0 & 0 \\ 0 & 0 & 0 & 0 & 11 & 0 & 0 & 0 & 0 & 0 \\ 0 & 0 & 0 & 0 & 0 & 13 & 0 & 0 & 0 & 0 \\ 0 & 0 & 0 & 0 & 0 & 0 & 17 & 0 & 0 & 0 \\ 0 & 0 & 0 & 0 & 0 & 0 & 0 & 19 & 0 & 0 \\ 0 & 0 & 0 & 0 & 0 & 0 & 0 & 0 & 23 & 0 \\ 0 & 0 & 0 & 0 & 0 & 0 & 0 & 0 & 0 & 29 \end{pmatrix}$$

12.5   Construct a 5×5 upper triangular matrix of 1s with 0s below the main diagonal.

SOLUTION

```
m = Table[If[i < = j, 1, 0], {i, 5}, {j, 5}];
m // MatrixForm
```

$$\begin{pmatrix} 1 & 1 & 1 & 1 & 1 \\ 0 & 1 & 1 & 1 & 1 \\ 0 & 0 & 1 & 1 & 1 \\ 0 & 0 & 0 & 1 & 1 \\ 0 & 0 & 0 & 0 & 1 \end{pmatrix}$$

12.6   Construct a 7×7 tridiagonal matrix with 2s on the main diagonal, 1s on the diagonals adjacent to the main diagonal, and 0s elsewhere.

SOLUTION

```
m = Table[If[Abs[i - j] == 1, 1, If[i == j, 2, 0]], {i, 1, 7}, {j, 1, 7}];
m // MatrixForm
```

$$\begin{pmatrix} 2 & 1 & 0 & 0 & 0 & 0 & 0 \\ 1 & 2 & 1 & 0 & 0 & 0 & 0 \\ 0 & 1 & 2 & 1 & 0 & 0 & 0 \\ 0 & 0 & 1 & 2 & 1 & 0 & 0 \\ 0 & 0 & 0 & 1 & 2 & 1 & 0 \\ 0 & 0 & 0 & 0 & 1 & 2 & 1 \\ 0 & 0 & 0 & 0 & 0 & 1 & 2 \end{pmatrix}$$

12.7   Let M be the 6×6 matrix containing the integers 1 through 36. Construct a 3×3 matrix consisting of the elements in the odd rows and even columns of M.

SOLUTION

```
m = Table[6i + j, {i, 0, 5}, {j, 1, 6}];
m // MatrixForm
```

$$\begin{pmatrix} 1 & 2 & 3 & 4 & 5 & 6 \\ 7 & 8 & 9 & 10 & 11 & 12 \\ 13 & 14 & 15 & 16 & 17 & 18 \\ 19 & 20 & 21 & 22 & 23 & 24 \\ 25 & 26 & 27 & 28 & 29 & 30 \\ 31 & 32 & 33 & 34 & 35 & 36 \end{pmatrix}$$

```
m[[{1, 3, 5}, {2, 4, 6}]] // MatrixForm
```

$$\begin{pmatrix} 2 & 4 & 6 \\ 14 & 16 & 18 \\ 26 & 28 & 30 \end{pmatrix}$$

12.2   Matrix Operations

Since vectors and matrices are stored as lists in *Mathematica*, all list operations described in Chapter 3 apply. In addition, there are some specialized commands that are applicable specifically to matrices. Since n-dimensional vectors can be considered to be $n \times 1$ matrices, many of these commands apply to vectors as well. In the following descriptions, m, m1, and m2 denote matrices and v1 and v2 denote vectors.

- **m1 + m2** computes the sum of two matrices.
- **m1 – m2** computes the difference of two matrices.
- **c m** multiplies each element of m by the scalar c.
- **m1 . m2** computes the matrix product of m1 and m2. **v1 . v2** computes the dot product of v1 and v2. For matrices, the operation returns a list; for vectors, a single number is returned.
- **Cross[v1, v2]** or **v1 × v2** returns the cross product of v1 and v2. (This applies to three-dimensional vectors only.) The cross product symbol, ×, can be inserted into the calculation by typing (without spaces) the key sequence [ESC]c-r-o-s-s[ESC] . (Do not confuse this with the × on the Basic Math Input palette. The latter represents simple multiplication.)

EXAMPLE 8   First we generate two 3×3 "random" matrices as lists.

```
m1 = Table[RandomInteger[9], {i, 1, 3}, {j, 1, 3}]
```
```
{{9, 4, 2}, {2, 9, 3}, {0, 1, 4}}
```
```
m2 = Table[RandomInteger[9], {i, 1, 3}, {j, 1, 3}]
```
```
{{2, 8, 1}, {8, 3, 4}, {6, 4, 0}}
```

Now we look at them in matrix form.

```
m1 // MatrixForm
```

$$\begin{pmatrix} 9 & 4 & 2 \\ 2 & 9 & 3 \\ 0 & 1 & 4 \end{pmatrix}$$

```
m2 // MatrixForm
```

$$\begin{pmatrix} 2 & 8 & 1 \\ 8 & 3 & 4 \\ 6 & 4 & 0 \end{pmatrix}$$

The next operation multiplies each element of m1 by 5.

5 m1

$$\begin{pmatrix} 45 & 20 & 10 \\ 10 & 45 & 14 \\ 0 & 5 & 20 \end{pmatrix}$$

Next compute their sum, difference, and product.

m1 + m2 // MatrixForm

$$\begin{pmatrix} 11 & 12 & 3 \\ 10 & 12 & 7 \\ 6 & 5 & 4 \end{pmatrix}$$

m1 - m2 // MatrixForm

$$\begin{pmatrix} 7 & -4 & 1 \\ -6 & 6 & -1 \\ -6 & -3 & 4 \end{pmatrix}$$

m1. m2 // MatrixForm

$$\begin{pmatrix} 62 & 92 & 25 \\ 94 & 55 & 38 \\ 32 & 19 & 4 \end{pmatrix}$$

Care must be taken not to use ∗ between the matrices to be multiplied, as this simply multiplies corresponding entries of the matrices, in accordance with list conventions.

m1 ∗ m2 // MatrixForm

$$\begin{pmatrix} 18 & 32 & 2 \\ 16 & 27 & 12 \\ 0 & 4 & 0 \end{pmatrix}$$

EXAMPLE 9

```
v1 = {1, 2, 3};
v2 = {4, 5, 6};
v1.v2
```

32   ← *Mathematica* expresses the dot product as a number rather
                                  than as a list containing a single entry.

Cross [v1, v2]

$\{-3, 6, -3\}$

Mathematica makes no distinction between row and column vectors. Therefore, if v is an *n*-dimensional vector and m is an $n \times n$ matrix, both v.m and m.v are defined (although they generally yield different results). Furthermore, if v1 is an $n \times 1$ matrix (row vector) and v2 is a $1 \times n$ matrix (column vector), v1.v2 should be an $n \times n$ matrix, but *Mathematica* still computes a dot product. The command **Outer** can be used to compute the "outer" product of two vectors.

- **Outer [Times, v1, v2]** computes the outer product of v1 and v2. (**Outer** is much more general and can be used for other purposes. See the Documentation Center for additional information.)

EXAMPLE 10

```
v1 = {1, 2, 3};
v2 = {4, 5, 6};
m = {{1, 2, 2}, {2, 3, 3}, {3, 1, 2}};
m //MatrixForm
```

$$\begin{pmatrix} 1 & 2 & 2 \\ 2 & 3 & 3 \\ 3 & 1 & 2 \end{pmatrix}$$

```
m.v1
{11, 17, 11}
```

$$\begin{pmatrix} 1 & 2 & 2 \\ 2 & 3 & 3 \\ 3 & 1 & 2 \end{pmatrix} \begin{pmatrix} 1 \\ 2 \\ 3 \end{pmatrix} = \begin{pmatrix} 11 \\ 17 \\ 11 \end{pmatrix}$$

```
v1.m
{14, 11, 14}
```

$$(1 \quad 2 \quad 3) \begin{pmatrix} 1 & 2 & 2 \\ 2 & 3 & 3 \\ 3 & 1 & 2 \end{pmatrix} = \begin{pmatrix} 14 \\ 11 \\ 14 \end{pmatrix}$$

```
Outer[Times, v1, v2] //MatrixForm
```

$$\begin{pmatrix} 4 & 5 & 6 \\ 8 & 10 & 12 \\ 12 & 15 & 18 \end{pmatrix}$$

$$\begin{pmatrix} 1 \\ 2 \\ 3 \end{pmatrix} (4 \quad 5 \quad 6) = \begin{pmatrix} 4 & 5 & 6 \\ 8 & 10 & 12 \\ 12 & 15 & 18 \end{pmatrix}$$

- **Inverse [*matrix*]** computes the inverse of *matrix*.
- **Det [*matrix*]** computes the determinant of *matrix*.
- **Transpose [*matrix*]** computes the transpose of *matrix*.
- **Tr [*matrix*]** computes the trace of *matrix*.
- **MatrixPower [*matrix*, n]** computes the nth power of *matrix*.
- **Minors [*matrix*]** produces a matrix whose (i, j)th entry is the determinant of the submatrix obtained from *matrix* by deleting row $n - i + 1$ and column $n - j + 1$. (*matrix* must be square.)
- **Minors [*matrix*, k]** produces the matrix whose entries are the determinants of all possible k × k submatrices of *matrix*. (*matrix* need not be square.)

EXAMPLE 11

$$m1 = \begin{pmatrix} 1 & 2 & 2 \\ 2 & 3 & 3 \\ 3 & 4 & 5 \end{pmatrix};$$

$$m2 = \begin{pmatrix} 1 & 2 & 3 & 4 \\ 5 & 6 & 7 & 8 \\ 9 & 10 & 11 & 12 \end{pmatrix};$$

> These, and subsequent examples, were created using Create Table/Matrix in the Insert menu.

```
Inverse[m1] //MatrixForm
```

$$\begin{pmatrix} -3 & 2 & 0 \\ 1 & 1 & -1 \\ 1 & -2 & 1 \end{pmatrix}$$

```
Tr[m1]
9
MatrixPower[m1, 3] // MatrixForm
```

$$\begin{pmatrix} 97 & 142 & 160 \\ 151 & 221 & 249 \\ 231 & 338 & 381 \end{pmatrix}$$

```
Transpose[m2] //MatrixForm
```

$$\begin{pmatrix} 1 & 5 & 9 \\ 2 & 6 & 10 \\ 3 & 7 & 11 \\ 4 & 8 & 12 \end{pmatrix}$$

```
Tr[m2]
```

18

> The matrix does not have to be square in order for its trace to be defined.

EXAMPLE 12

```
m = Table[a[i, j], {i, 1, 3}, {j, 1, 3}];
m //MatrixForm
```

$$\begin{pmatrix} a[1,1] & a[1,2] & a[1,3] \\ a[2,1] & a[2,2] & a[2,3] \\ a[3,1] & a[3,2] & a[3,3] \end{pmatrix}$$

```
Minors[m] //MatrixForm
```

$$\begin{pmatrix} -a[1,2]a[2,1]+a[1,1]a[2,2] & -a[1,3]a[2,1]+a[1,1]a[2,3] & -a[1,3]a[2,2]+a[1,2]a[2,3] \\ -a[1,2]a[3,1]+a[1,1]a[3,2] & -a[1,3]a[3,1]+a[1,1]a[3,3] & -a[1,3]a[3,2]+a[1,2]a[3,3] \\ -a[2,2]a[3,1]+a[2,1]a[3,2] & -a[2,3]a[3,1]+a[2,1]a[3,3] & -a[2,3]a[3,2]+a[2,2]a[3,3] \end{pmatrix}$$

SOLVED PROBLEMS

12.8 The (Euclidean) norm of a vector is the square root of the sum of the squares of its components. Compute the norm of the vector $(1, 3, 5, 7, 9, 11, 13, 15)$.

SOLUTION

```
v = Table[2k - 1, {k, 1, 8}]
```
$\{1, 3, 5, 7, 9, 11, 13, 15\}$

```
norm = √v.v
```

$2\sqrt{170}$

12.9 Prove that the cross product of two vectors in $\mathbb{R}^3$ is orthogonal to each of the vectors that form it.

SOLUTION

Let $\mathbf{u} = (a, b, c)$ and $\mathbf{v} = (d, e, f)$ and compute $\mathbf{w} = \mathbf{u} \times \mathbf{v}$. Then verify that $\mathbf{w}\perp\mathbf{u}$ and $\mathbf{w}\perp\mathbf{v}$. Two vectors are orthogonal ($\perp$) if their dot product is 0.

```
u = {u1, u2, u3};
v = {v1, v2, v3};
w = Cross[u, v]
```
$\{- u3\ v2 + u2\ v3,\ u3\ v1 - u1\ v3,\ - u2\ v1 + u1\ v2\}$

```
w.u // Expand
```
0

```
w.v // Expand
```
0

12.10 It can be shown that the volume of a parallelepiped formed by **u**, **v**, and **w** is $|\mathbf{u} \bullet (\mathbf{v} \times \mathbf{w})|$. Compute the volume of the parallelepiped formed by $\mathbf{i} + 2\mathbf{j} - 3\mathbf{k}$, $2\mathbf{i} - 5\mathbf{j} + \mathbf{k}$, and $3\mathbf{i} + \mathbf{j} + 2\mathbf{k}$. (The quantity $\mathbf{u} \bullet (\mathbf{v} \times \mathbf{w})$ is called the scalar triple product.)

SOLUTION

```
u = {1, 2, -3};
v = {2, -5, 1};
w = {3, 1, 2};
volume = Abs[u.Cross[v, w]]
64
```

12.11 Let $\mathbf{u} = (u1, u2, u3)$, $\mathbf{v} = (v1, v2, v3)$, and $\mathbf{w} = (w1, w2, w3)$. Prove that the scalar triple product,

$$\mathbf{u} \bullet (\mathbf{v} \times \mathbf{w}) = \begin{vmatrix} u1 & u2 & u3 \\ v1 & v2 & v3 \\ w1 & w2 & w3 \end{vmatrix}.$$

SOLUTION

```
u = {u1, u2, u3};
v = {v1, v2, v3};
w = {w1, w2, w3};
matrix = {{u1, u2, u3}, {v1, v2, v3}, {w1, w2, w3}};
lhs = u.Cross[v, w] // Expand;
rhs = Det[matrix] // Expand;
lhs == rhs
True
```

12.12 Construct the Hilbert matrix of order 6, and compute its determinant and its inverse.

SOLUTION

```
hilbert = HilbertMatrix[6];
hilbert // MatrixForm
```

$$\begin{pmatrix} 1 & \frac{1}{2} & \frac{1}{3} & \frac{1}{4} & \frac{1}{5} & \frac{1}{6} \\ \frac{1}{2} & \frac{1}{3} & \frac{1}{4} & \frac{1}{5} & \frac{1}{6} & \frac{1}{7} \\ \frac{1}{3} & \frac{1}{4} & \frac{1}{5} & \frac{1}{6} & \frac{1}{7} & \frac{1}{8} \\ \frac{1}{4} & \frac{1}{5} & \frac{1}{6} & \frac{1}{7} & \frac{1}{8} & \frac{1}{9} \\ \frac{1}{5} & \frac{1}{6} & \frac{1}{7} & \frac{1}{8} & \frac{1}{9} & \frac{1}{10} \\ \frac{1}{6} & \frac{1}{7} & \frac{1}{8} & \frac{1}{9} & \frac{1}{10} & \frac{1}{11} \end{pmatrix}$$

```
Det[hilbert]
```

$$\frac{1}{186\,313\,420\,339\,200\,000}$$

```
MatrixForm[Inverse[hilbert], TableAlignments → Right]
```

$$\begin{pmatrix} 36 & -630 & 3\,360 & -7\,560 & 7\,560 & -2\,772 \\ -630 & 14\,700 & -88\,200 & 211\,680 & -220\,500 & 83\,160 \\ 3\,360 & -88\,200 & 564\,480 & -1\,411\,200 & 1\,512\,000 & -582\,120 \\ -7\,560 & 211\,680 & -1\,411\,200 & 3\,628\,800 & -3\,969\,000 & 1\,552\,320 \\ 7\,560 & -220\,500 & 1\,512\,000 & -3\,969\,000 & 4\,410\,000 & -1\,746\,360 \\ -2\,772 & 83\,160 & -582\,120 & 1\,552\,320 & -1\,746\,360 & 698\,544 \end{pmatrix}$$

12.13 Construct a table that shows the determinant of the Hilbert matrices of orders 1 through 10.

SOLUTION

```
TableForm[Table[{k, Det[HilbertMatrix[k]] // N}, {k, 1, 10}],
        TableSpacing → {1, 5}, TableHeadings → {None, {"k", "determinant"}}]
```

k	determinant
1	1.
2	0.0833333
3	0.000462963
4	1.65344×10^{-7}
5	3.7493×10^{-12}
6	5.3673×10^{-18}
7	4.8358×10^{-25}
8	2.73705×10^{-33}
9	9.72023×10^{-43}
10	2.16418×10^{-53}

Since the determinants are nonzero, each Hilbert matrix is invertible. The Hilbert matrix is a classic example of an *ill-conditioned* matrix.

12.14 Let $M = \begin{pmatrix} \frac{1}{10} & \frac{2}{10} & \frac{7}{10} \\ \frac{3}{10} & \frac{3}{10} & \frac{4}{10} \\ \frac{5}{10} & \frac{4}{10} & \frac{1}{10} \end{pmatrix}$. Compute $\lim_{n\to\infty} M^n$. (M is a *stochastic* matrix.)

SOLUTION

$$m = \frac{1}{10} \begin{pmatrix} 1 & 2 & 7 \\ 3 & 3 & 4 \\ 5 & 4 & 1 \end{pmatrix};$$

```
Limit[MatrixPower[m, n], n → ∞] // MatrixForm
```

$$\begin{pmatrix} \frac{47}{150} & \frac{23}{75} & \frac{19}{50} \\ \frac{47}{150} & \frac{23}{75} & \frac{19}{50} \\ \frac{47}{150} & \frac{23}{75} & \frac{19}{50} \end{pmatrix}$$

12.15 Let $A = \begin{pmatrix} 1 & 2 & -1 & -2 & 3 \\ 2 & 1 & 2 & -2 & 0 \\ 0 & 1 & -2 & 3 & -1 \\ 1 & -1 & 1 & 2 & -3 \\ -2 & -2 & 1 & 1 & 2 \end{pmatrix}$ and $f(x) = x^5 + 2x^4 - x^3 + x^2 - 3x + 2$. Compute $f(A)$.

SOLUTION

$$a = \begin{pmatrix} 1 & 2 & -1 & -2 & 3 \\ 2 & 1 & 2 & -2 & 0 \\ 0 & 1 & -2 & 3 & -1 \\ 1 & -1 & 1 & 2 & -3 \\ -2 & -2 & 1 & 1 & 2 \end{pmatrix};$$

```
m = MatrixPower[a, 5] + 2 MatrixPower[a, 4] - MatrixPower[a, 3] +
    MatrixPower[a, 2] - 3 a + 2 IdentityMatrix[5] ;

MatrixForm[m, TableAlignments → Right]
```

$$\begin{pmatrix} -496 & -948 & -189 & 1776 & -1695 \\ -726 & -862 & 288 & 714 & -66 \\ -117 & 399 & -103 & -648 & 1233 \\ -174 & 324 & 315 & -1216 & 1875 \\ 1419 & 1068 & -267 & -702 & -1069 \end{pmatrix}$$

12.16 It can be shown that the complex number $a + bi$ and the matrix $\begin{pmatrix} a & b \\ -b & a \end{pmatrix}$ have the same algebraic properties. Compute $(2 + 3i)^5$ using matrices and verify using complex arithmetic that this value is correct.

SOLUTION

```
a = ( 2   3 ) ;
    ( -3  2 )
```

```
MatrixPower[a, 5];
```

$$\begin{pmatrix} 122 & -597 \\ 597 & 122 \end{pmatrix}$$ ← This represents the number $122 - 597\,i$.

```
(2 + 3I)^5
```

```
122 - 597i
```

12.17 Compute the determinants

$$\begin{vmatrix} 1 & 1 & 1 & . & . & 1 \\ x_1 & x_2 & x_3 & . & . & x_n \\ x_1^2 & x_2^2 & x_3^2 & . & . & x_n^2 \\ . & . & . & . & . & . \\ . & . & . & . & . & . \\ x_1^{n-1} & x_2^{n-1} & x_3^{n-1} & . & . & x_n^{n-1} \end{vmatrix}$$

for $n = 2, 3, 4,$ and 5. Can you determine a pattern? These are known as *Vandermonde determinants*.

SOLUTION

```
m[n_] := Table[x[i]^j, {j, 0, n - 1}, {i, 1, n}];
m[2] //MatrixForm
```

$$\begin{pmatrix} 1 & 1 \\ x[1] & x[2] \end{pmatrix}$$

```
m[3] //MatrixForm
```

$$\begin{pmatrix} 1 & 1 & 1 \\ x[1] & x[2] & x[3] \\ x[1]^2 & x[2]^2 & x[3]^2 \end{pmatrix}$$

```
Det[m[2]] // Factor
```

```
-x[1] + x[2]
```

```
Det[m[3]] // Factor
```

```
-(x[1] - x[2]) (x[1] - x[3]) (x[2] - x[3])
```

```
Det[m[4]] // Factor
```

```
(x[1] - x[2]) (x[1] - x[3]) (x[2] - x[3]) (x[1] - x[4]) (x[2] - x[4]) (x[3] - x[4])
```

```
Det[m[5]]  // Factor
```

$(x[1]-x[2])(x[1]-x[3])(x[2]-x[3])(x[1]-x[4])$
$(x[2]-x[4])(x[3]-x[4])(x[1]-x[5])(x[2]-x[5])$
$(x[3]-x[5])(x[4]-x[5])$

In general $\text{Det}[m[n]] = \prod_{i>j}(x[i]-x[j])$.

12.18   A theorem of linear algebra says that the determinant of a matrix is the sum of the products of each entry of any row or column by its corresponding cofactor. (The cofactor, C_{ij}, of a_{ij} is $(-1)^{i+j} M_{ij}$ where M_{ij} is the corresponding minor.) Use this to compute the determinant of a randomly generated 5×5 matrix and verify its value.

SOLUTION

```
n = 5;
a = TableRandomInteger[9], {i, 1, n}, {j, 1, n}];
a // MatrixForm
```

$$\begin{pmatrix} 4 & 6 & 5 & 3 & 3 \\ 5 & 3 & 0 & 5 & 6 \\ 1 & 6 & 9 & 7 & 7 \\ 7 & 9 & 7 & 1 & 2 \\ 0 & 9 & 4 & 5 & 6 \end{pmatrix}$$

```
matrixofminors = Minors[a];
MatrixForm[matrixofminors, TableAlignments → Right]
```

$$\begin{pmatrix} 171 & 159 & -174 & -283 & -216 \\ -1350 & -1584 & -270 & -140 & -48 \\ 669 & 549 & 342 & 293 & -78 \\ -561 & -339 & 96 & -143 & -168 \\ -3231 & -3333 & -438 & -497 & 246 \end{pmatrix}$$

```
signs = Table[(-1)^(i + j), {i, 1, n}, {j, 1, n}];
cofactors = matrixofminors * signs;
MatrixForm[cofactors, TableAlignments → Right]
```

$$\begin{pmatrix} 171 & -159 & -174 & 283 & -216 \\ 1350 & -1584 & 270 & -140 & 48 \\ 669 & -549 & 342 & -293 & -78 \\ 561 & -339 & -96 & -143 & 168 \\ -3231 & 3333 & -438 & 497 & 246 \end{pmatrix}$$

```
i=3    (* we expand using the third row *)
```

$$\text{determinant} = \sum_{j=1}^{n} a[[n-i+1,n-j+1]] * \text{cofactors}[[i,j]]$$

2082

```
Det[a]
```

2082

> The (i, j)th element of `Minors[a]` gives the determinant of the matrix obtained by deleting row $n-i+1$ and column $n-j+1$.

12.19 Let $\mathbf{x} = \begin{bmatrix} 1 \\ 2 \\ 3 \\ 4 \\ 5 \end{bmatrix}$. Compute $\mathbf{x^Tx} = \begin{bmatrix} 1 & 2 & 3 & 4 & 5 \end{bmatrix} \begin{bmatrix} 1 \\ 2 \\ 3 \\ 4 \\ 5 \end{bmatrix}$ and $\mathbf{xx^T} = \begin{bmatrix} 1 \\ 2 \\ 3 \\ 4 \\ 5 \end{bmatrix} \begin{bmatrix} 1 & 2 & 3 & 4 & 5 \end{bmatrix}$.

SOLUTION

$\mathbf{x^Tx}$ is the dot product of the vector $\mathbf{x}$ with itself. $\mathbf{xx^T}$, however, is a 5×5 matrix.

```
x.x
```
```
55
```
```
Outer[Times, x, x] // MatrixForm
```

$$\begin{pmatrix} 1 & 2 & 3 & 4 & 5 \\ 2 & 4 & 6 & 8 & 10 \\ 3 & 6 & 9 & 12 & 15 \\ 4 & 8 & 12 & 16 & 20 \\ 5 & 10 & 15 & 20 & 25 \end{pmatrix}$$

12.3   Matrix Manipulation

Mathematica offers a variety of matrix manipulation commands that are quite useful when working problems in linear algebra. Since matrices are actually lists, many of the commands are the same as described in Chapter 3.

- **Join[***list1*, *list2***]** combines the two lists *list1* and *list2* into one list consisting of the elements from *list1* and from *list2*. For matrices, this has the effect of placing the rows of *list2* under the rows of *list1*.
- **Join[***list1*, *list2*, *n***]** joins the objects at level *n* in each *list*. If *n* = 2, this has the effect of placing the columns of *list2* to the right of the columns of *list1*.
- **ArrayFlatten[***{{m*$_{11}$, *m*$_{12}$, ...}, {*m*$_{21}$, *m*$_{22}$, ...}, ...}***]** creates a single flattened matrix from a matrix of matrices *m*$_{ij}$. All the matrices in the same row must have the same first dimension, and all the matrices in the same column must have the same second dimension.

EXAMPLE 13   The following examples illustrate the commands described previously. To see their effects more clearly, the matrices are shown as lists.

$$m1 = \begin{pmatrix} a & b & c \\ d & e & f \\ g & h & i \end{pmatrix}$$

```
{{a, b, c}, {d, e, f}, {g, h, i}}
```

$$m2 = \begin{pmatrix} aa & bb & cc \\ dd & ee & ff \\ gg & hh & ii \end{pmatrix}$$

```
{{aa, bb, cc}, {dd, ee, ff}, {gg, hh, ii}}
```

```
Join[m1, m2]
```
```
{{a, b, c}, {d, e, f}, {g, h, i}, {aa, bb, cc}, {dd, ee, ff}, {gg, hh, ii}}
```

```
% //MatrixForm
```

$$\begin{pmatrix} a & b & c \\ d & e & f \\ g & h & i \\ aa & bb & cc \\ dd & ee & ff \\ gg & hh & ii \end{pmatrix}$$

```
Join[m1, m2, 2]
```

{{a, b, c, aa, bb, cc}, {d, e, f, dd, ee, ff}, {g, h, i, gg, hh, ii}}

```
% //MatrixForm
```

$$\begin{pmatrix} a & b & c & aa & bb & cc \\ d & e & f & dd & ee & ff \\ g & h & i & gg & hh & ii \end{pmatrix}$$

```
ArrayFlatten[{{m1, m2}, {m2, m1}}]
```

{{a, b, c, aa, bb, cc}, {d, e, f, dd, ee, ff}, {g, h, i, gg, hh, ii},
 {aa, bb, cc, a, b, c}, {dd, ee, ff, d, e, f}, {gg, hh, ii, g, h, i}}

```
% //MatrixForm
```

$$\begin{pmatrix} a & b & c & aa & bb & cc \\ d & e & f & dd & ee & ff \\ g & h & i & gg & hh & ii \\ aa & bb & cc & a & b & c \\ dd & ee & ff & d & e & f \\ gg & hh & ii & g & h & i \end{pmatrix}$$

The arrangement of the lists indicates that the two blocks on top, left to right, are **m1** and **m2**. The bottom blocks are **m2** and **m1**.

The following commands can be used to form submatrices:

- **Take[***matrix***, n]** returns the first n rows of *matrix*.
- **Take[***matrix***, –n]** returns the last n rows of *matrix*.
- **Take[***matrix***, {m, n}]** returns rows m through n of *matrix*.
- **Take[***matrix***, m, n]** returns a submatrix containing rows 1 through m and columns 1 through n of *matrix*.
- **Take[***matrix***, {m, n}, {p, q}]** returns rows m through n and colums p through q of *matrix*.
- **Drop[***matrix***, n]** returns *matrix* with its first n rows deleted.
- **Drop[***matrix***, –n]** returns *matrix* with its last n rows deleted.
- **Drop[***matrix***, {n}]** returns *matrix* with its nth row deleted.
- **Drop[***matrix***, {–n}]** returns *matrix* with the nth row from the end deleted.
- **Drop[***matrix***, {m, n}]** returns *matrix* with rows m through n deleted.
- **Drop[***matrix***, m, n]** returns *matrix* with rows 1 through m and columns 1 through n deleted.
- **Drop[***matrix***, {m}, {n}]** returns *matrix* with row m and column n deleted.
- **Drop[***matrix***, {m, n}, {p, q}]** returns *matrix* with rows m through n and columns p through q deleted.
- **Delete[***matrix***, n]** deletes the nth row of *matrix*.
- **Delete[***matrix***, –n]** deletes the nth from the last row of *matrix*.
- **Delete[***matrix***, {{p$_1$}, {p$_2$}, ...}]** deletes rows p$_1$, p$_2$ ...

EXAMPLE 14

```
m = Partition[Range[20],5];
m //MatrixForm
```

$$\begin{pmatrix} 1 & 2 & 3 & 4 & 5 \\ 6 & 7 & 8 & 9 & 10 \\ 11 & 12 & 13 & 14 & 15 \\ 16 & 17 & 18 & 19 & 20 \end{pmatrix}$$

```
Take[m, 3] //MatrixForm
```

$$\begin{pmatrix} 1 & 2 & 3 & 4 & 5 \\ 6 & 7 & 8 & 9 & 10 \\ 11 & 12 & 13 & 14 & 15 \end{pmatrix}$$

```
Take[m, {1, 4}, {3, 5}] //MatrixForm
```

$$\begin{pmatrix} 3 & 4 & 5 \\ 8 & 9 & 10 \\ 13 & 14 & 15 \\ 18 & 19 & 20 \end{pmatrix}$$

```
Take[m, {2, 3}] // MatrixForm
```

$$\begin{pmatrix} 6 & 7 & 8 & 9 & 10 \\ 11 & 12 & 13 & 14 & 15 \end{pmatrix}$$

```
Take[m, 3, {2, 3}] //MatrixForm
```

$$\begin{pmatrix} 2 & 3 \\ 7 & 8 \\ 12 & 13 \end{pmatrix}$$

```
Take[m, {2, 3}, {2, 4}] //MatrixForm
```

$$\begin{pmatrix} 7 & 8 & 9 \\ 12 & 13 & 14 \end{pmatrix}$$

SOLVED PROBLEMS

12.20 Construct a 10×10 upper triangular matrix whose nonzero entries are random digits. Show that its determinant is equal to the product of the entries on its main diagonal.

SOLUTION

```
f[i_ , j_ ]:= Random[Integer,{0,9}] /; i < j
f[i_ , j_ ]:= 0/; i ≥ j
m = Array[f, {10, 10}];
m //MatrixForm
```

$$\begin{pmatrix} 3 & 1 & 6 & 7 & 0 & 6 & 4 & 8 & 9 & 1 \\ 0 & 1 & 7 & 5 & 1 & 9 & 9 & 0 & 5 & 8 \\ 0 & 0 & 5 & 6 & 9 & 5 & 3 & 3 & 3 & 5 \\ 0 & 0 & 0 & 1 & 5 & 7 & 2 & 5 & 3 & 4 \\ 0 & 0 & 0 & 0 & 6 & 8 & 4 & 0 & 9 & 0 \\ 0 & 0 & 0 & 0 & 0 & 5 & 2 & 0 & 5 & 0 \\ 0 & 0 & 0 & 0 & 0 & 0 & 3 & 5 & 5 & 7 \\ 0 & 0 & 0 & 0 & 0 & 0 & 0 & 3 & 6 & 8 \\ 0 & 0 & 0 & 0 & 0 & 0 & 0 & 0 & 3 & 7 \\ 0 & 0 & 0 & 0 & 0 & 0 & 0 & 0 & 0 & 4 \end{pmatrix}$$

```
det[m] == ∏_{i=1}^{10} m[[i,i]]
```
```
True
```

12.21 Construct a 9×9 block diagonal matrix with a 2×2 block of 2s, a 3×3 block of 3s and a 4×4 block of 4s. (A block diagonal matrix is a square partitioned matrix whose diagonal matrices are square and all others are zero matrices.)

SOLUTION

```
m2 = Table[2, {2}, {2}];
m3 = Table[3, {3}, {3}];
```

```
m4 = Table[4, {4}, {4}];
z27 = ConstantArray[0, {2, 7}];        ← z27 is a 2 × 7 array of zeros, etc.
z32 = ConstantArray[0, {3, 2}];
z34 = ConstantArray[0, {3, 4}];
z45 = ConstantArray[0, {4, 5}];
top = ArrayFlatten[{{m2, z27}}];
middle = ArrayFlatten[{{z32, m3, z34}}];
bottom = ArrayFlatten[{{z45, m4}}];
ArrayFlatten[{{top}, {middle}, {bottom}}] //MatrixForm
```

$$
\begin{pmatrix}
2 & 2 & 0 & 0 & 0 & 0 & 0 & 0 & 0 \\
2 & 2 & 0 & 0 & 0 & 0 & 0 & 0 & 0 \\
0 & 0 & 3 & 3 & 3 & 0 & 0 & 0 & 0 \\
0 & 0 & 3 & 3 & 3 & 0 & 0 & 0 & 0 \\
0 & 0 & 3 & 3 & 3 & 0 & 0 & 0 & 0 \\
0 & 0 & 0 & 0 & 0 & 4 & 4 & 4 & 4 \\
0 & 0 & 0 & 0 & 0 & 4 & 4 & 4 & 4 \\
0 & 0 & 0 & 0 & 0 & 4 & 4 & 4 & 4 \\
0 & 0 & 0 & 0 & 0 & 4 & 4 & 4 & 4
\end{pmatrix}
$$

12.22 Let $M = \begin{pmatrix} a & b & c & d & e \\ f & g & h & i & j \\ k & l & m & n & o \\ p & q & r & s & t \\ u & v & w & x & y \end{pmatrix}$.

Find the matrix P obtained from M by deleting its fourth row and third column.

SOLUTION

```
temp = CharacterRange["a", "y"];          ← Generates a list of alphabet letters.
m = Partition[temp, 5];                    ← Forms five sublists of five letters each.
m //MatrixForm
```

$$
\begin{pmatrix}
a & b & c & d & e \\
f & g & h & i & j \\
k & l & m & n & o \\
p & q & r & s & t \\
u & v & w & x & y
\end{pmatrix}
$$

```
Drop[m, {4}, {3}] //MatrixForm
```

$$
\begin{pmatrix}
a & b & d & e \\
f & g & i & j \\
k & l & n & o \\
u & v & x & y
\end{pmatrix}
$$

12.4   Linear Systems of Equations

Mathematica offers a number of ways to solve systems of linear equations. **Solve**, discussed in Chapter 6, offers one alternative, but it is somewhat clumsy and inefficient for use on large systems. In this section we discuss a number of other procedures for solving systems of linear equations.

- **LinearSolve[a, b]** produces vectors, x, such that a . x = b.
- **LinearSolve[a]** produces a LinearSolveFunction that can be used to solve a . x = b for different vectors b.

Here a is the matrix of coefficients of the unknowns, and b is the "right-hand side" of the linear system. If a is invertible, **LinearSolve** will produce a unique solution to the linear system. If a is singular, either no solution exists or there are an infinite number of solutions.

If a system has a unique solution, *Mathematica* returns the solution. If no solution exists, *Mathematica* returns an error message.

EXAMPLE 15   The system $2x + y + z = 7$, $x - 4y + 3z = 2$, $3x + 2y + 2z = 13$ has a unique solution.

The system $2x + y + z = 7$, $x - 4y + 3z = 2$, $3x - 3y + 4z = 13$ has no solution.

$$a1 = \begin{pmatrix} 2 & 1 & 1 \\ 1 & -4 & 3 \\ 3 & 2 & 2 \end{pmatrix};$$

$$a2 = \begin{pmatrix} 2 & 1 & 1 \\ 1 & -4 & 3 \\ 3 & -3 & 4 \end{pmatrix};$$

b = {7, 2, 13};

LinearSolve[a1, b]

{1, 2, 3}

LinearSolve[a2, b]

LinearSolve::nosol : Linear equation encountered that has no solution. »

LinearSolve[{{2, 1, 1}, {1, -4, 3}, {3, -3, 4}}, {{7}, {2}, {13}}]

If the system a . x = b has an infinite number of solutions, the treatment is a bit more complicated. In this case, *Mathematica* returns one solution, known as a *particular solution*. The full set of solutions is constructed by adding to the particular solution the set of all solutions of the corresponding homogeneous system, a . x = 0.

The set of all vectors, x, such that a . x = 0, is called the *null space* of a and is easily determined by the command **NullSpace**.

- **NullSpace[a]** returns the basis vectors of the null space of a.

The nullity of a, the dimension of the null space of a, can be found by computing **Length[NullSpace[a]]**. The rank of a may be computed as **n - Length[NullSpace[a]]** where n represents the number of columns of a.

EXAMPLE 16 $2x + y + z = 7$, $x - 4y + 3z = 2$, $3x - 3y + 4z = 9$ has an infinite number of solutions.

$$a = \begin{pmatrix} 2 & 1 & 1 \\ 1 & -4 & 3 \\ 3 & -3 & 4 \end{pmatrix}; \qquad b = \{7, 2, 9\};$$

nullspacebasis = NullSpace[a]

{{-7, 5, 9}}

particular = LinearSolve[a, b]

| Because the null space contains a nonzero vector, there is no unique solution. |

$\left\{ \dfrac{10}{3}, \dfrac{1}{3}, 0 \right\}$ ← This is a particular solution.

The full set of solutions to the system is of the form `t*nullspacebasis + particular` where t is an arbitrary parameter. However, to express as a single list, we must first flatten `nullspacebasis`.

generalsolution = t*Flatten[nullspacebasis] + particular

$$\left\{ \frac{10}{3} - 7t,\ \frac{1}{3} + 5t,\ 9t \right\}$$

As a check, we substitute our general solution back into the original system.

a.generalsolution // Expand

$\{7, 2, 9\}$

The Gauss-Jordan method for solving the linear system `a.x = b` is based upon the reduction of the augmented matrix `[a|b]` into *reduced row echelon form* by a series of *elementary row operations*. The three basic elementary row operations are:

1. interchanging two rows
2. multiplying a row by a non-zero constant
3. replacing one row by itself plus a multiple of another row

It is easily seen that elementary row operations have no effect upon the solution of the system.

A matrix is said to be in *reduced row echelon form* if

1. all zero rows are placed at the bottom of the matrix
2. each leading nonzero entry is 1 (called a leading 1)
3. each entry above and below a leading 1 is 0
4. if two rows have leading 1s, the lower row has its leading 1 farther to the right

To solve a linear system, we use elementary row operations to reduce the augmented matrix to reduced row echelon form. The solution(s) of the system, or the fact that no solution exists, may then be easily determined.

Every student of linear algebra knows that row reduction is a time-consuming, tedious process that is highly prone to error. However, the *Mathematica* command **RowReduce** quickly reduces any matrix to reduced row echelon form.

■ **RowReduce[*matrix*]** reduces *matrix* to reduced row echelon form.

EXAMPLE 17　Determine the reduced row echelon form of the 4×5 matrix whose general entry $a_{i,j} = |i - j|$.

a = Table[Abs[i - j], {i, 1, 4}, {j, 1, 5}];

a // MatrixForm

$$\begin{pmatrix} 0 & 1 & 2 & 3 & 4 \\ 1 & 0 & 1 & 2 & 3 \\ 2 & 1 & 0 & 1 & 2 \\ 3 & 2 & 1 & 0 & 1 \end{pmatrix}$$

RowReduce[a]

$$\begin{pmatrix} 1 & 0 & 0 & 0 & \frac{1}{3} \\ 0 & 1 & 0 & 0 & 0 \\ 0 & 0 & 1 & 0 & 0 \\ 0 & 0 & 0 & 1 & \frac{4}{3} \end{pmatrix}$$

We now illustrate how row reduction can be used to solve a linear system. For comparison purposes we use the three examples previously considered in Examples 15 and 16.

EXAMPLE 18

(a) $2x + y + z = 7$, $x - 4y + 3z = 2$, $3x + 2y + 2z = 13$ (unique solution)
(b) $2x + y + z = 7$, $x - 4y + 3z = 2$, $3x - 3y + 4z = 13$ (no solution)
(c) $2x + y + z = 7$, $x - 4y + 3z = 2$, $3x - 3y + 4z = 9$ (infinite number of solutions)

We find the augmented matrix for each of the three systems.

$$a1 = \begin{pmatrix} 2 & 1 & 1 & 7 \\ 1 & -4 & 3 & 2 \\ 3 & 2 & 2 & 13 \end{pmatrix}; \quad a2 = \begin{pmatrix} 2 & 1 & 1 & 7 \\ 1 & -4 & 3 & 2 \\ 3 & -3 & 4 & 13 \end{pmatrix}; \quad a3 = \begin{pmatrix} 2 & 1 & 1 & 7 \\ 1 & -4 & 3 & 2 \\ 3 & -3 & 4 & 9 \end{pmatrix};$$

`RowReduce[a1] //MatrixForm`

$$\begin{pmatrix} 1 & 0 & 0 & 1 \\ 0 & 1 & 0 & 2 \\ 0 & 0 & 1 & 3 \end{pmatrix}$$

This reduced matrix, when interpreted as a system of equations, reads: $x = 1$, $y = 2$, $z = 3$.

`RowReduce[a2] //MatrixForm`

$$\begin{pmatrix} 1 & 0 & \frac{7}{9} & 0 \\ 0 & 1 & -\frac{5}{9} & 0 \\ 0 & 0 & 0 & 1 \end{pmatrix}$$

The bottom row reads $0x + 0y + 0z = 1$, which, of course, is impossible. This contradiction (a row of 0s and a final 1) reveals that no solution is possible.

`RowReduce[a3] //MatrixForm`

$$\begin{pmatrix} 1 & 0 & \frac{7}{9} & \frac{10}{3} \\ 0 & 1 & -\frac{5}{9} & \frac{1}{3} \\ 0 & 0 & 0 & 0 \end{pmatrix}$$

The bottom row of 0s is not a contradiction. However, there cannot be a unique solution. If we let $z = t$, an independent parameter, the solution may be put into the form

$$x = \frac{10}{3} - \frac{7}{9}t, \qquad y = \frac{1}{3} + \frac{5}{9}t, \qquad z = t.$$

Note: Although the solution looks slightly different than the solution obtained previously, it is equivalent in the sense that it describes precisely the same solution set.

Another popular method, *LU* decomposition, is useful, particularly if you have many systems, all having the same coefficient matrix. The idea behind the method is simple.

If A is a square matrix, it may be possible to factor $A = LU$ where L is lower triangular with 1s on the main diagonal and U is upper triangular. The system $A\mathbf{x} = \mathbf{b}$ then reads $(LU)\mathbf{x} = \mathbf{b}$, which can be written $L(U\mathbf{x}) = \mathbf{b}$. If we let $\mathbf{y} = U\mathbf{x}$, we can solve $L\mathbf{y} = \mathbf{b}$ for $\mathbf{y}$. Once we have determined $\mathbf{y}$, we solve $U\mathbf{x} = \mathbf{y}$ for $\mathbf{x}$.

Even though the solution of a system by *LU* decomposition involves solving two systems of equations, each involves a triangular matrix so the computation is efficient.

Thus, there are two steps to solving a system of equations by *LU* decomposition: factorization and back substitution. The corresponding *Mathematica* commands are `LUDecomposition` and `LUBackSubstitution`.

- `LUDecomposition[`*matrix*`]` finds the *LU* decomposition of *matrix*.
- `LUBackSubstitution[`*data*`, b]` uses the output of `LUDecomposition[`*matrix*`]` to solve the system *matrix*.x = b.

The output of `LUDecomposition` consists of three parts: (1) the matrices L and U "packed" as a single matrix, (2) a permutation vector, and (3) the L^∞ condition number of the matrix. The output of `LUDecomposition`, *data*, is fed into `LUBackSubstitution` to solve the system.

The permutation vector rearranges the rows in order to ensure a maximum degree of numerical stability. The condition number will be of no concern to us in this chapter.

`LUDecomposition` and `LUBackSubstitution` cannot be used on systems that possess an infiite number of solutions.

EXAMPLE 19   To solve the system $2x + y + z = 7$, $x - 4y + 3z = 2$, $3x + 2y + 2z = 13$ using *LU* decomposition, we must first obtain the matrix factorization of the coefficient matrix.

$$a = \begin{pmatrix} 2 & 1 & 1 \\ 1 & -4 & 3 \\ 3 & 2 & 2 \end{pmatrix}; \quad b = \{7, 2, 13\};$$

data = LUDecomposition[a]

$$\left\{ \left\{ \{1, -4, 3\}, \{2, 9, -5\}, \left\{3, \frac{14}{9}, \frac{7}{9}\right\} \right\}, \{2, 1, 3\}, 1 \right\}$$

LUBackSubstitution[data, b]

$\{1, 2, 3\}$

The next two examples illustrate the structure of *data*.

EXAMPLE 20

$$m = \begin{pmatrix} 2 & 3 & 4 \\ 4 & 11 & 14 \\ 6 & 29 & 43 \end{pmatrix};$$

{lu, p, cond} = LUDecomposition[m]

$\{\{\{2, 3, 4\}, \{2, 5, 6\}, \{3, 4, 7\}\}, \{1, 2, 3\}, 1\}$

In this example, no rearrangement of the rows was performed because the permutation vector, p, is $\{1, 2, 3\}$.

The first part of **LUDecomposition[m]** is given in a "packed" format. Since *LU* is known to be of the form $\begin{pmatrix} 1 & 0 & 0 \\ x & 1 & 0 \\ x & x & 1 \end{pmatrix} \begin{pmatrix} x & x & x \\ 0 & x & x \\ 0 & 0 & x \end{pmatrix}$, only nine entries (represented by x) need be specified. The first part of **LUDecomposition[m]** specifies these nine numbers as a single matrix.

lu //MatrixForm

$$\begin{pmatrix} 2 & 3 & 4 \\ 2 & 5 & 6 \\ 3 & 4 & 7 \end{pmatrix}$$

The numbers, although combined into one matrix, are in their correct positions.

$$l = \begin{pmatrix} 1 & 0 & 0 \\ 2 & 1 & 0 \\ 3 & 4 & 1 \end{pmatrix} \quad \text{and} \quad u = \begin{pmatrix} 2 & 3 & 4 \\ 0 & 5 & 6 \\ 0 & 0 & 7 \end{pmatrix}$$

EXAMPLE 21

$$m = \begin{pmatrix} 2 & 1 & 1 \\ 1 & -4 & 3 \\ 3 & 2 & 2 \end{pmatrix};$$

{lu, p, cond} = LUDecomposition[m]

$$\left\{ \left\{ \{1, -4, 3\}, \{2, 9, -5\}, \left\{3, \frac{14}{9}, \frac{7}{9}\right\} \right\}, \{2, 1, 3\}, 1 \right\}$$

lu //MatrixForm

$$\begin{pmatrix} 1 & -4 & 3 \\ 2 & 9 & -5 \\ 3 & \frac{14}{9} & \frac{7}{9} \end{pmatrix}$$

If we proceed as in the previous example, we would be tempted to say that

$$1 = \begin{pmatrix} 1 & 0 & 0 \\ 2 & 1 & 0 \\ 3 & \frac{14}{9} & 1 \end{pmatrix} \quad \text{and} \quad u = \begin{pmatrix} 1 & -4 & 3 \\ 0 & 9 & -5 \\ 0 & 0 & \frac{7}{9} \end{pmatrix}$$

However, multiplying **1** by **u** does *not* give back the original matrix:

l.u //MatrixForm

$$\begin{pmatrix} 1 & -4 & 3 \\ 2 & 1 & 1 \\ 3 & 2 & 2 \end{pmatrix}$$

The permutation vector, p = {2, 1, 3}, indicates that the rows of the matrix have been interchanged. Indeed rows 1 and 2 have been switched. If we permute the rows of **1**, we should get back our original matrix upon multiplication by u.

$$1 = \begin{pmatrix} 2 & 1 & 0 \\ 1 & 0 & 0 \\ 3 & \frac{14}{9} & 1 \end{pmatrix} ; \quad u = \begin{pmatrix} 1 & -4 & 3 \\ 0 & 9 & -5 \\ 0 & 0 & \frac{7}{9} \end{pmatrix} ;$$

l.u //MatrixForm

$$\begin{pmatrix} 2 & 1 & 1 \\ 1 & -4 & 3 \\ 3 & 2 & 2 \end{pmatrix}$$

SOLVED PROBLEMS

12.23 Describe the set of vectors, *S*, spanned by (1, 2, 1, 2, 1), (1, 3, 2, 4, 2), and (1, 4, 3, 6, 3).

SOLUTION

```
a = {1, 2, 1, 2, 1};
b = {1, 3, 2, 4, 2};
c = {1, 4, 3, 6, 3};
m = {a, b, c};
m // MatrixForm
```

$$\begin{pmatrix} 1 & 2 & 1 & 2 & 1 \\ 1 & 3 & 2 & 4 & 2 \\ 1 & 4 & 3 & 6 & 3 \end{pmatrix}$$

Form a matrix, **m**, using the given vectors as rows. The row space of **m** is the space spanned by **a**, **b**, and **c**. Then reduce the matrix to reduced row echelon form. The non-zero rows form a basis for the row space. Every vector in *S* is a linear combination of its basis vectors.

```
rref = RowReduce [m] ;
rref // MatrixForm
```

$$\begin{pmatrix} 1 & 0 & -1 & -2 & -1 \\ 0 & 1 & 1 & 2 & 1 \\ 0 & 0 & 0 & 0 & 0 \end{pmatrix}$$

```
rref [[1]]
{1, 0, -1, -2, -1}
rref [[2]]
{0, 1, 1, 2, 1}
s * rref [[1]] + t * rref [[2]]
{s, t, -s + t, -2 s + 2 t, -s + t}        ← Every vector in S is of this form.
```

12.24 A theorem of linear algebra says that every vector in the row space of A is orthogonal to every vector in the null space of A. Verify this result for

$$A = \begin{pmatrix} 1 & 2 & 3 & 4 & 5 \\ 6 & 7 & 8 & 9 & 10 \\ 11 & 12 & 13 & 14 & 15 \\ 16 & 17 & 18 & 19 & 20 \\ 21 & 22 & 23 & 24 & 25 \end{pmatrix}$$

SOLUTION

It suffices to show that each basis vector of the row space is orthogonal to every basis vector in the null space.

```
a = Partition[Range[25], 5];

a // MatrixForm
```

$$\begin{pmatrix} 1 & 2 & 3 & 4 & 5 \\ 6 & 7 & 8 & 9 & 10 \\ 11 & 12 & 13 & 14 & 15 \\ 16 & 17 & 18 & 19 & 20 \\ 21 & 22 & 23 & 24 & 25 \end{pmatrix}$$

```
rowspacebasis = RowReduce[a];

rowspacebasis // MatrixForm
```

$$\begin{pmatrix} 1 & 0 & -1 & -2 & -3 \\ 0 & 1 & 2 & 3 & 4 \\ 0 & 0 & 0 & 0 & 0 \\ 0 & 0 & 0 & 0 & 0 \\ 0 & 0 & 0 & 0 & 0 \end{pmatrix}$$

```
nullspacebasis = NullSpace[a];

nullspacebasis // MatrixForm
```

$$\begin{pmatrix} 3 & -4 & 0 & 0 & 1 \\ 2 & -3 & 0 & 1 & 0 \\ 1 & -2 & 1 & 0 & 0 \end{pmatrix}$$

> To show that each vector in rowspacebasis is orthogonal to every vector in nullspacebasis, we must show that every dot product is 0. The easiest way is to multiply rowspacebasis by the *transpose* of nullspacebasis. The bottom three rows of zeros may be ignored.

```
rowspacebasis.Transpose[nullspacebasis] // MatrixForm
```

$$\begin{pmatrix} 0 & 0 & 0 \\ 0 & 0 & 0 \\ 0 & 0 & 0 \\ 0 & 0 & 0 \\ 0 & 0 & 0 \end{pmatrix}$$

12.25 Construct a 5×5 matrix of random digits, a, and a 5×1 matrix of random digits, b, and solve the linear system ax = b using **LinearSolve**. Then verify that your solution is correct.

SOLUTION

```
a = Table[RandomInteger[9], {i, 1, 5}, {j, 1, 5}];

a // MatrixForm
```

$$\begin{pmatrix} 0 & 9 & 5 & 0 & 2 \\ 7 & 7 & 5 & 9 & 3 \\ 0 & 6 & 6 & 2 & 6 \\ 5 & 8 & 4 & 3 & 9 \\ 3 & 0 & 8 & 1 & 5 \end{pmatrix}$$

```
Det[a] == 0
```

False $\leftarrow$ Since the determinant $\neq 0$, the system has a unique solution.

```
b = Table[RandomInteger[9], {i, 1, 5}];
b // MatrixForm
```

$$\begin{pmatrix} 1 \\ 2 \\ 4 \\ 6 \\ 1 \end{pmatrix}$$

```
x = LinearSolve[a, b]
```

$$\left\{ -\frac{58}{217}, \frac{65}{651}, -\frac{187}{651}, \frac{111}{434}, \frac{143}{186} \right\}$$

```
a.x == b
```
True

12.26 Solve the system $A\mathbf{x} = \mathbf{b}$ where $A = \begin{pmatrix} 1 & 2 & 3 \\ 2 & -1 & 4 \\ 3 & -4 & 5 \end{pmatrix}$ and $\mathbf{b} = \begin{pmatrix} 14 \\ 12 \\ 13 \end{pmatrix}, \begin{pmatrix} 9 \\ 17 \\ 28 \end{pmatrix}$, and $\begin{pmatrix} 10 \\ 22 \\ 38 \end{pmatrix}$.

SOLUTION

$$a = \begin{pmatrix} 1 & 2 & 3 \\ 2 & -1 & 4 \\ 3 & -4 & 6 \end{pmatrix};$$

```
f = LinearSolve[a]
LinearSolveFunction[{3,3}, <>]
f[{14, 12, 13}]
{1, 2, 3}
f[{9, 17, 28}]
{2, -1, 3}
f[{10, 22, 38}]
{2, -2, 4}
```

12.27 Find the general solution of the system $\begin{cases} w + 2x + 3y + 3z = \ \ 9 \\ 2w + \ \ x + 2y + 5z = \ 10 \\ 2w + 2x + \ \ y + 2z = \ \ 7 \\ 2w - \ \ x - 3y + \ \ z = -1 \end{cases}$

SOLUTION

$$a = \begin{pmatrix} 1 & 2 & 3 & 3 \\ 2 & 1 & 2 & 5 \\ 2 & 2 & 1 & 2 \\ 2 & -1 & -3 & 1 \end{pmatrix}; \qquad b = \{9, 10, 7, -1\};$$

```
Det[a] == 0
```
← Since the determinant is 0, we anticipate either
no solution or an infinite number of solutions.
True

```
nullspacebasis = NullSpace[a]
```
← Since the null space contains a non-zero vector,
there will be an infinite number of solutions.
$\{\{-13, 11, -10, 7\}\}$

```
particular = LinearSolve[a, b]
```

$$\left\{ \frac{20}{7}, -\frac{4}{7}, \frac{17}{7}, 0 \right\}$$

```
generalsolution = t * Flatten[nullspacebasis] + particular
```

$$\left\{ \frac{20}{7} - 13t, -\frac{4}{7} + 11t, \frac{17}{7} - 10t, 7t \right\}$$

12.28 Find the general solution of the system $\begin{cases} w + 2x + 3y + 3z = 9 \\ 3w + 4x + 4y + 5z = 16 \\ 2w + 2x + y + 2z = 7 \\ 4w + 6x + 7y + 8z = 25 \end{cases}$

SOLUTION

```
a = ⎛1 2 3 3⎞;    b = {9, 16, 7, 25};
    ⎜3 4 4 5⎟
    ⎜2 2 1 2⎟
    ⎝4 6 7 8⎠

Det[a] == 0
True
nullspacebasis = NullSpace[a]
{{1, -2, 0, 1}, {4, -5, 2, 0}}
particular = LinearSolve[a, b]
```

$\left\{-2, \dfrac{11}{2}, 0, 0\right\}$

```
generalsolution = s * nullspacebasis[[1]] +
        t * nullspacebasis[[2]] + particular
```

$\left\{-2 + s + 4t, \dfrac{11}{2} - 2s - 5t, 2t, s\right\}$

12.29 Let A be a 7×7 tridiagonal matrix having 3s on the main diagonal and -1s on the diagonals adjacent to the main diagonal. Let e_i be a 7-dimensional vector having 1 in the ith position and 0s elsewhere. Solve $Ax = e_i$, $i = 1, \ldots, 7$.

SOLUTION

```
a = Table[If[Abs[i - j] == 1, -1, If[i == j, 3, 0]], {i, 1, 7}, {j, 1, 7}];
a // MatrixForm
```

$$\begin{pmatrix} 3 & -1 & 0 & 0 & 0 & 0 & 0 \\ -1 & 3 & -1 & 0 & 0 & 0 & 0 \\ 0 & -1 & 3 & -1 & 0 & 0 & 0 \\ 0 & 0 & -1 & 3 & -1 & 0 & 0 \\ 0 & 0 & 0 & -1 & 3 & -1 & 0 \\ 0 & 0 & 0 & 0 & -1 & 3 & -1 \\ 0 & 0 & 0 & 0 & 0 & -1 & 3 \end{pmatrix}$$

> KroneckerDelta[i, j] = 1 if i = j and 0 otherwise.

```
ludata = LUDecomposition[a];
b = Table[KroneckerDelta[i, j], {i, 1, 7}, {j, 1, 7}];
LUBackSubstitution[ludata, b]  // TableForm
```

$\dfrac{377}{987}$	$\dfrac{48}{329}$	$\dfrac{55}{987}$	$\dfrac{1}{47}$	$\dfrac{8}{987}$	$\dfrac{1}{329}$	$\dfrac{1}{987}$
$\dfrac{48}{329}$	$\dfrac{144}{329}$	$\dfrac{55}{329}$	$\dfrac{3}{47}$	$\dfrac{8}{329}$	$\dfrac{3}{329}$	$\dfrac{1}{329}$
$\dfrac{55}{987}$	$\dfrac{55}{329}$	$\dfrac{440}{987}$	$\dfrac{8}{47}$	$\dfrac{64}{987}$	$\dfrac{8}{329}$	$\dfrac{8}{987}$
$\dfrac{1}{47}$	$\dfrac{3}{47}$	$\dfrac{8}{47}$	$\dfrac{21}{47}$	$\dfrac{8}{47}$	$\dfrac{3}{47}$	$\dfrac{1}{47}$
$\dfrac{8}{987}$	$\dfrac{8}{329}$	$\dfrac{64}{987}$	$\dfrac{8}{47}$	$\dfrac{440}{987}$	$\dfrac{55}{329}$	$\dfrac{55}{987}$
$\dfrac{1}{329}$	$\dfrac{3}{329}$	$\dfrac{8}{329}$	$\dfrac{3}{47}$	$\dfrac{55}{329}$	$\dfrac{144}{329}$	$\dfrac{48}{329}$
$\dfrac{1}{987}$	$\dfrac{1}{329}$	$\dfrac{8}{987}$	$\dfrac{1}{47}$	$\dfrac{55}{987}$	$\dfrac{48}{329}$	$\dfrac{377}{987}$

The ith column of the table represents the solution of $Ax = e_i$.

12.5 Orthogonality

Two vectors are *orthogonal* if their inner product is 0. Orthogonal vectors possess useful properties that make working with them convenient. For example, if **u** and **v** are orthogonal, they satisfy the (generalized) Theorem of Pythagoras: $\|\mathbf{u} + \mathbf{v}\|^2 = \|\mathbf{u}\|^2 + \|\mathbf{v}\|^2$.

Orthogonality also allows us to introduce the concept of projection. In $\mathbb{R}^2$ it is easy to visualize what projection means. If $\mathbf{a} = \overrightarrow{PQ}$ and $\mathbf{b} = \overrightarrow{PR}$ are two vectors with the same initial point P, then if S is the foot of the perpendicular from R to $\overrightarrow{PQ}$, the projection of **b** onto **a** is the vector $\overrightarrow{PS}$. This vector is often represented as $\text{proj}_\mathbf{a}\mathbf{b}$.

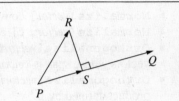

The projection vector can be computed using the *Mathematica* command **Projection**.

- **Projection[*vector1*, *vector2*]** returns the orthogonal projection of *vector1* onto *vector2*.

EXAMPLE 22 Compute the projection of $(1, 2, 3)$ onto $(-2, 3, -1)$.

```
a = {1, 2, 3};
b = {-2, 3, -1};
Projection[a, b]
```

$$\left\{-\frac{1}{7}, \frac{3}{14}, -\frac{1}{14}\right\}$$

The concept of orthogonality depends upon the definition of inner product for the space under consideration. By default, *Mathematica* uses the Euclidean inner product (dot product) in linear algebra commands. However, this can be changed by including an alternate definition in the third argument of **Projection**.

- **Projection[*vector1*, *vector2*, *f*]** returns the orthogonal projection of *vector1* onto *vector2* with respect to an inner product defined by *f*.

It can be shown that if $c1$, $c2$, and $c3$ are positive real numbers, then

$$\langle\mathbf{a}, \mathbf{b}\rangle = c_1 a_1 b_1 + c_2 a_2 b_2 + c_3 a_3 b_3$$

defines an inner product on $\mathbb{R}^3$. To compute the orthogonal projection of $(1, 2, 3)$ onto $(-2, 3, -1)$ using this inner product, we must define an appropriate function describing the inner product. To do this, we compute $\mathbf{a} * \mathbf{b}$ and then take the dot product with $\mathbf{c} = (c_1, c_2, c_3)$.

EXAMPLE 23 Compute the projection of $(1, 2, 3)$ onto $(-2, 3, -1)$ using the inner product $\langle\mathbf{a}, \mathbf{b}\rangle = 2a_1 b_1 + 3a_2 b_2 + 4a_3 b_3$.

```
a = {1, 2, 3};
b = {-2, 3, -1};
f[a_, b_]:= {2, 3, 4}.(a * b)          ← Note: It is important to use := here.
Projection[a, b, f]
```

$$\left\{-\frac{4}{39}, \frac{2}{13}, -\frac{2}{39}\right\}$$

EXAMPLE 24 A useful inner product often used in function spaces is $\langle f, g\rangle = \int_{-1}^{1} f(x)\,g(x)\,dx$. Using this inner product, compute the projection of x^2 on $x^3 + 1$.

```
a = x^2;
b = x^3 + 1;
f[p1_, p2_]:= ∫_{-1}^{1} p1 p2 dx
Projection[a, b, f]
```

$$\frac{7}{24}(1+x^3)$$

A finite dimensional vector space, by definition, has a finite basis. However, except for the trivial vector space that contains only the zero vector, an infinite number of different bases are possible.

The most convenient basis for any vector space is an orthonormal basis. The *Gram-Schmidt orthogonalization process* provides a "recipe" for converting any basis into an orthonormal basis.

- **Normalize[*vector*]** converts *vector* into a unit vector.
- **Normalize[*vector*, f]** converts *vector* into a unit vector with respect to the norm function f.
- **Orthogonalize[*vectorlist*]** uses the Gram-Schmidt method to produce an orthonormal set of vectors whose span is *vectorlist*.
- **Orthogonalize[*vectorlist*, f]** produces an orthonormal set of vectors with respect to the inner product defined by f.
- **Norm[v]** returns the Euclidean norm of **v**. $\| \mathbf{v} \| = \sqrt{\sum_{i=1}^{n} v_i^2}$.

EXAMPLE 25 To normalize (3, 4, 12) with respect to the Euclidean inner product, we type

```
Normalize[{3, 4, 12}]
```

$$\left\{ \frac{3}{13}, \frac{4}{13}, \frac{12}{13} \right\}$$

To normalize with respect to the norm $\| \mathbf{v} \| = \sqrt{2v_1^2 + 3v_2^2 + 4v_3^2}$, we define

```
f[v_] = √{2,3,4}.(v * v)
```

```
Normalize[{3, 4, 12}, f]
```

$$\left\{ \sqrt{\frac{3}{214}}, 2\sqrt{\frac{2}{321}}, 2\sqrt{\frac{6}{107}} \right\}$$

EXAMPLE 26 Find an orthonormal basis for the space spanned by (1, 1, 1, 0, 0), (0, 1, 1, 1, 0), and (0, 0, 1, 1, 1), and verify that the result is correct.

```
v = {{1, 1, 1, 0, 0}, {0, 1, 1, 1, 0}, {0, 0, 1, 1, 1}};
w = Orthogonalize[v]
```

$$\left\{ \left\{ \frac{1}{\sqrt{3}}, \frac{1}{\sqrt{3}}, \frac{1}{\sqrt{3}}, 0, 0 \right\}, \left\{ -\frac{2}{\sqrt{15}}, \frac{1}{\sqrt{15}}, \frac{1}{\sqrt{15}}, \sqrt{\frac{3}{5}}, 0 \right\}, \left\{ \frac{1}{2\sqrt{10}}, -\frac{3}{2\sqrt{10}}, \frac{1}{\sqrt{10}}, \frac{1}{2\sqrt{10}}, \frac{\sqrt{\frac{5}{2}}}{2} \right\} \right\}$$

To verify that the result is correct, we compute six dot products.

```
w[[1]].w[[1]]
```
1
```
w[[2]].w[[2]]
```
1
```
w[[3]].w[[3]]
```
1

> The vectors are all unit length.

```
w[[1]].w[[2]]
```
0
```
w[[1]].w[[3]]
```
0
```
w[[2]].w[[3]]
```
0

> The vectors are mutually orthogonal.

SOLVED PROBLEMS

12.30 Compute the norm of the vector (1, 2, 3, 4, 5) with respect to (a) the Euclidean inner product and (b) $\langle \mathbf{u}, \mathbf{v} \rangle = 2u_1v_1 + 3u_2v_2 + u_3v_3 + 3u_4v_4 + 2u_5v_5$.

SOLUTION

(a) ```
u = {1,2,3,4,5};

norm = √u.u

√55
```

(b) ```
u = {1,2,3,4,5};

c = {2,3,1,3,2};

norm = √c.(u*u)

11
```

12.31 Find the projection of the vector (3, 4, 5) onto each of the coordinate axes.

SOLUTION

```
v = {3, 4, 5};
Projection[v, {1, 0, 0}]
{3, 0, 0}
Projection[v, {0, 1, 0}]
{0, 4, 0}
Projection[v, {0, 0, 1}]
{0, 0, 5}
```

12.32 Find a unit vector having the same direction as (1, –2, 2, –3).

SOLUTION

```
Normalize[{1, -2, 2, -3}]
```

$$\left\{ \frac{1}{3\sqrt{2}}, -\frac{\sqrt{2}}{3}, \frac{\sqrt{2}}{3}, -\frac{1}{\sqrt{2}} \right\}$$

12.33 If $\mathbf{a} = (1, 2, 3)$ and $\mathbf{b} = (1, –2, 5)$, compute the length of the vector $\mathbf{v}$ shown in the diagram.

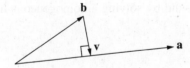

SOLUTION

Since $\mathbf{b} + \mathbf{v} = \text{proj}_a\mathbf{b}$, it follows that $\mathbf{v} = \text{proj}_a\mathbf{b} - \mathbf{b}$.

```
a = {1, 2, 3}; b = {1, -2, 5};
v = Projection[b, a] - b;
Norm[v]
```

$$\sqrt{\frac{138}{7}}$$

12.34 Find an orthonormal basis for the space spanned by (1, 2, 1, 3), (2, 2, 2, 2), (1, –1, 1, –1), and (3, 4, 3, 5).

SOLUTION

```
v1 = {1, 2, 1, 3};
v2 = {2, 2, 2, 2};
```

```
v3 = {1, -1, 1, -1};
v4 = {3, 4, 3, 5};
v = {v1, v2, v3, v4};
w = Orthogonalize[v]
```

$$\left\{\left\{\frac{1}{\sqrt{15}}, \frac{2}{\sqrt{15}}, \frac{1}{\sqrt{15}}, \sqrt{\frac{3}{5}}\right\}, \left\{\frac{8}{\sqrt{165}}, \frac{1}{\sqrt{165}}, \frac{8}{\sqrt{165}}, -2\sqrt{\frac{3}{55}}\right\},\right.$$

$$\left.\left\{\frac{1}{\sqrt{22}}, -2\sqrt{\frac{2}{11}}, \frac{1}{\sqrt{22}}, \sqrt{\frac{2}{11}}\right\}, \{0,0,0,0\}\right\}$$

> The set **v** is linearly dependent, so **Orthogonalize** can only produce three basis vectors. **(0, 0, 0, 0)** can be disregarded.

12.35 Construct an orthonormal basis for P_5, the set of all polynomials of degree ≤ 5 with respect to the inner product $<p, q> = \int_0^1 p(x)q(x)dx$.

SOLUTION

One basis for P_5 is the set $\mathbf{v} = \{1, x, x^2, x^3, x^4, x^5\}$. They comprise a linearly independent set that spans P_5.

```
v = {1, x, x², x³, x⁴, x⁵};

f[p_, q_] := ∫₀¹ p q dx

Orthogonalize[v, f] //Simplify
```

$$\{1, \sqrt{3}(-1+2x), \sqrt{5}(1-6x+6x^2), \sqrt{7}(-1+12x-30x^2+20x^3),$$

$$3(1-20x+90x^2-140x^3+70x^4), \sqrt{11}(-1+30x-210x^2+560x^3-630x^4+252x^5)\}$$

12.6   Eigenvalues and Eigenvectors

λ is said to be an eigenvalue of a square matrix, A, if there exists a non-zero vector, $\mathbf{x}$, such that $A\mathbf{x} = \lambda\mathbf{x}$. As powerful as the eigenvalue concept is in linear algebra, the computation of eigenvalues and their corresponding eigenvectors can be extremely difficult if the matrix is large.

One way to determine the eigenvalues of a matrix is to solve the characteristic equation $\det(A - \lambda I) = 0$. Once the eigenvalues are determined, the eigenvectors can be found by solving a homogeneous linear system.

EXAMPLE 27

$$a = \begin{pmatrix} 4 & 1 & -1 \\ 2 & 5 & -2 \\ 1 & 1 & 2 \end{pmatrix};$$

```
length = Length[a];
Solve[Det[a - λ IdentityMatrix[length]] == 0, λ]
{{λ → 3}, {{λ → 3}, {λ → 5}}
```

The eigenvalues are 3 (with multiplicity 2) and 5. To find the eigenvectors, we look at the null space of $A - \lambda I$:

```
NullSpace[a - 3 IdentityMatrix[length]]
{{1, 0, 1}, {-1, 1, 0}}
NullSpace[a - 5 IdentityMatrix[length]]
{{1, 2, 1}}
```

Of course, as one might expect, *Mathematica* contains commands that automatically compute eigenvalues, eigenvectors, and some other related items.

- **CharacteristicPolynomial[*matrix, var*]** returns the characteristic polynomial of *matrix* expressed in terms of variable *var*.
- **Eigenvalues[*matrix*]** returns a list of the eigenvalues of *matrix*.
- **Eigenvectors[*matrix*]** returns a list of the eigenvectors of *matrix*.
- **Eigensystem[*matrix*]** returns a list of the form {*eigenvalues, eigenvectors*}.

EXAMPLE 28

$$a = \begin{pmatrix} 4 & 1 & -1 \\ 2 & 5 & -2 \\ 1 & 1 & 2 \end{pmatrix};$$

CharacteristicPolynomial[a, x]

$45 - 39x + 11x^2 - x^3$

Eigenvalues[a]

{5, 3, 3}

Eigenvectors[a]

{{1, 2, 1}, {1, 0, 1}, {-1, 1, 0}}

Eigensystem[a]

{{5, 3, 3}, {{1, 2, 1}, {1, 0, 1}, {-1, 1, 0}}}

If the entries of the matrix are expressed exactly, i.e., in non-decimal form, *Mathematica* tries to determine the eigenvalues and eigenvectors exactly. If any of the entries of the matrix are expressed in decimal form, *Mathematica* returns decimal approximations. Alternatively, one can use **N[*matrix*]** as the argument of **CharacteristicPolynomial**, **Eigenvalues**, **Eigenvectors**, and **Eigensystem** to force *Mathematica* to return decimal eigenvalues and eigenvectors. If *k* digit precision is desired, **N[*matrix*, k]** will return k significant digits.

EXAMPLE 29

$$a = \begin{pmatrix} 1 & 1 \\ 1 & 3 \end{pmatrix};$$

Eigenvalues[a]

$\left\{ 2 + \sqrt{2}, 2 - \sqrt{2} \right\}$

Eigenvectors[a]

$\left\{ \left\{ -1 + \sqrt{2}, 1 \right\}, \left\{ -1 - \sqrt{2}, 1 \right\} \right\}$

Eigenvalues[N[a]]

{3.41421, 0.585786}

Eigenvectors[N[a]]

{{0.382683, 0.92388}, {-0.92388, 0.382683}}

Eigenvalues[N[a, 20]]

{3.4142135623730950488, 0.58578643762690495120}

Eigenvectors[N[a, 20]]

{{-0.38268343236508977173, -0.92387953251128675613}, {-0.92387953251128675613, 0.38268343236508977173}}

Eigensystem[N[a]]

{{3.41421, 0.585786}, {{0.382683, 0.92388}, {-0.92388, 0.382683}}}

Note: Because different algorithms are used for computing numerical eigenvalues, they sometimes emerge in a different order. Furthermore, since eigenvectors are not uniquely determined, the numerical eigenvectors may appear to be multiples or linear combinations of those obtained previously.

SOLVED PROBLEMS

12.36 What is the characteristic polynomial of the matrix A, whose entries are the first 25 consecutive integers?

SOLUTION

```
a = Partition[Range[25], 5];
a // MatrixForm
```

$$\begin{pmatrix} 1 & 2 & 3 & 4 & 5 \\ 6 & 7 & 8 & 9 & 10 \\ 11 & 12 & 13 & 14 & 15 \\ 16 & 17 & 18 & 19 & 20 \\ 21 & 22 & 23 & 24 & 25 \end{pmatrix}$$

```
CharacteristicPolynomial[a, x]
```

$250 x^3 + 65 x^4 - x^5$

12.37 Consider the tridiagonal 5×5 matrix whose main diagonal entries are 4, with 1s on the adjacent diagonals. Show the eigenvalues and corresponding eigenvectors in a clear, unambiguous manner.

SOLUTION

```
m = Table[If[Abs[i - j] == 1, 1, If[i == j, 4, 0]], {i, 1, 5}, {j, 1, 5}];
m // MatrixForm
```

$$\begin{pmatrix} 4 & 1 & 0 & 0 & 0 \\ 1 & 4 & 1 & 0 & 0 \\ 0 & 1 & 4 & 1 & 0 \\ 0 & 0 & 1 & 4 & 1 \\ 0 & 0 & 0 & 1 & 4 \end{pmatrix}$$

```
data = Eigensystem[m]
```

$\{\{4 + \sqrt{3}, 5, 4, 3, 4, -\sqrt{3}\}, \{\{1, \sqrt{3}, 2, \sqrt{3}, 1\}, \{-1, -1, 0, 1, 1\},$

$\{1, 0, -1, 0, 1\}, \{-1, 1, 0, -1, 1\}, \{1, -\sqrt{3}, 2, -\sqrt{3}, 1\}\}\}$

```
Do[Print["eigenvalue #", k, "is", data[[1, k]],
"with corresponding eigenvector:", data[[2, k]]], {k, 1, 5}]
```

eigenvalue #1 is $4 + \sqrt{3}$ with corresponding eigenvector: $\{1, \sqrt{3}, 2, \sqrt{3}, 1\}$

eigenvalue #2 is 5 with corresponding eigenvector: $\{-1, -1, 0, 1, 1\}$

eigenvalue #3 is 4 with corresponding eigenvector: $\{1, 0, -1, 0, 1\}$

eigenvalue #4 is 3 with corresponding eigenvector: $\{-1, 1, 0, -1, 1\}$

eigenvalue #5 is $4 - \sqrt{3}$ with corresponding eigenvector: $\{1, -\sqrt{3}, 2, -\sqrt{3}, 1\}$

12.38 An important theorem in linear algebra, the *Cayley-Hamilton theorem*, says that every square matrix satisfies its characteristic equation. Verify the Cayley-Hamilton theorem for

$$A = \begin{pmatrix} 1 & 2 & 1 & 2 \\ 3 & -1 & 3 & -1 \\ 2 & 5 & 7 & 1 \\ 1 & 2 & 3 & 6 \end{pmatrix}$$

SOLUTION

```
a = ( 1   2   1   2
      3  -1   3  -1   ;
      2   5   7   1
      1   2   3   6 )
```

```
CharacteristicPolynomial[a, x]
```

$$-196 + 161\,x + 15\,x^2 - 13\,x^3 + x^4$$

```
-196 IdentityMatrix[4] + 161 a + 15 MatrixPower[a, 2] - 13 MatrixPower[a, 3] +
    MatrixPower[a, 4]  // MatrixForm
```

$$\begin{pmatrix} 0 & 0 & 0 & 0 \\ 0 & 0 & 0 & 0 \\ 0 & 0 & 0 & 0 \\ 0 & 0 & 0 & 0 \end{pmatrix}$$

12.39 Approximate the eigenvalues of the 10×10 Hilbert matrix: $h_{ij} = \dfrac{1}{i+j-1}$.

SOLUTION

```
hilbert = HilbertMatrix[10];
hilbert // MatrixForm
```

$$\begin{pmatrix}
1 & \frac{1}{2} & \frac{1}{3} & \frac{1}{4} & \frac{1}{5} & \frac{1}{6} & \frac{1}{7} & \frac{1}{8} & \frac{1}{9} & \frac{1}{10} \\
\frac{1}{2} & \frac{1}{3} & \frac{1}{4} & \frac{1}{5} & \frac{1}{6} & \frac{1}{7} & \frac{1}{8} & \frac{1}{9} & \frac{1}{10} & \frac{1}{11} \\
\frac{1}{3} & \frac{1}{4} & \frac{1}{5} & \frac{1}{6} & \frac{1}{7} & \frac{1}{8} & \frac{1}{9} & \frac{1}{10} & \frac{1}{11} & \frac{1}{12} \\
\frac{1}{4} & \frac{1}{5} & \frac{1}{6} & \frac{1}{7} & \frac{1}{8} & \frac{1}{9} & \frac{1}{10} & \frac{1}{11} & \frac{1}{12} & \frac{1}{13} \\
\frac{1}{5} & \frac{1}{6} & \frac{1}{7} & \frac{1}{8} & \frac{1}{9} & \frac{1}{10} & \frac{1}{11} & \frac{1}{12} & \frac{1}{13} & \frac{1}{14} \\
\frac{1}{6} & \frac{1}{7} & \frac{1}{8} & \frac{1}{9} & \frac{1}{10} & \frac{1}{11} & \frac{1}{12} & \frac{1}{13} & \frac{1}{14} & \frac{1}{15} \\
\frac{1}{7} & \frac{1}{8} & \frac{1}{9} & \frac{1}{10} & \frac{1}{11} & \frac{1}{12} & \frac{1}{13} & \frac{1}{14} & \frac{1}{15} & \frac{1}{16} \\
\frac{1}{8} & \frac{1}{9} & \frac{1}{10} & \frac{1}{11} & \frac{1}{12} & \frac{1}{13} & \frac{1}{14} & \frac{1}{15} & \frac{1}{16} & \frac{1}{17} \\
\frac{1}{9} & \frac{1}{10} & \frac{1}{11} & \frac{1}{12} & \frac{1}{13} & \frac{1}{14} & \frac{1}{15} & \frac{1}{16} & \frac{1}{17} & \frac{1}{18} \\
\frac{1}{10} & \frac{1}{11} & \frac{1}{12} & \frac{1}{13} & \frac{1}{14} & \frac{1}{15} & \frac{1}{16} & \frac{1}{17} & \frac{1}{18} & \frac{1}{19}
\end{pmatrix}$$

```
Eigenvalues[N[hilbert]]
```

$\{1.75192, 0.34293, 0.0357418, 0.00253089, 0.00012875, 4.72969 \times 10^{-6}, 1.22897 \times 10^{-7}, 2.14744 \times 10^{-9}, 2.26675 \times 10^{-11}, 1.09287 \times 10^{-13}\}$

12.40 Approximate the eigenvalues of the 10×10 matrix A such that $a_{i,j} = \begin{cases} i+j-1 & \text{if } i+j \leq 11 \\ 21-i-j & \text{if } i+j > 11 \end{cases}$

SOLUTION

```
f[i_, j_]:= i + j - 1 /; i + j ≤ 11
f[i_, j_]:= 21 - i - j /; i + j > 11
a = Array[f, {10, 10}];
a // MatrixForm
```

$$\begin{pmatrix}
1 & 2 & 3 & 4 & 5 & 6 & 7 & 8 & 9 & 10 \\
2 & 3 & 4 & 5 & 6 & 7 & 8 & 9 & 10 & 9 \\
3 & 4 & 5 & 6 & 7 & 8 & 9 & 10 & 9 & 8 \\
4 & 5 & 6 & 7 & 8 & 9 & 10 & 9 & 8 & 7 \\
5 & 6 & 7 & 8 & 9 & 10 & 9 & 8 & 7 & 6 \\
6 & 7 & 8 & 9 & 10 & 9 & 8 & 7 & 6 & 5 \\
7 & 8 & 9 & 10 & 9 & 8 & 7 & 6 & 5 & 4 \\
8 & 9 & 10 & 9 & 8 & 7 & 6 & 5 & 4 & 3 \\
9 & 10 & 9 & 8 & 7 & 6 & 5 & 4 & 3 & 2 \\
10 & 9 & 8 & 7 & 6 & 5 & 4 & 3 & 2 & 1
\end{pmatrix}$$

```
Eigenvalues[N[a]]
```

{67.8404, − 20.4317, 4.45599, − 2.42592, 1.39587, − 1., 0.756101,
  − 0.629808, 0.55164, − 0.512543}

12.7 Diagonalization and Jordan Canonical Form

Given an $n \times n$ matrix, A, if there exists an invertible matrix, P, such that $A = PDP^{-1}$, where D is a diagonal matrix, we say that A is diagonalizable.

Not every matrix is diagonalizable. However, it can be shown that if A has a set of n linearly independent eigenvectors, then A is diagonalizable. P is the matrix whose columns are the eigenvectors of A, and D is the diagonal matrix whose main diagonal entries are their respective eigenvalues.

EXAMPLE 30

$$a = \begin{pmatrix}
18 & -51 & 27 & -15 \\
8 & -24 & 14 & -8 \\
15 & -48 & 28 & -15 \\
15 & -47 & 25 & -12
\end{pmatrix};$$

```
Eigenvalues[a]
```
{4, 3, 2, 1}   ← Since the eigenvalues are distinct, the
                                                       corresponding eigenvectors will be
```
Eigenvectors[a]
```                                                    linearly independent.
{{3, 1, 2, 3}, {1, 0, 0, 1}, {0, 1, 3, 2}, {3, 2, 3, 2}}
```
p = Transpose[Eigenvectors[a]]
```                                                    ← The transpose makes the eigenvectors
{{3, 1, 0, 3}, {1, 0, 1, 2}, {2, 0, 3, 3}, {3, 1, 2, 2}}   columns.
```
d = DiagonalMatrix[Eigenvalues[a]]
```
{{4, 0, 0, 0}, {0, 3, 0, 0}, {0, 0, 2, 0}, {0, 0, 0, 1}}
```
p.d.Inverse[p] // MatrixForm
```

$$\begin{pmatrix}
18 & -51 & 27 & -15 \\
8 & -24 & 14 & -8 \\
15 & -48 & 28 & -15 \\
15 & -47 & 25 & -12
\end{pmatrix}$$

← p.d.Inverse[p] = a

To summarize,

`MatrixForm[a] == MatrixForm[p].MatrixForm[d].MatrixForm[Inverse[p]]`

$$
\begin{pmatrix}
18 & -51 & 27 & -15 \\
8 & -24 & 14 & -8 \\
15 & -48 & 28 & -15 \\
15 & -47 & 25 & -12
\end{pmatrix}
=
\begin{pmatrix}
3 & 0 & 1 & 3 \\
1 & 0 & 1 & 2 \\
2 & 0 & 3 & 3 \\
3 & 1 & 2 & 2
\end{pmatrix}
\cdot
\begin{pmatrix}
4 & 0 & 0 & 0 \\
0 & 3 & 0 & 0 \\
0 & 0 & 2 & 0 \\
0 & 0 & 0 & 1
\end{pmatrix}
\cdot
\begin{pmatrix}
3 & -9 & 5 & -3 \\
-5 & 15 & -9 & 6 \\
-1 & 2 & -1 & 1 \\
-1 & 4 & -2 & 1
\end{pmatrix}
$$

Unfortunately, not every matrix can be diagonalized. However, there is a standard form, called *Jordan canonical form*, that every matrix possesses.

A Jordan block is a square matrix whose elements are zero except for the main diagonal, where all numbers are equal, and the superdiagonal, where all values are 1:

$$
\begin{pmatrix}
\lambda & 1 & 0 & 0 & .. & 0 & 0 \\
0 & \lambda & 1 & 0 & .. & 0 & 0 \\
0 & 0 & \lambda & 1 & .. & 0 & 0 \\
. & . & . & & .. & 1 & 0 \\
0 & 0 & 0 & 0 & .. & \lambda & 1 \\
0 & 0 & 0 & 0 & .. & 0 & \lambda
\end{pmatrix}
$$

If A is any $n \times n$ matrix, there exists a matrix Q such that $A = QJQ^{-1}$ and

$$
J =
\begin{pmatrix}
J_1 & 0 & 0 & .. & 0 \\
0 & J_2 & 0 & .. & 0 \\
0 & 0 & J_3 & .. & 0 \\
. & . & . & .. & . \\
0 & 0 & 0 & .. & J_k
\end{pmatrix}
$$

The J_is are Jordan blocks. The same eigenvalue may occur in different blocks. The number of distinct blocks corresponding to a given eigenvalue is equal to the number of independent eigenvectors belonging to that eigenvalue.

EXAMPLE 31

$$
a =
\begin{pmatrix}
5 & 4 & 3 \\
-1 & 0 & -3 \\
1 & -2 & 1
\end{pmatrix};
$$

`Eigensystem[a]`

`{{4, 4, -2}, {{1, -1, 1}, {0, 0, 0}, {-1, 1, 1}}}`

The eigenvalues are –2 and 4 with eigenvectors, respectively, (–1, 1, 1) and (1, –1, 1). The vector {0, 0, 0} is not an eigenvector; its presence simply indicates that a third linearly independent eigenvector cannot be found. To construct Q, the standard procedure is to find a vector **x** such that $(A - 4\,I)\,\mathbf{x} = (1, -1, 1)$.

`LinearSolve[a - 4 IdentityMatrix[3], {1, -1, 1}]`

`{1, 0, 0}`

The matrices Q and J may now be constructed:

$$
q =
\begin{pmatrix}
-1 & 1 & 1 \\
1 & -1 & 0 \\
1 & 1 & 0
\end{pmatrix};
$$

$$
j =
\begin{pmatrix}
-2 & 0 & 0 \\
0 & 4 & 1 \\
0 & 0 & 4
\end{pmatrix};
$$

```
q.j.Inverse[q] // MatrixForm
```

$$\begin{pmatrix} 5 & 4 & 3 \\ -1 & 0 & -3 \\ 1 & -2 & 1 \end{pmatrix}$$

- **JordanDecomposition[*matrix*]** computes the Jordan canonical form of *matrix*. The output is a list $\{q, j\}$ where q and j correspond to Q and J as described previously.

EXAMPLE 32   (Continuation of Example 31)

$$a = \begin{pmatrix} 5 & 4 & 3 \\ -1 & 0 & -3 \\ 1 & -2 & 1 \end{pmatrix}$$

```
{q, j} = JordanDecomposition[a];
q // MatrixForm
```

$$\begin{pmatrix} -1 & 1 & 1 \\ 1 & -1 & 0 \\ 1 & 1 & 0 \end{pmatrix}$$

```
j // MatrixForm
```

$$\begin{pmatrix} -2 & 0 & 0 \\ 0 & 4 & 1 \\ 0 & 0 & 4 \end{pmatrix}$$

> Of course this agrees with the results of Example 31.

EXAMPLE 33

$$a = \begin{pmatrix} 65 & 88 & -129 & -23 & -1 & -97 & -19 \\ 86 & 124 & -180 & -32 & -4 & -134 & -21 \\ 29 & 39 & -54 & -11 & 2 & -43 & -13 \\ 36 & 50 & -77 & -11 & -3 & -56 & -5 \\ 63 & 88 & -131 & -24 & 0 & -97 & -16 \\ 58 & 85 & -126 & -21 & -6 & -91 & -9 \\ 63 & 87 & -129 & -24 & -1 & -96 & -16 \end{pmatrix} ;$$

```
{q, j} = JordanDecomposition[a];
MatrixForm[a] == MatrixForm[q].MatrixForm[j].MatrixForm[Inverse[q]]
```

$$\begin{pmatrix} 65 & 88 & -129 & -23 & -1 & -97 & -19 \\ 86 & 124 & -180 & -32 & -4 & -134 & -21 \\ 29 & 39 & -54 & -11 & 2 & -43 & -13 \\ 36 & 50 & -77 & -11 & -3 & -56 & -5 \\ 63 & 88 & -131 & -24 & 0 & -97 & -16 \\ 58 & 85 & -126 & -21 & -6 & -91 & -9 \\ 63 & 87 & -129 & -24 & -1 & -96 & -16 \end{pmatrix} ==$$

$$\begin{pmatrix} 4 & 2 & -2 & -2 & 1 & 0 & 0 \\ 5 & \frac{11}{2} & -\frac{3}{4} & -\frac{29}{8} & 3 & -1 & -1 \\ 0 & -4 & -2 & 0 & & -1 & 1 & 1 \\ 3 & \frac{13}{2} & \frac{11}{4} & -\frac{19}{8} & 2 & -1 & 1 \\ 3 & \frac{5}{2} & \frac{3}{4} & -\frac{3}{8} & 2 & -1 & 0 \\ 6 & 10 & 0 & -4 & 4 & -2 & -2 \\ 2 & 0 & 0 & 0 & 1 & 0 & 0 \end{pmatrix} \cdot \begin{pmatrix} 2 & 1 & 0 & 0 & 0 & 0 & 0 \\ 0 & 2 & 1 & 0 & 0 & 0 & 0 \\ 0 & 0 & 2 & 1 & 0 & 0 & 0 \\ 0 & 0 & 0 & 2 & 0 & 0 & 0 \\ 0 & 0 & 0 & 0 & 3 & 1 & 0 \\ 0 & 0 & 0 & 0 & 0 & 3 & 1 \\ 0 & 0 & 0 & 0 & 0 & 0 & 3 \end{pmatrix} \cdot \begin{pmatrix} -\frac{17}{2} & -\frac{25}{2} & 18 & \frac{7}{2} & 0 & \frac{27}{2} & 3 \\ \frac{19}{2} & \frac{25}{2} & -19 & -\frac{7}{2} & 0 & -14 & -3 \\ -\frac{9}{2} & -6 & 9 & 2 & 0 & \frac{13}{2} & \frac{3}{2} \\ 5 & 6 & -10 & -2 & 0 & -7 & -1 \\ 17 & 25 & -36 & -7 & 0 & -27 & -5 \\ 27 & 37 & -55 & -10 & -1 & -41 & -7 \\ 19 & 26 & -38 & -7 & 1 & -29 & -7 \end{pmatrix}$$

Other decompositions such as QR decomposition and Schur decomposition are available in *Mathematica*, but shall not be discussed in this book. Their respective command names are **QRDecomposition** and **SchurDecomposition**.

SOLVED PROBLEMS

12.41 Construct a 5×5 matrix whose eigenvalues are $-2, -1, 0, 1, 2$ with respective eigenvectors $(1, 1, 0, 0, 0)$, $(0, 1, 1, 0, 0)$, $(0, 0, 1, 1, 0)$, $(0, 0, 0, 1, 1)$, and $(1, 0, 0, 0, 1)$.

SOLUTION

```
d = DiagonalMatrix[{-2, -1, 0, 1, 2}];
p = Transpose[{{1, 1, 0, 0, 0}, {0, 1, 1, 0, 0}, {0, 0, 1, 1, 0}, {0, 0, 0, 1, 1},
               {1, 0, 0, 0, 1}}];
a = p.d.Inverse[p];
a // MatrixForm
```

$$\begin{pmatrix} 0 & -2 & 2 & -2 & 2 \\ -\frac{1}{2} & -\frac{3}{2} & \frac{1}{2} & -\frac{1}{2} & \frac{1}{2} \\ \frac{1}{2} & -\frac{1}{2} & -\frac{1}{2} & \frac{1}{2} & -\frac{1}{2} \\ -\frac{1}{2} & \frac{1}{2} & -\frac{1}{2} & \frac{1}{2} & \frac{1}{2} \\ \frac{1}{2} & -\frac{1}{2} & \frac{1}{2} & -\frac{1}{2} & \frac{3}{2} \end{pmatrix}$$

12.42 Show that the matrix $\begin{pmatrix} a & b \\ c & d \end{pmatrix}$ has real eigenvalues if and only if $a^2 + 4bc - 2ad + d^2 \geq 0$.

SOLUTION

```
m = {{a, b}, {c, d}};
Eigenvalues[m]
```

$$\left\{ \frac{1}{2}\left(a + d - \sqrt{a^2 + 4bc - 2ad + d^2}\right), \frac{1}{2}\left(a + d + \sqrt{a^2 + 4bc - 2ad + d^2}\right) \right\}$$

The eigenvalues will be real if and only if the expression inside the radical symbol is non-negative.

12.43 Construct a 7×7 matrix of random digits and show that the sum of its eigenvalues is equal to its trace and the product of its eigenvalues is equal to its determinant.

SOLUTION

```
a = Table[RandomInteger[9], {i, 1, 7}, {j, 1, 7}];
a // MatrixForm
```

$$\begin{pmatrix} 2 & 9 & 8 & 0 & 7 & 6 & 2 \\ 2 & 3 & 5 & 1 & 9 & 5 & 1 \\ 3 & 8 & 5 & 0 & 8 & 3 & 1 \\ 6 & 3 & 9 & 3 & 6 & 5 & 9 \\ 4 & 2 & 9 & 9 & 6 & 9 & 4 \\ 4 & 1 & 7 & 6 & 7 & 3 & 7 \\ 1 & 5 & 8 & 3 & 9 & 9 & 0 \end{pmatrix}$$

```
eigenvalues = Eigenvalues[N[a]]
```

{34.6689, −1.97195 + 6.25152i, −1.97195 − 6.25152i, −6.08883, −2.35886, −0.138652 + 0.334612i, −0.138652 − 0.334612i}

$$\sum_{i=1}^{7} \text{eigenvalues}[[i]] \quad \text{or} \quad \textbf{Total[eigenvalues]}$$

22. + 0. i

```
Tr[a]
```
22

$$\prod_{i=1}^{7} \text{eigenvalues[[i]]} \quad \text{or} \quad \text{Product[eigenvalues[[i]], \{i, 1, 7\}]}$$

```
2807. + 1.42585×10⁻¹³ i
```

```
Det[a]
```
2807

12.44 A matrix, P, is said to be orthogonal if $P^TP = I$. If it is possible to find an orthogonal matrix P that diagonalizes A, then A is said to be *orthogonally diagonalizable*. However, only symmetric matrices are orthogonally diagonalizable. (A matrix is symmetric if $A^T = A$. If A is symmetric, it can be shown that the eigenvectors corresponding to distinct eigenvalues are orthogonal.) Find a matrix P that orthogonally diagonalizes

$$A = \begin{pmatrix} 3 & 1 & 0 & 0 & 0 \\ 1 & 3 & 0 & 0 & 0 \\ 0 & 0 & 2 & 1 & 1 \\ 0 & 0 & 1 & 2 & 1 \\ 0 & 0 & 1 & 1 & 2 \end{pmatrix}$$

SOLUTION

$$a = \begin{pmatrix} 3 & 1 & 0 & 0 & 0 \\ 1 & 3 & 0 & 0 & 0 \\ 0 & 0 & 2 & 1 & 1 \\ 0 & 0 & 1 & 2 & 1 \\ 0 & 0 & 1 & 1 & 2 \end{pmatrix};$$

```
{values, vectors} = Eigensystem[a]
{{4, 4, 2, 1, 1}, {{0, 0, 1, 1, 1}, {1, 1, 0, 0, 0}, {-1, 1, 0, 0, 0},
  {0, 0, -1, 0, 1}, {0, 0, -1, 1, 0}}}
eigenspace1 = {vectors[[1]], vectors[[2]]};
eigenspace2 = {vectors[[3]]};
eigenspace3 = {vectors[[4]], vectors[[5]]};
v1 = Orthogonalize[eigenspace1];
v2 = Orthogonalize[eigenspace2];
v3 = Orthogonalize[eigenspace3];
p = Transpose[Join[v1, v2, v3]];
p // MatrixForm
```

> There are five eigenvalues, two of which have multiplicity 2. We group their eigenvectors into three eigenspaces and apply the Gram-Schmidt process to each. The orthogonal matrix is the matrix whose columns are the vectors from **Orthogonalize**.

$$\begin{pmatrix} 0 & \dfrac{1}{\sqrt{2}} & -\dfrac{1}{\sqrt{2}} & 0 & 0 \\ 0 & \dfrac{1}{\sqrt{2}} & \dfrac{1}{\sqrt{2}} & 0 & 0 \\ \dfrac{1}{\sqrt{3}} & 0 & 0 & -\dfrac{1}{\sqrt{2}} & -\dfrac{1}{\sqrt{6}} \\ \dfrac{1}{\sqrt{3}} & 0 & 0 & 0 & \sqrt{\dfrac{2}{3}} \\ \dfrac{1}{\sqrt{3}} & 0 & 0 & \dfrac{1}{\sqrt{2}} & -\dfrac{1}{\sqrt{6}} \end{pmatrix}$$

```
Transpose[p].p // MatrixForm
```

$$\begin{pmatrix} 1 & 0 & 0 & 0 & 0 \\ 0 & 1 & 0 & 0 & 0 \\ 0 & 0 & 1 & 0 & 0 \\ 0 & 0 & 0 & 1 & 0 \\ 0 & 0 & 0 & 0 & 1 \end{pmatrix}$$ ←Just a check to see if the matrix is orthogonal.

```
d = DiagonalMatrix[values];
p.d.Transpose[p]  // MatrixForm
```

$$\begin{pmatrix} 3 & 1 & 0 & 0 & 0 \\ 1 & 3 & 0 & 0 & 0 \\ 0 & 0 & 2 & 1 & 1 \\ 0 & 0 & 1 & 2 & 1 \\ 0 & 0 & 1 & 1 & 2 \end{pmatrix}$$ ← This gives us back our original matrix.

```
MatrixForm[a] == MatrixForm[p] . MatrixForm[d] . MatrixForm[Transpose[p]]
```

$$\begin{pmatrix} 3 & 1 & 0 & 0 & 0 \\ 1 & 3 & 0 & 0 & 0 \\ 0 & 0 & 2 & 1 & 1 \\ 0 & 0 & 1 & 2 & 1 \\ 0 & 0 & 1 & 1 & 2 \end{pmatrix} ==$$

$$\begin{pmatrix} 0 & \frac{1}{\sqrt{2}} & -\frac{1}{\sqrt{2}} & 0 & 0 \\ 0 & \frac{1}{\sqrt{2}} & \frac{1}{\sqrt{2}} & 0 & 0 \\ \frac{1}{\sqrt{3}} & 0 & 0 & -\frac{1}{\sqrt{2}} & -\frac{1}{\sqrt{6}} \\ \frac{1}{\sqrt{3}} & 0 & 0 & 0 & \sqrt{\frac{2}{3}} \\ \frac{1}{\sqrt{3}} & 0 & 0 & \frac{1}{\sqrt{2}} & -\frac{1}{\sqrt{6}} \end{pmatrix} \cdot \begin{pmatrix} 4 & 0 & 0 & 0 & 0 \\ 0 & 4 & 0 & 0 & 0 \\ 0 & 0 & 2 & 0 & 0 \\ 0 & 0 & 0 & 1 & 0 \\ 0 & 0 & 0 & 0 & 1 \end{pmatrix} \cdot \begin{pmatrix} 0 & 0 & \frac{1}{\sqrt{3}} & \frac{1}{\sqrt{3}} & \frac{1}{\sqrt{3}} \\ \frac{1}{\sqrt{2}} & \frac{1}{\sqrt{2}} & 0 & 0 & 0 \\ -\frac{1}{\sqrt{2}} & \frac{1}{\sqrt{2}} & 0 & 0 & 0 \\ 0 & 0 & -\frac{1}{\sqrt{2}} & 0 & \frac{1}{\sqrt{2}} \\ 0 & 0 & -\frac{1}{\sqrt{6}} & \sqrt{\frac{2}{3}} & -\frac{1}{\sqrt{6}} \end{pmatrix}$$

Appendix

A.1 Pure Functions

A function is a correspondence between two sets of numbers *A* and *B* such that for each number in *A* there corresponds a unique number in *B*. For example, the "squaring" function: For each real number there corresponds a unique non-negative real number called its square.

While it is customary to write $f(x) = x^2$, one must understand that there is no special significance to the letter *x*. It is the process of *squaring* that defines the function.

Although *Mathematica* allows a function to be defined in terms of a variable, as in f [x_] = x^2, the variable x acts as a "dummy" and is insignificant. The function would be the same had we used y, z, or any other symbol.

A "pure" function is defined without reference to any specific variable. Its arguments are labeled #1, #2, #3, and so forth. To distinguish a pure function from any other *Mathematica* construct, an ampersand, &, is used at the end of its definition. Once defined, we can deal with a pure function as we would any other function.

Although the concept of a pure function is a natural one, it is possible to use *Mathematica* and never be concerned with it. Occasionally, however, *Mathematica* will express an answer as a pure function and it is therefore worthy of a brief mention. The interested reader can find more information in *Mathematica*'s Documentation Center.

EXAMPLE 1

```
f = #1² &;
f [3]
9
f [x]
x²
f [a + b]
(a + b)²
```

EXAMPLE 2

```
g = #1 #2² + 3&;
g [3, 4]
51
g [u, v]
3 + u v²
```

Another way of specifying a pure function is by use of *Mathematica*'s **Function** command.

- **Function[x,** *body***]** is a pure function with a single parameter x.
- **Function[{x₁, x₂, ...},** *body***]** is a pure function with a list of parameters x_1, x_2, . . .

EXAMPLE 3 Express the solution of the differential equation

$$\frac{d^2y}{dx^2} + y = 0; \quad y'(0) = y(0) = 1$$

as a pure function and evaluate it for $x = \pi/4$.

```
DSolve[{y''[x]+y[x]== 0, y[0]== 1, y'[0]=1}, y, x]
{{y→ Function[{x}, Cos[x] +Sin[x]]}}
Function[{x}, Cos[x] + Sin[x]] [π/4]
```
$\sqrt{2}$

SOLVED PROBLEMS

A.1 Express as a pure function the process of adding the square of a number to its square root and compute its value at 9.

SOLUTION

```
f = #1^2 + Sqrt[#1] &;
f[9]
84
```

A.2 A number is formed from two other numbers by adding the square of their sum to the sum of their squares. Express this operation as a pure function and compute its value for the numbers 3 and 4.

SOLUTION

```
g = (#1 + #2)^2 + #1^2 + #2^2 &;
g[3, 4]
74
```

A.3 Express the derivative of the function **Sin** as a pure function and compute its vale at $\pi/6$.

SOLUTION

```
f = Sin'
Cos[#1] &
f[π/6]
```
$\dfrac{\sqrt{3}}{2}$

A.4 Define $f(x) = (1 + x + x^2)^5$ and express its second derivative as a pure function.

SOLUTION

```
f[x_] = (1 + x + x²)⁵;
f''
20 (1 + 2 #1)² (1 + #1 + #1²)³ + 10 (1 + #1 + #1²)⁴ &
```

A.2 Patterns

You have certainly noticed the use of the underscore (_) character when defining functions in *Mathematica*. The use of the underscore is an important concept in *Mathematica* called *pattern matching*.

A pattern is an expression such as `x_` that contains an underscore character. The pattern can stand for any expression. Thus, `f[x_]` specifies how the function `f` should be applied to *any* argument. When you define a function such as `f[x_] = x²`, you are telling *Mathematica* to automatically apply the transformation rule `f[x_] → x²` whenever possible.

In contrast, a transformation rule for `f[x]` without an underscore specifies only how the literal expression `f[x]` should be transformed, and does not say anything about the transformation of `f[y]`, `f[z]`, etc.

EXAMPLE 4

```
Clear[f]
f[x_] = x²;
f[x]
x²
f[y]
y²
f[a + b]
(a + b)²
```

x_ is matched by any expression.

```
Clear[f]
f[x] = x²;
f[x]
x²
f[y]
f[y]
f[a + b]
f[a + b]
```

x is matched only by x.

EXAMPLE 5

```
1 + xᵖ + xᑫ /. x^q_ → Log[q]
1 + Log[p] + Log[q]                ← All exponentials are transformed to Log.
1 + xᵖ + xᑫ /. xᑫ → Log[q]
1 + xᵖ + Log[q]                    ← Only xᑫ is transformed.
```

Patterns can specify the *type* of an expression as well as its format. For example, **`_Integer`** stands for an integer pattern. Similarly, **`_Rational`**, **`_Real`**, and **`_Complex`** are acceptable patterns representing other types of numbers.

EXAMPLE 6 The *Mathematica* function **`Factorial[n]`** computes `n!` if `n` is a positive integer and `Γ[1+n]` if `n` is a positive real number. For certain applications, it might be useful to leave the factorial of a non-integer undefined.

```
fact[n_Integer] = Factorial[n];
fact[n_Real] = "undefined";
fact[5]
120
fact[5.5]
undefined
```

```
Factorial[5]
120
Factorial[5.5]
287.885
```

EXAMPLE 7 This example defines the function

$$f(x,y)=\begin{cases} xy & \text{if both } x \text{ and } y \text{ are integers} \\ x+y & \text{if both } x \text{ and } y \text{ are real} \\ x-y & \text{if } x \text{ or } y \text{ is an integer and the other is real} \end{cases}$$

```
f[a_Integer, b_Integer] = a b;
f[a_Real, b_Real] = a + b;
f[a_Real, b_Integer] = f[a_Integer, b_Real] = a - b;
f[2, 3]
6
f[2., 3.]
5.
f[2., 3]
-1.
f[2, 3.]
-1.
f[I, 1]
f[i,1]                          ← f is undefined for complex arguments.
```

A.3 Contexts

It is common practice to define symbols using names that are reminiscent of the symbol's purpose. Sometimes, however, the names get unwieldy and cumbersome to work with. *Contexts* are used as a tool to help organize the symbols used in a *Mathematica* session.

The complete name of a symbol is divided into two parts, a context and a shorter name, separated by a backquote (`) character. Used for this purpose the backquote is called a *context mark*.

EXAMPLE 8 `atomicnumber`au and `atomicweight`au are two distinct symbols with a common short name, au. (Au is the chemical symbol for gold.)

```
atomicnumber`au = 79;

atomicweight`au = 196.967;

atomicnumber`au

79

atomicweight`au

196.967
```

When you begin a *Mathematica* session, the default context is `Global` . Thus, for example, the symbol `object` is equivalent to `Global`object`. The default can be changed by redefining the symbol `$Context`.

- ▪ **$Context** is the current default context.
- ▪ **Context [*symbol*]** returns the context of *symbol*.

EXAMPLE 9

```
atomicnumber`au = 79;

atomicweight`au = 196.967;

$Context = "atomicweight`" ;          ← Context names are strings; quotes are important.

au

196.967

$Context = "atomicnumber`" ;

au

79
```

"Built-in" *Mathematica* symbols have context `System`.

EXAMPLE 10

```
Context[Pi]
System`
```

It is common for symbols in different contexts to have the same short name. If only the short name is referenced, *Mathematica* decides which is called by its position in a list called **$ContextPath**.

- **$ContextPath** is the current search path.

EXAMPLE 11

```
au = "gold";
atomicnumber`au = 79;
atomicweight`au = 196.967;
$ContextPath
{PacletManager`, WebServices`, System`, Global`}        ← Default context path.
$ContextPath = Join[$ContextPath, {"atomicweight`"}, {"atomicnumber`"}]
{PacletManager`, WebServices`, System`, Global`, atomicweight`, atomicnumber`}
au
gold                        ← Global` comes before atomicweight` and atomicnumber`
Remove[Global`au]             in $ContextPath.
au                          ← atomicweight is now the first element of $ContextPath in which
196.967                       au appears.
Remove[atomicweight`au]
au                          ← atomicnumber is now the first element of $ContextPath in
79                            which au appears.
```

A.4   Modules

Mathematica, by default, assumes that all objects are *global*. This means, for example, that if you define x to have a value of 3, x will remain 3 until its value is changed. In contrast, a *local* object has a limited scope valid only within a certain group of instructions.

Modules allow you to define local variables whose values are defined only within the module. Outside of the module, the object may either be undefined or have a completely different value.

- **Module[{var1, var2, ...}, body]** defines a module with local variables *var1, var2, ...*
- **Module[{var1 = v1, var2 = v2, ...}, body]** defines a module with local variables *var1, var2, ...* initialized to *v1, v2, . . .*, respectively.

EXAMPLE 12

```
x = 3;                      ← Global variable x is set to 3.
Module[{x = 8}, x + 1]      ← Module is defined with local variable x initialized to 8.
9                           ← x is incremented.
x                           ← Global x is called.
3                           ← Original value of x is returned.
```

It is often useful to group several commands into one unit to be executed as a group. This is especially true if complicated structures involving loops are involved. Several commands may be incorporated within *body* if they are separated by semicolons.

EXAMPLE 13

```
Module[{x = 1, y = 2}, x = x + 3; y = y + 4; Print[x y]]
24
```

| *x* and *y* are initialized to 1 and 2, respectively; *x* is incremented by 3, *y* is incremented by 4, and the two are multiplied. |

It is often convenient to define a function whose value is a module. This allows considerably more flexibility when dealing with functions whose definitions are complicated. When defining a function in this manner, it is important that the delayed assignment, := , be used.

EXAMPLE 14   The following defines the factorial function. The value of x0 is assumed to be a non-negative integer. The variables fact and x, which are initialized to be 1 and x0, respectively, are local so there is no conflict with any variables of the same name elsewhere in the program. x0 is a "dummy" variable.

```
f[x0_] := Module[{fact = 1, x = x0}, While[x > 1, fact = x * fact; x = x - 1];
               Print[fact]]
f[0]
1
f[5]
120
f[10]
3 628 800
```

To clarify how a module works, consider the next example. Although the same module is executed three times, the variable, which appears as x, is actually assigned three different local names. Because of this clever "bookkeeping," all three are independent and none will conflict with global variable x.

EXAMPLE 15

```
x = 3
3
Module[{x}, Print[x]]
x$342
Module[{x}, Print[x]]
x$344
Module[{x}, Print[x]]
x$346
x
3
```

> Since all three values of x are given different internal names, there can be no conflict.

SOLVED PROBLEMS

A.5   Write a module that will take an integer and return all its factors.

SOLUTION

```
factorlist[x0_] := Module[{x = 1},
                    While[x ≤ x0, If[Mod[x0, x] == 0, Print[x]]; x++]]
factorlist[1]
1
factorlist[10]
1
2
5
10
factorlist[11]
1
11
```

```
factorlist[90]
```

```
1
2
3
5
6
9
10
15
18
30
45
90
```

A.6　A very crude way of determining the position of a prime within the sequence of primes is to examine the list of all primes up to and including the prime in question and determine its position in the list. If the number is not in the list, then the number is not prime. Construct a module that will determine whether a number is prime, and if so, determine its position. If not, return a message indicating that it is not prime.

SOLUTION

```
pos[x0_]:= Module[{x = 1, prm}, prm = False;
While[Prime[x] ≤ x0 && Not[prm],
        If[Prime[x] == x0, prm = True]; x ++];
        If[prm, Print[x – 1], Print["Not a Prime"]]]
```

```
pos[1]
```

Not a Prime

```
pos[2]
```

1

```
pos[3]
```

2

```
pos[101]
```

26

```
pos[1001]
```

Not a Prime

A.7　A famous conjecture asserts that if you start with a positive integer, n, and replace it by $n/2$ if n is even and by $3n + 1$ if n is odd, and repeat the process over and over in an iterative manner, then you will *always* wind up with 1. (This conjecture has never been proven or disproved.) Construct a module that simulates this iterative process.

SOLUTION

We first define a function, **successor**, that will define one iteration step.

```
successor[n_]:= If[EvenQ[n], n/2, 3n + 1]
```

Next we introduce a module, **allvalues**, that will produce a list of all successors, starting with the successor of n.

```
allvalues[n_] := Module[{m = n}, While[m ≠ 1, m = successor[m]; Print[m]]]
allvalues[6]
```

3

10

5

16

8

4

2

1

Since this list might be long if *n* is large, and all we are really interested in is the final value and the number of iterations it takes to get there, another module might be more appropriate.

```
finalvalue[n_] := Module[{m = n, k = 0}, While[m ≠ 1, m = successor[m]; k++];
              Print["final value = ", m, ", # iterations = ", k]]
```

finalvalue lists the final value of the process, together with the number of iterations needed to reach the final value.

```
finalvalue[6]
```
final value = 1, # iterations = 8
```
finalvalue[100]
```
final value = 1, # iterations = 25
```
finalvalue[1000]
```
final value = 1, # iterations = 111

A.5   Commands Used in This Book
Options are not included in this list. Please refer to the index.

- **$Context** is the current default context.
- **$ContextPath** is the current search path.
- **Abs[x]** returns x if $x \geq 0$ and $-x$ if $x < 0$.
- **Accumulate[*list*]** returns a list having the same length as *list* containing the successive partial sums of *list*.
- **AddTo[x,y]** or **x + = y** adds y to x and returns the new value of x.
- **AffineShape[*object*, {*xscale*, *yscale*, *zscale*}]** scales the *x*-, *y*-, and *z*-coordinates by *xscale*, *yscale*, and *zscale*, respectively.
- **And[p, q]** or **p && q** or **p ∧ q** is True if both p and q are True; False otherwise.
- **Animate[*expression*, {k, m, n, i}]** displays several different graphics images rapidly in succession, producing the illusion of movement.
- **Animate[*expression*, {k1, m1, n1, i1}, {k2, m2, n2, i2}, ...,]** allows multiple parameters which can be independently controlled.
- **Apart[*fraction*]** writes *fraction* as a sum of partial fractions.
- **Append[*list*, x]** returns *list* with x inserted to the right of its last element.
- **ArcSin, ArcCos, ArcTan, ArcSec, ArcCsc,** and **ArcCot** are the inverse trigonometric functions. Only the *principal values*, expressed in radians, are returned by these functions.
- **ArcSinh, ArcCosh, ArcTanh, ArcSech, ArcCsch,** and **ArcCoth** are the inverse hyperbolic functions.
- **Array[f, n]** generates a list consisting of n values, f[1], f[2], ..., f[n].
- **Array[f, n, r]** generates a list consisting of n values, f[i] starting with f[r], i.e., f[r], f[r+1], ..., f[r + n − 1].

- **Array[f,{m, n}]** generates a nested list consisting of an array of m elements, each of which is an array of n elements, whose values are f[i,j] as j goes from 1 to n and i goes from 1 to m. Here f is a function of two variables. The second index varies most rapidly.
- **Array[f,{m, n},{r, s}]** generates a nested list consisting of an array of m elements, each of which is an array of n elements. The first element of the first sublist is f[r, s].
- **ArrayFlatten[{{m_{11}, m_{12},...}, {m_{21}, m_{22},...},...}]** creates a single flattened matrix from a matrix of matrices m_{ij}. All the matrices in the same row must have the same first dimension, and all the matrices in the same column must have the same second dimension.
- **BarChart[*datalist*]** draws a simple bar graph. *datalist* is a set of numbers enclosed within braces.
- **BarChart[{*datalist1*, *datalist2*,...}]** draws a bar graph containing data from multiple data sets. Each data list is a set of numbers enclosed within braces.
- **BarChart3D[*datalist*]** draws a 3-D bar graph corresponding to the numbers in *datalist*.
- **BarChart3D[{*datalist1*, *datalist2*,...}]** draws a bar graph containing data from multiple data sets.
- **Cancel[*fraction*]** cancels out common factors in the numerator and denominator of *fraction*. The option **Extension → Automatic** allows operations to be performed on algebraic numbers that appear in *fraction*.
- **CartesianProduct[*list1*, *list2*]** returns the Cartesian product of *list1* and *list2*.
- **Catalan** is Catalan's constant and is approximately 0.915966. It is used in the theory of combinatorial functions.
- **Ceiling[x]** returns the smallest integer not less than x. Many textbooks represent this by $\lceil x \rceil$.
- **CharacteristicPolynomial[*matrix*, *var*]** returns the characteristic polynomial of *matrix* expressed in terms of variable *var*.
- **CharacterRange["*char1*", "*char2*"]** produces a list of characters from *char1* to *char2*, based upon their standard ASCII values (assuming an American English alphabet).
- **Characters[*string*]** produces a list of characters in *string*.
- **ChebyshevT[n, x]** gives the Chebyshev polynomial (of the first kind) of degree n.
- **Clear[*symbol*]** clears *symbol*'s definition and values, but does not clear its attributes, messages, or defaults. *symbol* remains in *Mathematica*'s symbol list. Typing *symbol* = . will also clear the definition of *symbol*.
- **Coefficient[*polynomial*, *form*]** gives the coefficient of *form* in *polynomial*.
- **Coefficient[*polynomial*, *form*, n]** gives the coefficient of *form* to the nth power in *polynomial*.
- **CoefficientList[*polynomial*, *variable*]** gives a list of the coefficients of powers of *variable* in *polynomial*, starting with the 0th power.
- **Collect[*poly*, *var*]** takes a polynomial having two or more variables and expresses it as a polynomial in *var*.
- **ColumnForm[*list*]** presents *list* as a single column of objects.
- **ColumnForm[list, *horizontal*]** specifies the horizontal alignment of each row. Acceptable values of *horizontal* are **Left** (default), **Center**, and **Right**.
- **ColumnForm[list, *horizontal*, *vertical*]** allows vertical alignment of the column. Acceptable values of *vertical* are **Above**, **Center**, and **Below** (default).
- **Complement[*universe*, *list*]** returns a sorted list consisting of those elements of *universe* that are not in *list*. In this context, universe represents the universal set.
- **Complement[*universe*, *list1*, *list2*]** returns a sorted list consisting of those elements of *universe* that are not in *list1* or *list2*. This command extends in a natural way to more than two sets.
- **Composition[f1, f2, f3,...]** constructs the composition f1 ∘ f2 ∘ f3...
- **ConstantArray[c, {m, n}]** generates an m × n array, each element of which is c.
- **Context[*symbol*]** returns the context of *symbol*.
- **ContourPlot[*equation*, {x, xmin, xmax}, {y, ymin, ymax}]** plots *equation* by treating the equation as a function in three-dimensional space, and generates a contour of the equation cutting through the plane where z equals zero.
- **ContourPlot[{*equation1*, *equation2*,...}, {x, xmin, xmax}, {y, ymin, ymax}]** plots several implicitly defined curves.
- **ContourPlot[f[x, y], {x, xmin, xmax}, {y, ymin, ymax}]** draws a contour plot of $f(x, y)$ in a rectangle determined by xmin, xmax, ymin, and ymax.
- **ContourPlot3D[f[x, y, z], {x, xmin, xmax}, {y, ymin, ymax}, {z, zmin, zmax}]** draws a three-dimensional contour plot of the level surface $f(x, y, z) = 0$ in a box determined by xmin, xmax, ymin, ymax, zmin, and zmax.

- **Cross[v1, v2]** returns the cross product of **v1** and **v2**. (This applies to three-dimensional vectors only.) The cross product symbol, ×, can be inserted into the calculation by typing (without spaces) the key sequence **[ESC]c-r-o-s-s[ESC]**.
- **D[f[x], x]** returns the derivative of f with respect to x.
- **D[f[x], {x, n}]** returns the nth derivative of f with respect to x.
- **D[f, x]** or $\partial_x f$ (on the **Basic Math Input** palette) returns $\partial f / \partial x$, the partial derivative of f with respect to x.
- **D[f, {x, n}]** or $\partial_{\{x, n\}} f$ returns $\partial^n f / \partial x^n$, the nth order partial derivative of f with respect to x.
- **D[f, x₁, x₂, ..., xₖ]** or $\partial_{x_1, x_2, \ldots, x_k} f$ returns the "mixed" partial derivative $\dfrac{\partial^k f}{\partial x_1 \, \partial x_2 \, \ldots \, \partial x_k}$.
- **D[f, {x₁, n₁}, {x₂, n₂}, ..., {xₖ, nₖ}]** or $\partial_{\{x_1, n_1\}, \{x_2, n_2\}, \ldots, \{x_k, n_k\}} f$ returns the partial derivative $\dfrac{\partial^n f [x_1, x_2, \ldots, x_k]}{\partial_{x_1}^{n_1} \partial_{x_2}^{n_2} \ldots \partial_{x_k}^{n_k}}$ where $n_1 + n_2 + \ldots + n_k = n$.
- **Decrement[x]** or **x --** decreases the value of x by 1 but returns the *old* value of x.
- **Degree** is equal to **Pi/180** and is used to convert degrees to radians.
- **Delete[list, n]** deletes the element in the nth position of list.
- **Delete[list, -n]** deletes the element in the nth position from the end of list.
- **Delete[list, {{p₁}, {p₂}, ...}]** deletes the elements in positions p_1, p_2, ...
- **Delete[list, {p, q}]** deletes the element in position q of part p.
- **Delete[list, {{p₁, q₁}, {p₂, q₂}, . . .]** deletes the elements in position q_1 of part p_1, position q_2 of part p_2, ...
- **Denominator[*fraction*]** returns the denominator of *fraction*.
- **DensityPlot[f[x, y], {x, xmin, xmax}, {y, ymin, ymax}]** draws a density plot of $f(x, y)$ in a rectangle determined by xmin, xmax, ymin, and ymax.
- **Depth[*list*]** returns *one more* than the number of levels in the list structure. Raw objects, i.e., objects that are not lists, have a depth of 1.
- **Derivative[n]** is a *functional operator* that acts on a function to produce a new function, namely, its nth derivative. **Derivative[n][f]** gives the nth derivative of f as a *pure* function and **Derivative[n][f][x]** will compute the nth derivative of f at x.
- **Derivative[n₁, n₂, ..., nₖ][f]** gives the partial derivative $\dfrac{\partial^n f}{\partial_{x_1}^{n_1} \partial_{x_2}^{n_2} \ldots \partial_{x_k}^{n_k}}$ where $n_1 + n_2 + \ldots + n_k = n$. It returns a pure function that may then be evaluated at $[x_1, x_2, \ldots, x_k]$.
- **Det[*matrix*]** computes the determinant of *matrix*.
- **DiagonalMatrix[*list*]** creates a diagonal matrix whose diagonal entries are the elements of *list*.
- **DiracDelta[t]** returns $\delta(t)$, the Dirac delta function that satisfies $\delta(t) = 0$ if $t \neq 0$, $\int_{-\infty}^{\infty} \delta(t) dt = 1$.
- **Divide[a, b]** computes the quotient of a and b. Only two arguments are permitted. **Divide[a, b]** is equivalent to **a/b**.
- **DivideBy[x, y]** or **x /= y** divides x by y and returns the new value of x.
- **Do[*expression*, {k}]** evaluates *expression* precisely k times.
- **Do[*expression*, {i, imax}]** evaluates *expression* imax times with the value of i changing from 1 to imax in increments of 1.
- **Do[*expression*, {i, imin, imax}]** evaluates *expression* with the value of i changing from imin to imax in increments of 1.
- **Do[*expression*, {i, imin, imax, increment}]** evaluates *expression* with the value of i changing from imin to imax in increments of increment.
- **Do[*expression*, {i, imin, imax}, {j, jmin, jmax}]** evaluates *expression* with the value of i changing from imin to imax and j changing from jmin to jmax in increments of 1. The variable i changes by 1 for each cycle of j. This is known as a nested **Do** loop.
- **Do[*expression*, {i, imin, imax, i_increment}, {j, jmin, jmax, j_increment}, ...,]** forms a nested **Do** loop allowing for incrimination values other than 1.
- **Drop[*list*, n]** returns *list* with its first n objects deleted.
- **Drop[*list*, -n]** returns *list* with its last n objects deleted.
- **Drop[*list*, {n}]** returns *list* with its nth object deleted.
- **Drop[*list*, {-n}]** returns *list* with the nth object from the end deleted.
- **Drop[*list*, {m, n}]** returns *list* with objects m through n deleted.

- **Drop** [*list*, {**m, n, k**}] returns *list* with objects m through n in increments of k deleted.
- **DSolve** [*equation*, **y** [**x**], **x**] gives the general solution, y [x], of the differential equation, *equation*, whose independent variable is x.
- **DSolve** [*equation*, **y, x**] gives the general solution, y, of the differential equation expressed as a "pure" function within a list. **ReplaceAll** (**/.**) may then be used to evaluate the solution. Alternatively, one may use **Part** or **[[ ]]** to extract the solution from the list.
- **Dt** [**f** [**x, y**]] returns the total differential of f [x, y].
- **Dt** [**f** [**x, y**], **x**] returns the total derivative of f [x, y] with respect to x.
- **E** or **e** is the base of the natural logarithm.
- **Eigensystem** [*matrix*] returns a list of the form {*eigenvalues, eigenvectors*}.
- **Eigenvalues** [*matrix*] returns a list of the eigenvalues of *matrix*.
- **Eigenvectors** [*matrix*] returns a list of the eigenvectors of *matrix*.
- **Eliminate** [*equations*, *variables*] eliminates *variables* from a set of simultaneous equations.
- **Equal** [**x, y**] or **x == y** is True if and only if x and y have the same value.
- **EulerGamma** is Euler's constant and is approximately 0.577216. It has applications in integration and asymptotic expansions.
- **Exp** [**x**] is the natural exponential function. Other equivalent forms are **E^x** and **E^x**. Lowercase e cannot be used, but the special symbol e from the Basic Math Input palette may be used instead. Exponential functions to the base b are computed by **b^x** or **b^x**.
- **Expand** [*poly*] expands products and powers, writing *poly* as a sum of individual terms.
- **ExpandAll** [*expression*] expands both numerator and denominator of *expression*, writing the result as a sum of fractions with a common denominator.
- **ExpandDenominator** [*expression*] expands the denominator of *expression* but leaves the numerator alone.
- **ExpandNumerator** [*expression*] expands the numerator of *expression* but leaves the denominator alone.
- **ExpToTrig** [*expression*] converts exponential functions to trigonometric and/or hyperbolic functions.
- **Factor** [*poly*] attempts to factor *poly* over the integers. If factoring is unsuccessful, *poly* is unchanged.
- **Factorial** [**n**] or n! gives the factorial of n if n is a positive integer and $\Gamma(n + 1)$ if n has a non-integer positive value.
- **FactorInteger** [**n**] gives the prime factors of n together with their respective exponents.
- **FactorTerms** [*poly*] factors out common constants that appear in the terms of *poly*.
- **FactorTerms** [*poly*, *var*] factors out any common monomials containing variables other than *var*.
- **Fibonacci** [**n**] returns the nth Fibonacci number.
- **FindMaximum** [**f** [**x**], {**x, x_0**}] finds the relative maximum of $f(x)$ near x_0.
- **FindMinimum** [**f** [**x**], {**x, x_0**}] finds the relative minimum of $f(x)$ near x_0.
- **FindRoot** [**lhs == rhs, {x, x0}**] solves the equation lhs = rhs using Newton's method with starting value x0.
- **FindRoot** [**lhs == rhs, {x, {x0, x1}}**] solves the equation lhs = rhs using (a variation of) the secant method with starting values x0 and x1.
- **FindRoot** [**lhs == rhs, {x, x0, xmin, xmax}**] attempts to solve the equation, but stops if the iteration goes outside the interval [xmin, xmax].
- **FindRoot** [**equations, {var1,a1},{var2,a2},...**] attempts to solve equations using initial values a1, a2, ... for var1, var2, ..., respectively. The equations are enclosed in a list: {equation1, equation2, ... }. Alternatively, the equations may be separated by **&&** (logical and).
- **First** [*list*] returns the element of *list* in the first position.
- **Flatten** [*list*] converts a nested list to a simple list containing the innermost objects of *list*.
- **Flatten** [*list*, **n**] flattens a nested list n times, each time removing the outermost level. The depth of each level is reduced by n or to a minimum level of 1.
- **FlattenAt** [*list*, **n**] flattens the sublist that is at the nth position of the list by one level. If n is negative, *Mathematica* counts backward, starting at the end of the list.

- **Floor[x]** returns the greatest integer which does not exceed x. This is sometimes known as the "greatest integer function" and is represented in many textbooks by $\lfloor x \rfloor$.
- **For[*initialization*, *test*, *increment*, *expression*]** executes *initialization*, then repeatedly evaluates *expression*, *increment*, and *test* until *test* becomes False.
- **FractionalPart[x]** gives the fractional portion of x (decimal point included).
- **FullForm[*expression*]** exhibits the internal form of *expression*.
- **FullSimplify[*expression*]** tries a wide range of transformations on *expression* involving elementary and special functions, and returns the simplest form it finds.
- **Function[x, *body*]** is a pure function with a single parameter x.
- **Function[{x_1, x_2,...}, *body*]** is a pure function with a list of parameters x_1, x_2, . . .
- **GCD[m, n]** returns the greatest common divisor of m and n.
- **GoldenRatio** has the value $(1+\sqrt{5})/2$ and has a special significance with respect to Fibonacci series. It is used in *Mathematica* as the default width-to-height ratio of two-dimensional plots.
- **Graphics[*primitives*]** creates a two-dimensional graphics object.
- **Graphics3D[*primitives*]** creates a three-dimensional graphics object.
- **GraphicsArray[{g1, g2, ...}]** plots a row of graphics objects.
- **GraphicsArray[{g11, g12, ...},{g21, g22, ...}}]** plots a two-dimensional array of graphics objects.
- **Greater[x, y]** or **x > y** is True if and only if x is numerically greater than y.
- **GreaterEqual[x, y]** or **x >= y** or $x \geq y$ is True if and only if x is numerically greater than y or equal to y.
- **HankelMatrix[n, *list*]** creates a Hankel matrix whose first row (and column) is *list*.
- **HankelMatrix[n]** creates a Hankel matrix whose first row (and column) is {1, 2, 3, . . . , n}.
- **HeavisideTheta[x]** returns a value of 0 if x < 0 and 1 if x > 0.
- **HilbertMatrix[m, n]** creates an m × n Hilbert matrix.
- **HilbertMatrix[n]** creates an n × n Hilbert matrix
- **IdentityMatrix[n]** creates an n × n identity matrix.
- **IdentityMatrix[n]** produces an n × n matrix with 1s on the main diagonal and 0s elsewhere.
- **If[*condition*, *true*, *false*]** evaluates *condition* and executes *true* if *condition* is True and executes *false* if *condition* is False.
- **If[*condition*, *true*, *false*, *neither*]** evaluates *condition* and executes *true* if *condition* is True, executes *false* if *condition* is False, and executes *neither* if *condition* is neither True nor False.
- **If[*condition*, *true*]** evaluates *condition* and executes *true* if *condition* is True. If *condition* is False no action is taken and Null is returned.
- **If[*condition*, , *false*]** evaluates *condition* and executes *false* if *condition* is False. If *condition* is True no action is taken and Null is returned. (Note the double comma.)
- **Implies[p, q]** or $p \Rightarrow q$ is False if p is True and q is False; True otherwise.
- **Increment[x]** or **x ++** increases the value of x by 1 but returns the *old* value of x.
- **Infinity** or ∞ is a constant with special properties. For example, $\infty + 1 = \infty$.
- **InputForm[*expression*]** prints *expression* in a form suitable for input to *Mathematica*.
- **Insert[*list*, x, n]** returns *list* with x inserted in position n.
- **Insert[*list*, x, −n]** returns *list* with x inserted in the nth position from the end.
- **Insert[*list*, x,{m, n}]** returns *list* with x inserted in the nth position of the mth entry in the outer level.
- **IntegerPart[x]** gives the integer portion of x (decimal point excluded).
- **Integrate[f[x], x]** computes the antiderivative (indefinite integral) $\int f(x)\,dx$.
- **Integrate[f[x], {x, a, b}]** computes, whenever possible, the exact value of $\int_a^b f(x)\,dx$. The symbol $\int_\square^\square \blacksquare d\square$ on the Basic Math Input palette may be used as well.
- **Integrate[f[x, y], {x, xmin, xmax}, {y, ymin, ymax}]** evaluates the double integral $\int_{xmin}^{xmax} \int_{ymin}^{ymax} f(x,y)\,dy\,dx$.
- **Integrate[f[x, y, z], {x, xmin, xmax}, {y, ymin, ymax}, {z, zmin, zmax}]** evaluates the triple integral $\int_{xmin}^{xmax} \int_{ymin}^{ymax} \int_{zmin}^{zmax} f(x,y,z)\,dz\,dy\,dx$.

- **InterpolateRoot[lhs == rhs, {x, a, b}]** solves the equation lhs = rhs using initial values a and b.
- **Intersection[*list1*, *list2*]** returns a sorted list of elements common to *list1* and *list2*. If *list1* and *list2* are disjoint, i.e., they have no common elements, the command returns the empty list, {}. *list1* ∩ *list2* is equivalent to **Intersection[*list1*, *list2*]**.
- **Inverse[*matrix*]** computes the inverse of *matrix*.
- **InverseLaplaceTransform[F[*var1*], *var1*, *var2*]** computes the inverse Laplace transform of the function F, with independent variable *var1*, and expresses it as a function of *var2*.
- **Join[*list1*, *list2*]** combines the two lists *list1* and *list2* into one list consisting of the elements from *list1* and *list2*.
- **JordanDecomposition[*matrix*]** computes the Jordan canonical form of *matrix*.
- **KSubsets[*list*, k]** returns a list containing all subsets of *list* of size k.
- **LaplaceTransform[f[*var1*], *var1*, *var2*]** computes the Laplace transform of the function f, with independent variable *var1*, and expresses it as a function of *var2*.
- **Last[*list*]** returns the element of *list* in the last position.
- **LCM[m, n]** returns the least common multiple of m and n.
- **Length[*list*]** returns the length of *list*, i.e., the number of elements in *list*.
- **Less[x, y]** or **x < y** is True if and only if x is numerically less than y.
- **LessEqual[x, y]** or **x <= y** or **x ≤ y** is True if and only if x is numerically less than or equal to y.
- **Level[*list*, {*levelspec*}]** returns a list consisting of those objects that are at level *levelspec* of *list*.
- **Level[*list*, *levelspec*]** returns a list consisting of those objects that are at or below level *levelspec* of *list*.
- **Limit[f[x], x → a]** computes the value of $\lim_{x \to a} f(x)$.
- **LinearSolve[a, b]** produces vectors x such that a.x = b.
- **LinearSolve[a]** produces a **LinearSolveFunction** that can be used to solve a.x = b for different vectors b.
- **List[*elements*]** represents a list of objects. *elements* represents the members of the list separated by commas. **List[*elements*]** is equivalent to {*elements*}.
- **ListContourPlot[*array*]** generates a contour plot from a two-dimensional *array* of numbers.
- **ListContourPlot3D[*array*]** draws a contour plot of the values in *array*, a three-dimensional array of numbers representing the values of a function.
- **ListDensityPlot[*array*]** generates a density plot from a two-dimensional *array* of numbers.
- **ListLinePlot[{y_1, y_2, ...}]** plots points whose y-coordinates are y_1, y_2, ... and connects them with line segments. The x-coordinates are taken to be the positive integers.
- **ListLinePlot[{{x_1, y_1}, {x_2, y_2}, ...,}]** plots the points (x_1, y_1), (x_2, y_2), ... and connects them with lines.
- **ListLinePlot[*list*$_1$, *list*$_2$, ...]** plots multiple lines through points defined by *list*$_1$, *list*$_2$, ...
- **ListPlot[{y1, y2, ...}]** plots points whose y-coordinates are y1, y2, ... The x-coordinates are taken to be the positive integers, 1, 2, ...
- **ListPlot[{{x1, y1}, {x2, y2}, ..., }]** plots the points (x1, y1), (x2, y2), ...
- **ListPlot3D[{{z_{11}, z_{12}, ...}, {z_{21}, z_{22}, ...}, ...}]** generates a three-dimensional surface based upon a given array of heights. The x- and y-coordinate values for each data point are taken to be consecutive integers beginning with 1.
- **ListPlot3D[{{x_1, y_1, z_1}, {x_2, y_2, z_2}, ...}]** generates a three-dimensional surface based upon a given array of heights z_j, which are the z-coordinates corresponding to the points {x_i, y_i}.
- **ListPointPlot3D[*list*]** plots the points in *list* in a three-dimensional box. *list* must be a list of sublists, each of which contains three numbers, representing the coordinates of points to be plotted.
- **ListSurfacePlot3D[*list*]** creates a three-dimensional polygonal mesh from the vertices specified in *list*, which should be of the form {{{x11, y11, z11}, {x12, y12, z12}, ...}, {{x21, y21, z21}, {x22, y22, z22}, ...}, ...}
- **Log[x]** represents the natural logarithm. If a base, *b*, other than *e* is required, the appropriate form is **Log[b, x]**.

- **LogicalExpand[*expression*]** applies the distributive laws for logical operations to *expression* and puts it into disjunctive normal form.
- **LUBackSubstitution[*data*, b]** uses the output of **LUDecomposition** [*matrix*] to solve the system matrix.x = b.
- **LUDecomposition[*matrix*]** finds the *LU* decomposition of *matrix*.
- **Manipulate[*expression*, {k, m, n, i}]** works very much the same way as **Animate** except it allows the user to control the parameterdirectly with a slider.
- **Manipulate[*expression*, {k1, m1, n1, i1}, {k2, m2, n2, i2}, ...]** allows multiple parameters which can be independently controlled.
- **MatrixForm[*list*]** prints double nested lists as a rectangular array enclosed within parentheses. The innermost lists are printed as rows. Single nested lists are printed as columns enclosed within parentheses.
- **MatrixPower[*matrix*, n]** computes the nth power of *matrix*.
- **Max[*list*]** returns the largest number in *list*.
- **Min[*list*]** returns the smallest number in *list*.
- **Minors[*matrix*]** produces a matrix whose (i, j)th entry is the determinant of the submatrix obtained from *matrix* by deleting row $n - i + 1$ and column $n - j + 1$.
- **Minors[*matrix*, k]** produces the matrix whose entries are the determinants of all possible k × k submatrices of *matrix* (*matrix* need not be square).
- **Minus[a]** produces the additive inverse (negative) of a. **Minus[a]** is equivalent to -a.
- **Mod[m, n]** returns the remainder when m is divided by n.
- **Module[{*var1*, *var2*, ...}, *body*]** defines a module with local variables *var1*, *var2*, ...
- **Module[{*var1* = *v1*, *var2* = *v2*, ...}, *body*]** defines a module with local variables *var1*, *var2*, ... initialized to *v1*, *v2*, ..., respectively.
- **N[*expression*]** gives the numerical approximation of *expression* to six significant digits (*Mathematica*'s default).
- **N[*expression*, n]** attempts to give an approximation accurate to n significant digits.
- **NDSolve[*equations*, y, {x, xmin, xmax}]** gives a numerical approximation to the solution, y, of the differential equation with initial conditions, *equations*, whose independent variable, x, satisfies xmin ≤ x ≤ xmax.
- **Nest[f, *expression*, n]** applies f to *expression* successively n times.
- **NestList[f, *expression*, n]** applies f to *expression* successively n times and returns a list of all the intermediate calculations from 0 to n.
- **NIntegrate[f[x], {x, a, b}]** computes an approximation to the value of $\int_a^b f(x)\, dx$ using strictly numerical methods.
- **NIntegrate[f[x, y], {x, xmin, xmax}, {y, ymin, ymax}]** returns a numerical approximation of the value of the double integral $\int_{xmin}^{xmax} \int_{ymin}^{ymax} f(x, y)\, dy\, dx$.
- **NIntegrate[f[x, y, z], {x, xmin, xmax}, {y, ymin, ymax}, {z, zmin, zmax}]** returns a numerical approximation of the value of the triple integral $\int_{xmin}^{xmax} \int_{ymin}^{ymax} \int_{zmin}^{zmax} f(x, y, z)\, dz\, dy\, dx$.

- **Norm[v]** returns the Euclidean norm of **v**. $\| \mathbf{v} \| = \sqrt{\sum_{i=1}^{n} v_i^2}$.
- **Normal[*series*]** returns a polynomial representation of the SeriesData object *series* which can then be evaluated numerically. The O[x]^n term is omitted.
- **Normalize[*vector*]** converts *vector* into a unit vector.
- **Normalize[*vector*, f]** converts *vector* into a unit vector with respect to the norm function f.
- **Not[p]** or !p or ¬p is True if p is False and False if p is True.
- **NProduct**, returns numerical approximations to each of the products described in **Product**.
- **NRoots[lhs == rhs, *variable*]** produces numerical approximations of the solutions of a polynomial equation.
- **NSolve[*equations*, *variables*]** solves *equations* numerically for *variables*.
- **NSolve[*equations*, *variables*, n]** solves *equations* numerically for *variables* to n digits of precision.
- **NSum**, returns numerical approximations to each of the sums described in **Sum**.

- **NullSpace[a]** returns the basis vectors of the null space of a.
- **Numerator[*fraction*]** returns the numerator of *fraction*.
- **Opacity[a]** specifies the degree of transparency of a graphics object. The value of a must be between 0 and 1, with 0 representing perfect transparency and 1 representing complete opaqueness.
- **Or[p,q]** or **p || q** or **p ∨ q** is True if p or q (or both) are True; False otherwise.
- **Orthogonalize[*vectorlist*]** uses the Gram-Schmidt method to produce an orthonormal set of vectors whose span is *vectorlist*.
- **Orthogonalize[*vectorlist*, f]** produces an orthonormal set of vectors with respect to the inner product defined by f.
- **Outer[Times, v1, v2]** computes the outer product of v1 and v2.
- **PaddedForm[*expression*, {n, f}]** prints the value of *expression* leaving space for a total of n digits, f of which are to the right of the decimal point. The fractional portion of the number is rounded if any digits are deleted.
- **PaddedForm[*expression*, n]** prints the value of *expression* leaving space for a total of n digits. This form of the command can be used for integers or real number approximations. The decimal point is not counted as a position.
- **ParametricPlot[{x[t], y[t]}, {t, tmin, tmax}]** plots the parametric equations $x = x(t)$, $y = y(t)$ over the interval tmin ≤ t ≤ tmax.
- **ParametricPlot[{{x1[t], y1[t]}, {x2[t], y2[t]}, ...}, {t, tmin, tmax}]** plots several sets of parametric equations over tmin ≤ t ≤ tmax.
- **ParametricPlot3D[{x[t], y[t], z[t]}, {t, tmin, tmax}]** plots a space curve in three dimensions for tmin ≤ t ≤ tmax.
- **ParametricPlot3D[{x[s, t], y[s, t], z[s, t]}, {s, smin, smax}, {t, tmin, tmax}]** plots a surface in three dimensions.
- **Part[list, k]** or **list[[k]]** returns the kth element of list.
- **Part[list, -k]** or **list[[-k]]** returns the kth element from the end of list.
- **Part[list, m, n]** or **list[[m, n]]** returns the nth entry of the mth element of list, provided list has depth of at least 2.
- **Partition[*list*, k]** converts *list* into sublists of length k. If list contains k n + m elements, where m < k, **Partition** will create n sublists and the remaining m elements will be dropped.
- **Partition[*list*, k, d]** partitions *list* into sublists of length k offsetting each sublist from the previous sublist by d elements. In other words, each sublist (other than the first) begins with the d + 1st element of the previous sublist.
- **Pi** or π is the ratio of the circumference of a circle to its diameter.
- **PieChart[*datalist*]** draws a simple pie chart. *datalist* is a list of numbers enclosed within braces.
- **PieChart[{*datalist1*, *datalist2*, ...}]** draws a pie chart containing data from multiple data sets. Each data set is a list of numbers enclosed within braces.
- **Plot[f[x], {x, xmin, xmax}** plots a two-dimensional graph of the function f[x] on the interval xmin ≤ x ≤ xmax.
- **Plot[{f[x], g[x]}, {x, xmin, xmax}]** plots the graphs of f[x] and g[x] from xmin to xmax on the same set of axes. This command can be generalized in a natural way to plot three or more functions.
- **Plot3D[f[x, y], {x, xmin, xmax}, {y, ymin, ymax}]** plots a three-dimensional graph of the function f[x, y] above the rectangle xmin ≤ x ≤ xmax, ymin ≤ y ≤ ymax.
- **Plot3D[{f1[x, y], f2[x, y], ...}, {x, xmin, xmax}, {y, ymin, ymax}]** plots several three-dimensional surfaces on one set of axes.
- **Plus[a, b, ...]** computes the sum of a, b, ... **Plus[a, b]** is equivalent to **a + b**.
- **PolarPlot[f[θ], {θ, θ_{min}, θ_{max}}]** generates a plot of the polar equation $r = f(\theta)$ as θ varies from θ_{min} to θ_{max}.
- **PolarPlot[{f1[θ], f2[θ], ...}, {θ, θ_{min}, θ_{max}}]** plots several polar graphs on one set of axes.
- **PolynomialGCD[p1, p2, ...]** computes the greatest common divisor of the polynomials p1, p2, ...

- **PolynomialLCM[p1, p2,...]** computes the least common multiple of the polynomials p1, p2,...
- **PolynomialQ[*expression*, *variable*]** yields True if *expression* is a polynomial in *variable*, and False otherwise.
- **PolynomialQuotient[p, s, x]** gives the quotient upon division of p by s expressed as a function of x. Any remainder is ignored.
- **PolynomialRemainder[p, s, x]** returns the remainder when p is divided by s. The degree of the remainder is less than the degree of s.
- **Power[a, b]** computes a^b, **Power[a, b, c]** produces a^{b^c}, etc.
- **PowerExpand[*expression*]** expands nested powers, powers of products and quotients, roots of products and quotients, and their logarithms.
- **PreDecrement[x]** or **-- x** decreases the value of x by 1 and returns the *new* value of x.
- **PreIncrement[x]** or **++ x** increases the value of x by 1 and returns the *new* value of x.
- **Prepend[*list*, x]** returns *list* with x inserted to the left of its first element.
- **Prime[n]** returns the nth prime.
- **PrimeQ[*expression*]** yields True if *expression* is a prime number, and yields False otherwise.
- **Print[*expression*]** prints *expression*, followed by a line feed.
- **Print[*expression1*, *expression2*,...]** prints *expression1*, *expression2*, ... followed by a single line feed.
- **Product[a[i], {i, imax}]** or $\prod_{i=1}^{imax} a[i]$ evaluates the product $\prod_{i=1}^{imax} a_i$.
- **Product[a[i], {i, imin, imax}]** or $\prod_{i=imin}^{imax} a[i]$ evaluates the product $\prod_{i=imin}^{imax} a_i$.
- **Product[a[i], {i, imin, imax, increment}]** evaluates the product $\prod_{i=imin}^{imax} a_i$ in steps of increment.
- **Product[a[i, j], {i, imax}, {j, jmax}]** or $\prod_{i=1}^{imax} \prod_{j=1}^{jmax} a[i, j]$ evaluates the product $\prod_{i=1}^{imax} \prod_{j=1}^{jmax} a_{i,j}$.
- **Product[a[i,j], {i,imin,imax}, {j,jmin,jmax}]** or $\prod_{i=imin}^{imax} \prod_{j=jmin}^{jmax} a[i, j]$ evaluates the product $\prod_{i=imin}^{imax} \prod_{j=jmin}^{jmax} a_{i,j}$.
- **Product[a[i,j], {i,imin,imax,i_increment}, {j,jmin,jmax, j_increment}]** evaluates the product $\prod_{i=imin}^{imax} \prod_{j=jmin}^{jmax} a_{i,j}$ in steps of i_increment and j_increment.
- **Projection[*vector1*, *vector2*]** returns the orthogonal projection of *vector1* onto *vector2*.
- **Projection[*vector1*, *vector2*, *f*]** returns the orthogonal projection of *vector1* onto *vector2* with respect to an inner product defined by *f*.
- **Quotient[m, n]** returns the quotient when m is divided by n.
- **Random[ ]** gives a uniformly distributed, real, pseudorandom number in the interval [0, 1].
- **Random[*type*]** returns a uniformly distributed pseudorandom number of type *type*, which is either Integer, Real, or Complex. Its values are between 0 and 1, in the case of Integer or Real, and contained within the square determined by 0 and 1 + i, if *type* is Complex.
- **Random[*type*, *range*]** gives a uniformly distributed pseudorandom number in the interval or rectangle determined by *range*. *range* can be either a single number or a list of two numbers such as {a,b} or {a + b I, c + d I}. A single number m, is equivalent to {0,m}.
- **Random[*type*, *range*, n]** gives a uniformly distributed pseudorandom number to n significant digits in the interval or rectangle determined by *range*.
- **RandomComplex[]** returns a pseudorandom complex number lying within the rectangle whose opposite vertices are 0 and 1+I.
- **RandomComplex[zmax]** returns a pseudorandom complex number that lies in the rectangle whose opposite vertices are 0 and zmax.
- **RandomComplex[{zmin, zmax}]** returns a pseudorandom complex number that lies in the rectangle whose opposite vertices are zmin and zmax.
- **RandomComplex[{zmin, zmax},n]** returns a list of n pseudorandom complex numbers each of which lies in the rectangle whose opposite vertices are zmin and zmax.

- **RandomComplex[{zmin, zmax},{m, n}]** returns an m × n list of pseudorandom complex numbers each of which lies in the rectangle whose opposite vertices are zmin and zmax.
- **RandomInteger[ ]** returns 0 or 1 with equal probability.
- **RandomInteger[imax]** returns a pseudorandom integer between 0 and imax.
- **RandomInteger[{imin, imax}]** returns a pseudorandom integer between imin and imax.
- **RandomInteger[{imin, imax},n]** returns a list of n pseudorandom integers between xmin and xmax. This extends in a natural way to lists of higher dimension.
- **RandomInteger[{imin, imax},{m, n}]** returns an m × n list of pseudorandom integers between xmin and xmax. This extends in a natural way to lists of higher dimension.
- **RandomPrime[n]** returns a pseudorandom prime number between 2 and n.
- **RandomPrime[{m, n}]** returns a pseudorandom prime number between m and n.
- **RandomPrime[{m, n}, k]** returns a list of k pseudorandom primes, each between m and n.
- **RandomReal[ ]** returns a pseudorandom real number between 0 and 1.
- **RandomReal[xmax]** returns a pseudorandom real number between 0 and xmax.
- **RandomReal[{xmin, xmax}]** returns a pseudorandom real number between xmin and xmax.
- **RandomReal[{xmin, xmax},n]** returns a list of n pseudorandom real numbers between xmin and xmax.
- **RandomReal[{xmin, xmax},{m, n}]** returns an m × n list of pseudorandom real numbers between xmin and xmax. This extends in a natural way to lists of higher dimension.
- **RandomSample[{e_1, e_2, . . . , e_n}]** gives a pseudorandom permutation of the list of e_i.
- **RandomSample[{e_1, e_2, . . . , e_n}, k]** gives a pseudorandom sample of k of the e_i.
- **Range[n]** generates a list of the first n consecutive integers.
- **Range[m, n]** generates a list of numbers from m to n in unit increments.
- **Range[m, n, d]** generates a list of numbers from m through n in increments of d.
- **Reduce[*equations*, *variables*]** simplifies *equations*, attempting to solve for *variables*. If *equations* is an identity, **Reduce** returns the value True. If *equations* is a contradiction, the value False is returned.
- **Remove[*symbol*]** removes *symbol* completely. *symbol* will no longer be recognized unless it is redefined.
- **ReplacePart[*list*, x, n]** replaces the object in the nth position of *list* by x.
- **ReplacePart[*list*, x, -n]** replaces the object in the nth position from the end by x.
- **ReplacePart[*list*, i → *new*]** replaces the ith part of *list* with *new*.
- **ReplacePart[*list*, {i_1 → new_1, i_2 → new_2, . . . , i_n → new_n}]** replaces parts i_1, i_2, . . . , i_n with new_1, new_2, . . . , new_n, respectively.
- **ReplacePart[*list*, {{i_1}, {i_2}, . . . , {i_n}} → *new*]** replaces all elements in positions i_1, i_2, . . . , i_n with *new*.
- **ReplacePart[*list*, {i, j} → *new*]** replaces the element in position j of the ith outer level entry with *new*.
- **ReplacePart[*list*, {i_1, j_1} → new_1, {i_2, j_2} → new_2, . . . , {i_n, j_n} → new_n]** replaces the entries in positions j_k of entry i_k in the outer level with new_k.
- **ReplacePart[*list*, {{i_1, j_1}, {i_2, j_2}, . . . , {i_n, j_n}} → *new*]** replaces all entries in positions j_k of entry i_k in the outer level with *new*.
- **Rest[*list*]** returns list with its *first* element deleted.
- **Reverse[*list*]** reverses the order of the elements of *list*.
- **RevolutionPlot3D[f[x], {x, xmin, xmax}]** plots the surface generated by rotating the curve $z = f(x)$, xmin ≤ x ≤ xmax, completely around the z-axis.
- **RevolutionPlot3D[f[x], {x, xmin, xmax}, {θ, θmin, θmax}]** plots the surface generated by rotating the curve $z = f(x)$, xmin ≤ x ≤ xmax, around the z-axis for θmin ≤ θ ≤ θmax where θ is the angle measured counterclockwise from the positive x-axis.
- **RevolutionPlot3D[{f[t],g[t]}, {t, tmin, tmax}]** generates a plot of the surface generated by rotating the curve $x = f(t)$, $z = g(t)$, tmin ≤ t ≤ tmax, completely around the z-axis.
- **RevolutionPlot3D[{f[t],g[t]}, {t, tmin, tmax}, {θ, θmin, θmax}]** generates a plot of the surface generated by the curve $x = f(t)$, $z = g(t)$, tmin ≤ t ≤ tmax, around the z-axis for θmin ≤ θ ≤ θmax where θ is the angle measured counterclockwise from the positive x-axis.

- **RevolutionPlot3D[z[r, θ], {r, rmin, rmax}]** generates a plot of the surface $z = z(r, \theta)$, $\text{rmin} \leq r \leq \text{rmax}$, described in cylindrical coordinates.
- **RevolutionPlot3D[z[r, θ], {r, rmin, rmax}, {θ, θmin, θmax}]** generates a plot of the surface $z = z(r, \theta)$, $\text{rmin} \leq r \leq \text{rmax}$, $\theta\text{min} \leq \theta \leq \theta\text{max}$.
- **Roots[lhs == rhs, *variable*]** produces the solutions of a polynomial equation.
- **RotateLeft[*list*]** cycles each element of *list* one position to the left. The leftmost element is moved to the extreme right of the list.
- **RotateLeft[*list*, n]** cycles the elements of *list* precisely n positions to the left. The leftmost n elements are moved to the extreme right of the list in their same relative positions. If n is negative, rotation occurs to the right.
- **RotateRight[*list*]** cycles each element of *list* one position to the right. The rightmost element is moved to the extreme left of the list.
- **RotateRight[*list*, n]** cycles the elements of *list* precisely n positions to the right. The rightmost n elements are moved to the extreme left of the list in their same relative positions. If n is negative, rotation occurs to the left.
- **RotateShape[*object*, ϕ, θ, ψ]** rotates *object* using the Euler angles ϕ, θ, and ψ.
- **Round[x]** returns the integer closest to x. If x lies exactly between two integers (e.g., 5.5), **Round** returns the nearest even integer.
- **RowReduce[*matrix*]** reduces *matrix* to reduced row echelon form.
- **SeedRandom[n]** initializes the random number generator using n as a seed. This guarantees that sequences of random numbers generated with the same seed will be identical.
- **SeedRandom[ ]** initializes the random number generator using the time of day and other attributes of the current *Mathematica* session.
- **Series[f[x], {x, a, n}]** generates a **SeriesData** object representing the nth degree Taylor polynomial of $f(x)$ about *a*.
- **SeriesCoefficient[*series*, n]** returns the coefficient of the nth degree term of a **SeriesData** object.
- **Show[g1, g2, . . .]** plots several graphs on a common set of axes.
- **Sign[x]** returns the values –1, 0, 1 depending upon whether x is negative, 0, or positive, respectively.
- **Simplify[*expression*]** performs a sequence of transformations on *expression*, and returns the simplest form it finds.
- **Sin, Cos, Tan, Sec, Csc,** and **Cot** respectively represent the six basic trigonometric functions, sine, cosine, tangent, secant, cosecant and cotangent.
- **Sinh, Cosh, Tanh, Sech, Csch,** and **Coth** represent the six hyperbolic functions.
- **Solve[*equations*, *variables*]** attempts to solve *equations* for *variables*.
- **Sort[*list*]** sorts the list *list* in increasing order. Real numbers are ordered according to their numerical value. Letters are arranged lexicographically, with capital letters coming after lowercase.
- **SphericalPlot3D[ρ, ϕ, θ]** generates a complete plot of the surface whose spherical radius, ρ, is defined as a function of ϕ and θ.
- **SphericalPlot3D[[ρ, {ϕ, ϕmin, ϕmax}, {θ, θmin, θmax}]** generates a plot of the surface whose spherical radius, ρ, is defined as a function of ϕ and θ over the intervals $\phi\text{min} \leq \phi \leq \phi\text{max}$, $\theta\text{min} \leq \theta \leq \theta\text{max}$.
- **Sqrt[x]** or $\sqrt{x}$ gives the non-negative square root of x.
- **StringDrop[*string*, n]** returns *string* with its first n characters dropped.
- **StringDrop[*string*, –n]** returns *string* with its last n characters dropped.
- **StringDrop[*string*, {n}]** returns *string* with its nth character dropped.
- **StringDrop[*string*, {–n}]** returns *string* with the nth character from the end dropped.
- **StringDrop[*string*, {m, n}]** returns *string* with characters m through n dropped.
- **StringInsert[*string1*, *string2*, n]** yields a string with *string2* inserted starting at position n in *string1*.
- **StringInsert[*string1*, *string2*, –n]** yields a string with *string2* inserted starting at the nth position from the end of *string1*.
- **StringInsert[*string1*, *string2*, {n1, n2, . . .}]** inserts a copy of *string2* at each of the positions n1, n2, . . . of *string1*.

- **StringJoin [** *string1* **,** *string2* **, ...]** or *string1* **<>** *string2* **<>** ... concatenates two or more strings to form a new string whose length is equal to the sum of the individual string lengths.
- **StringLength [** *string* **]** returns the number of characters in *string*.
- **StringPosition [** *string* **,** *substring* **]** returns a list of the start and end positions of all occurrances of *substring* within *string*.
- **StringReplace [** *string* **,** *string1* → *newstring1* **]** replaces *string1* by *newstring1* whenever it appears in *string*.
- **StringReplace [** *string* **,** {*string1* → *newstring1* **,** *string2* → *newstring2* **, ...}]** replaces *string1* by *newstring1*, *string2* by *newstring2*, ... whenever they appear in *string*.
- **StringReverse [** *string* **]** reverses the characters in *string*.
- **StringTake [** *string* **, n]** returns the first n characters of *string*.
- **StringTake [** *string* **, −n]** returns the last n characters of *string*.
- **StringTake [** *string* **, {n}]** returns the nth character of *string*.
- **StringTake [** *string* **, {−n}]** returns the nth character from the end of *string*.
- **StringTake [** *string* **, {m, n}]** returns characters m through n of *string*.
- **Subsets [** *list* **]** returns a list containing all subsets of *list*, including the empty set, i.e., the power set of *list*.
- **Subtract [a, b]** computes the difference of a and b. Only two arguments are permitted. **Subtract [a, b]** is equivalent to **a − b**.
- **SubtractFrom [x, y]** or **x −= y** subtracts y from x and returns the new value of x.
- **Sum [a [i] , {i, imax}]** or $\sum_{i=1}^{imax} a[i]$ evaluates the sum $\sum_{i=1}^{imax} a_i$.
- **Sum [a [i] , {i, imin, imax}]** or $\sum_{i=imin}^{imax} a[i]$ evaluates the sum $\sum_{i=imin}^{imax} a_i$.
- **Sum [a [i] , {i, imin, imax, increment}]** evaluates the sum $\sum_{i=imin}^{imax} a_i$ in steps of **increment**. Summation continues as long as i ≤ imax.
- **Sum [a [i, j] , {i, imax} , {j, jmax}]** or $\sum_{i=1}^{imax} \sum_{j=1}^{jmax} a[i, j]$ evaluates the sum $\sum_{i=1}^{imax} \sum_{j=1}^{jmax} a_{i,j}$.
- **Sum [a [i, j] , {i, imin, imax} , {j, jmin, jmax}]** or $\sum_{i=imin}^{imax} \sum_{j=jmin}^{jmax} a[i, j]$ evaluates the sum $\sum_{i=imin}^{imax} \sum_{j=jmin}^{jmax} a_{i,j}$.
- **Sum [a [i, j] , {i, imin, imax, i_increment} , {j, jmin, jmax, j_increment}]** evaluates the sum $\sum_{i=imin}^{imax} \sum_{j=jmin}^{jmax} a_{i,j}$ in steps of **i_increment** and **j_increment**.
- **SurfaceOfRevolution [f [x] , {x, xmin, xmax}]** generates the surface of revolution obtained by rotating the curve $z = f(x)$ about the z-axis.
- **SurfaceOfRevolution [f [x] , {x, xmin, xmax}, {θ, θmin, θmax}]** generates the surface of revolution obtained by rotating the curve $z = f(x)$ about the z-axis, for $θmin ≤ θ ≤ θmax$.
- **SurfaceOfRevolution [{x [t] , z [t] }, {t, tmin, tmax}]** generates the surface of revolution obtained by rotating the curve defined parametrically by $x = x(t)$, $z = z(t)$, about the z-axis.
- **Table [** *expression* **, {n}]** generates a list containing n copies of the object *expression*.
- **Table [** *expression* **, {k, n}]** generates a list of the values of *expression* as k varies from 1 to n.
- **Table [** *expression* **, {k, m, n}]** generates a list of the values of *expression* as k varies from m to n.
- **Table [** *expression* **, {k, m, n, d}]** generates a list of the values of *expression* as k varies from m to n in steps of d.
- **Table [** *expression* **, {m}, {n}]** generates a two-dimensional list, each element of which is the object *expression*.
- **Table [** *expression* **, {i, m_i, n_i}, {j, m_j, n_j}]** generates a nested list whose values are *expression*, computed as j goes from m_j to n_j and as i goes from m_i to n_i. The index j varies most rapidly.
- **TableForm [** *list* **]** prints *list* the same way as **MatrixForm** except the surrounding parentheses are omitted.

- **TableForm[*list*, *options*]** allows the use of various formatting options in determining the appearance of a table.
- **Take[list, n]** returns a list consisting of the first n elements of list.
- **Take[list, -n]** returns a list consisting of the last n elements of list.
- **Take[list, {n}]** returns a list consisting of the nth element of list.
- **Take[list, {-n}]** returns a list consisting of the nth element from the end of list.
- **Take[list, {m, n}]** returns a list consisting of the elements of list in positions m through n inclusive.
- **Take[list, {m, n, k}]** returns a list consisting of the elements of list in positions m through n in increments of k.
- **Times[a, b, ...]** computes the product of a, b, ... **Times[a, b]** is equivalent to **a * b**.
- **TimesBy[x,y]** or **x *= y** multiplies x by y and returns the new value of x.
- **Timing[*expression*]** evaluates *expression*, and returns a list of time used, in seconds, together with the result obtained.
- **Together[*expression*]** combines the terms of *expression* using a common denominator. Any common factors in the numerator and denominator are cancelled.
- **Total[*list*]** gives the sum of the elements of *list*.
- **Tr[*matrix*]** computes the trace of *matrix*.
- **TraditionalForm[*expression*]** prints *expression* in a traditional mathematical format.
- **TranslateShape[*object*, {x, y, z}]** translates *object* by the vector {x, y, z}].
- **Transpose[*matrix*]** computes the transpose of *matrix*.
- **TrigExpand[*expression*]** expands *expression*, splitting up sums and multiples that appear in arguments of trigonometric functions and expanding out products of trigonometric functions into sums and powers, taking advantage of trigonometric identities whenever possible.
- **TrigFactor[*expression*]** converts *expression* into a factored expression of trigonometric functions of a single argument.
- **TrigReduce[*expression*]** rewrites products and powers of trig functions in *expression* as trigonometric expressions with combined arguments, reducing *expression* to a linear trig function (i.e., without powers or products).
- **TrigToExp[*expression*]** converts trigonometric and hyperbolic functions to exponential form.
- **Unequal[x, y]** or **x != y** or $x \neq y$ is True if and only if x and y have different values.
- **Union[*list1*, *list2*]** combines lists *list1* and *list2* into one sorted list, eliminating any duplicate elements. Although only two lists are presented in this description, any number of lists may be used. As a special case, **Union[*list*]** will eliminate duplicate elements in *list*. *list1* $\cup$ *list* is equivalent to **Union[*list1*, *list2*]**.
- **UnitStep[x]** returns a value of 0 if $x < 0$ and 1 if $x \geq 0$.
- **Variables[*polynomial*]** gives a list of all independent variables in *polynomial*.
- **VectorPlot[{Fx, Fy}, {x, xmin, xmax}, {y, ymin, ymax}]** produces a vector field plot of the two-dimensional vector function **F**, whose components are Fx and Fy.
- **While[*condition*, *expression*]** evaluates *condition*, then *expression*, repetitively, until *condition* is False.
- **WireFrame[*object*]** shows all polygons used in the construction of *object* as transparent. It may be used on any Graphics3D object that contains the primitives Polygon, Line, and Point.
- **Xor[p, q]** is True if p or q (but not both) are True; False otherwise.

Index